T0091649

GPU Pro 360

Guide to Geometry Manipulation

GPU Pro 360

Guide to Geometry Manipulation

Edited by Wolfgang Engel

CRC Press is an imprint of the
Taylor & Francis Group, an **Informa** business
AN A K PETERS BOOK

CRC Press
Taylor & Francis Group
6000 Broken Sound Parkway NW, Suite 300
Boca Raton, FL 33487-2742

Printed on acid-free paper

International Standard Book Number-13: 9781138568259 (Hardback)
International Standard Book Number-13: 9781138568242 (Paperback)

Library of Congress Cataloging-in-Publication Data

Names: Engel, Wolfgang F., editor.
Title: GPU pro 360 guide to geometry manipulation / [edited by] Wolfgang Engel.
Description: Boca Raton : Taylor & Francis, CRC Press, [2018] | Includes bibliographical references.
Identifiers: LCCN 2017059328| ISBN 9781138568242 (pbk. : alk. paper) | ISBN 9781138568259 (hardback : alk. paper)
Subjects: LCSH: Computer graphics. | Rendering (Computer graphics) | Graphics processing units--Programming. | Computer animation. | Shapes--Computer simulation.
Classification: LCC T385 .G6888 2018 | DDC 006.6–dc23
LC record available at https://lccn.loc.gov/2017059328

Visit the eResources: www.crcpress.com/9781138568242

Visit the Taylor & Francis Web site at
http://www.taylorandfrancis.com

and the CRC Press Web site at
http://www.crcpress.com

Contents

Introduction

"Geometry manipulation" focuses on the ability of *graphics processor units* (GPUs) to process and generate geometry in exciting and interesting ways.

Tamy Boubekeur covers Phong Tesselation in the chapter "As Simple as Possible Tessellation for Interactive Applications." This operator is simpler than GPU subdivision surfaces and their approximations but succeeds at hiding standard "polygonization" artifacts often encountered on mesh silhouettes and interior contours. *Phong Tessellation* can be implemented on today's GPU using either uniform or adaptive instanced tessellation (vertex shader), or on the geometry shader for low tessellation rates.

The chapter "Rule-Based Geometry Synthesis in Real-Time" by Milán Magdics and Gergely Klár presents a framework for synthesizing and rendering geometry described by a rule-based representation in real time. The representation is evaluated completely on the GPU; thus, the geometry synthesis can be very fast and there is no need to copy data between the CPU and the graphics card. By applying frustum culling and rule selection based on the distance from the camera during the synthesis, only what is required for rendering with dynamic level of detail is generated.

Graham Hemingway describes in the chapter "GPU-based NURBS Geometry Evaluation and Rendering" a method for using the GPU to calculate NURBS geometry. Compared to evaluation on the CPU, this method yields significant performance improvements without drawbacks in precision or flexibility.

"Polygonal-Functional Hybrids for Computer Animation and Games," by Denis Kravtsov et al., describes how to represent geometry with functions to overcome some of the challenges with polygons like produce animations involving dramatic change of the shape of the model and creating complex shapes with changing topology. They also cover the integration of existing polygonal models and functional representations.

The chapter "Terrain and Ocean Rendering" looks at the tessellation related stages of DirectX 11, explains a simple implementation of terrain rendering, and implements the techniques from the *ShaderX*^{6} article "Procedural Ocean Effects" by László Szécsi and Khashayar Arman.

Jorge Jimenez, Jose I. Echevarria, Christopher Oat, and Diego Gutierrez present a method to add expressive and animated wrinkles to characters in the chapter "Practical and Realistic Facial Wrinkles Animation." Their system allows the animator to independently blend multiple wrinkle maps across regions of a character's face. When combined with traditional blend-target morphing for facial animation, this technique can produce very compelling results that enable virtual characters to be much more expressive in both their actions and dialog.

The chapter "Procedural Content Generation on the GPU," by Aleksander Netzel and Pawel Rohleder, demonstrates the generating and rendering of infinite and deterministic heightmap-based terrain utilizing fractal Brownian noise calculated in real time on the GPU. Additionally it proposes a random tree distribution scheme that exploits previously generated terrain information. The authors use spectral synthesis to accumulate several layers of approximated fractal Brownian motion. They also show how to simulate erosion in real time.

The chapter "Vertex Shader Tessellation" by Holger Gruen presents a method to implement tessellation using only the vertex shader. It requires DirectX 10 and above to work. This method does not require any data in addition to the already available vertex data, in contrast to older techniques that were called "Instanced Tessellation." It relies solely on the delivery of `SV_VertexID` and uses the original vertex and index buffers as input shader resources.

In "Optimized Stadium Crowd Rendering," Alan Chambers describes in detail the design and methods used to reproduce a 80,000-seat stadium. This method was used in the game *Rugby Challenge* on XBOX 360, PS3, and PC. Chambers reveals several tricks used to achieve colored "writing" in the stands, ambient occlusion that darkens the upper echelons, and variable crowd density that can be controlled live in-game.

"Geometric Antialiasing Methods" is about replacing hardware multisample antialiasing (MSAA) with a software method that works in the postprocessing pipeline, which has been very popular since a multisample antialiasing (MLAA) solution on the GPU was presented in *GPU Pro*^{2}. Persson discusses two antialiasing methods that are driven by additional geometric data generated in the geometry shader or stored upfront in a dedicated geometry buffer that might be part of the G-Buffer.

The chapter "GPU Terrain Subdivision and Tessellation" presents a GPU-based algorithm to perform real-time terrain subdivision and rendering of vast detailed landscapes without preprocessing data on the CPU. It also achieves smooth level of detail transitions from any viewpoint.

"Introducing the Programmable Vertex Pulling Rendering Pipeline" discusses one of the bigger challenges in game development targeting PC platforms: the GPU driver overhead. By moving more tasks onto the quickly evolving GPUs, the number of draw calls per frame can be increased. The chapter gives also an in-depth view on the latest AMD GPUs.

"A WebGL Globe Rendering Pipeline" describes a globe rendering pipeline that integrates hierarchical levels of detail (HLOD) algorithms used to manage high resolution imagery streamed from standard map servers, such as Esri or OpenStreetMap.

The chapter "Dynamic GPU Terrain" by David Pangerl presents a GPU-based algorithm to dynamically modify terrain topology and synchronize the changes with a physics simulation.

The next chapter, "Bandwidth-Efficient Procedural Meshes in the GPU via Tessellation" by Gustavo Bastos Nunes and João Lucas Guberman Raza, covers the procedural generation of highly detailed meshes with the help of the hardware tessellator while integrating a geomorphic-enabled level-of-detail (LOD) scheme.

"Real-Time Deformation of Subdivision Surfaces on Object Collisions" by Henry Schäfer, Matthias Nießner, Benjamin Keinert, and Marc Stamminger shows how to mimic residuals such as scratches or impacts with soft materials like snow or sand by enabling automated fine-scale surface deformations resulting from object collisions. This is achieved by using dynamic displacement maps on the GPU.

"Realistic Volumetric Explosions in Games" by Alex Dunn covers a single-pass volumetric explosion effect with the help of ray marching, sphere tracing, and the hardware tessellation pipeline to generate a volumetric sphere.

The chapter by Anton Kai Michels and Peter Sikachev describes the procedural snow deformation rendering in *Rise of the Tomb Raider*. Their deferred deformation is used to render trails with depression at the center and elevation on the edges, allowing gradual refilling of the snow tracks, but it can also easily be extended to handle other procedural interactions with the environment. The technique is scalable and memory friendly and provides centimeter-accurate deformations. It decouples the deformation logic from the geometry that is actually affected and thus can handle dozens of NPCs and works on any type of terrain.

The last chapter in this book deals with Catmull-Clark subdivision surfaces widely used in film production and more recently also in video games because of their intuitive authoring and surfaces with nice properties. They are defined by bicubic B-spline patches obtained from a recursively subdivided control mesh of arbitrary topology. Wade Brainerd describes a real-time method for rendering such subdivision surfaces, which has been used for the key assets in *Call of Duty* on the Playstation 4 and runs at FullHD at 60 frames per second.

Web Materials

Example programs and source code to accompany some of the chapters are available on the CRC Press website: go to https://www.crcpress.com/9781138568242 and click on the "Downloads" tab.

The directory structure follows the book structure by using the chapter numbers as the name of the subdirectory.

General System Requirements

The material presented in this book was originally published between 2010 and 2016, and the most recent developments have the following system requirements:

- The DirectX June 2010 SDK (the latest SDK is installed with Visual Studio 2012).
- DirectX 11 or DirectX 12 capable GPU are required to run the examples. The chapter will mention the exact requirement.
- The OS should be Microsoft Windows 10, following the requirement of DirectX 11 or 12 capable GPUs.
- Visual Studio C++ 2012 (some examples might require older versions).
- 2GB RAM or more.
- The latest GPU driver.

As Simple as Possible Tessellation for Interactive Applications

Tamy Boubekeur

A *tessellation operator* increases the number of nodes in a mesh and can be understood as an upsampling method for polygonal surfaces.

Using real-time tessellation allows the application developer to specify a coarse mesh at the application level, but dynamically add vertices as needed at GPU level. This high resolution mesh may then be used to sample a (visually) smoother surface, eventually displaced using scalar or vector maps.

We describe (Section 1.1) and extend (Section 1.2) *Phong tessellation*, a simple operator generating curved surfaces over piecewise linear surface meshes. This operator is simpler than GPU subdivision surfaces [Shiue et al. 05] and their approximations [Vlachos et al. 01] [Boubekeur and Schlick 07b] [Loop and Schaefer 08]. However, for a minimal computational cost, it succeeds at hiding standard "polygonization" artifacts often encountered on mesh silhouettes and interior contours (see Figure 1.1). This operator is purely local (per-polygon computation, without neighborhood queries) and reproduces quadratic geometry variations. It naturally completes Phong Normal Interpolation.

Phong tessellation can be implemented on today's GPU using either *uniform* or *adaptive instanced tessellation* (vertex shader), or on the geometry shader for low tessellation rates. Upcoming graphics hardware and APIs such as Microsoft Direct3D 11 [Gee 08] will feature a standard tessellator unit with two new programmable stages: the *Hull Shader*, which defines the control network of the polygon to tessellate (in our case, this is the polygon itself together with its normal vectors and without any additional control points), and the *Domain Shader*, which evaluates the curved geometry (in our case, a code similar to Listing 1.1).

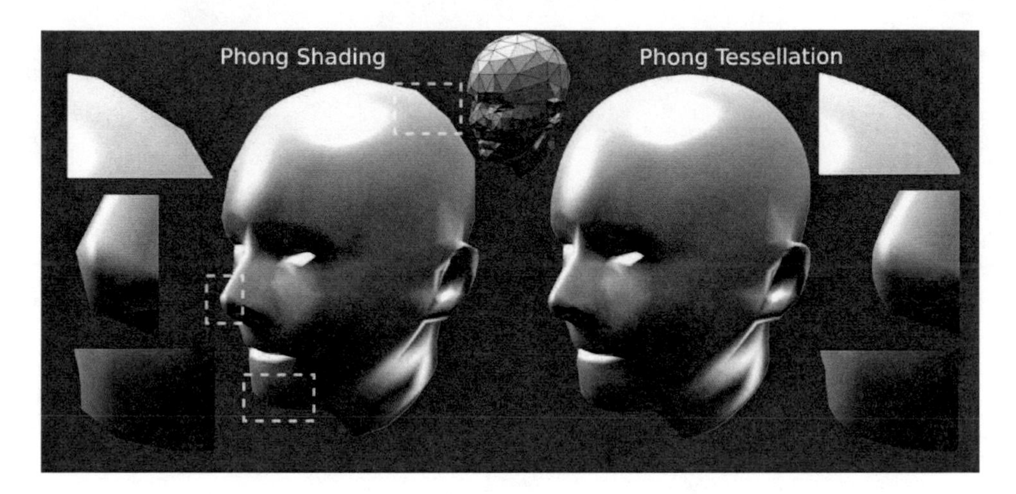

Figure 1.1. Phong tessellation (right) completes Phong shading (left).

1.1 Basic Phong Tessellation Operator

Consider a triangle $\mathbf{t}$ indexing three vertices $\{\mathbf{v}_i, \mathbf{v}_j, \mathbf{v}_k\}$, with $\mathbf{v} = \{\mathbf{p}, \mathbf{n}\}$, $\mathbf{p} \in \mathbb{R}^3$ the position and $\mathbf{n} \in S^2$ the normal vector. Generally, the tessellation of $\mathbf{t}$ generates a set of new vertices lying on a surface defined over $\mathbf{t}$. Linear tessellation simply generates vertices on the plane defined by $\mathbf{t}$. Each generated vertex $\mathbf{p}(u,v)$ in the linear tessellation corresponds to a barycentric coordinate $(u,v,w), u,v \in [0,1], w = 1-u-v$ defined by

$$\mathbf{p}(u,v) = (u,v,w)(\mathbf{p}_i, \mathbf{p}_j, \mathbf{p}_k)^{\mathsf{T}}.$$

Phong normal interpolation [Phong 75] uses the same process but normalizes the result in the end:

$$\mathbf{n}'(u,v) = (u,v,w)(\mathbf{n}_i, \mathbf{n}_j, \mathbf{n}_k)^{\mathsf{T}}, \quad \mathbf{n}(u,v) = \mathbf{n}'/\|\mathbf{n}'\|.$$

Phong tessellation [Boubekeur and Alexa 08] generalizes this idea to the generation of curved geometry (see Figure 1.2). Each vertex of the triangle defines a tangent plane (vertex position and normal), and the Phong tessellation interpolates between three points, which are the projections on each plane of a linear interpolation over $\mathbf{t}$:

$$\mathbf{p}^*{}_\alpha(u,v) = (1-\alpha)\mathbf{p}(u,v) + \alpha(u,v,w)\begin{pmatrix} \pi_i(\mathbf{p}(u,v)) \\ \pi_j(\mathbf{p}(u,v)) \\ \pi_k(\mathbf{p}(u,v)) \end{pmatrix}.$$

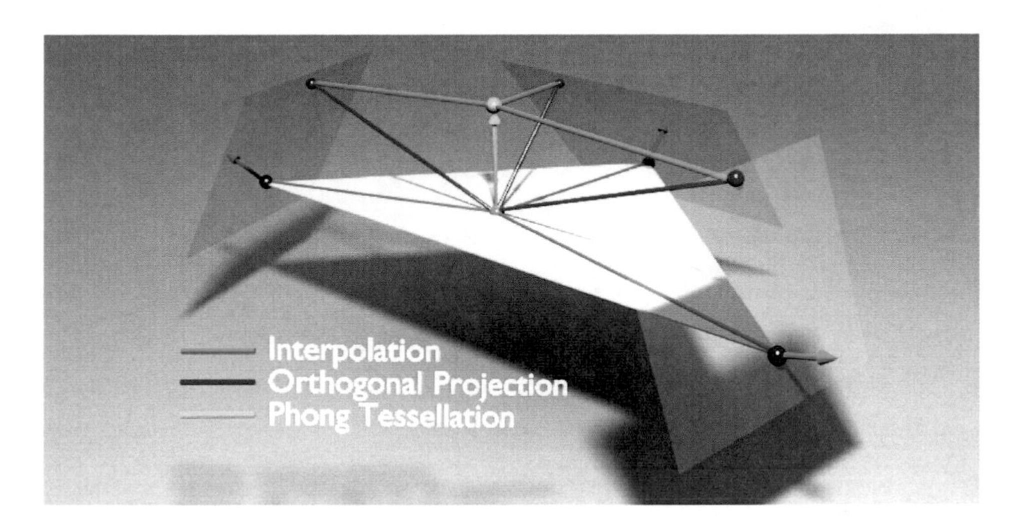

Figure 1.2. Phong tessellation performs in three steps: (1) Flat tessellation: vertices positions (red) are interpolated at a point (light blue); (2) Orthogonal projection: this point is orthogonally projected onto the three tangent planes defined at vertices (blue lines); and (3) Final displacement: these three points are interpolated again, defining the Phong tessellation displacement vector (in yellow). The initial interpolation (1) accounts for dynamic base mesh and standard tessellation methods, which starts from local coordinates expressed relatively to the current polygon.

Here $\pi_i(\mathbf{x})$ is the orthogonal projection of $\mathbf{x}$ on the plane defined by $\mathbf{v_i}$ and α acts as a simple shape factor interpolating between flat and curved triangle patch. We usually set it to 3/4. Note that this operator defines a quadratic patch implicitly. The coarse mesh to tessellate can simply be set on the GPU using extra vertex attributes, textures or, for the sake of simplicity, *uniform* variables in the vertex shader (on a per-triangle basis):

```
uniform vec3 p0, p1, p2;
```

The Phong tessellation vertex shader is given in Listing 1.1. At rendering time, when a triangle has to be tessellated, we use *Instanced Tessellation* [Boubekeur and Schlick 05] for generating a large number of vertices for each triangle and evaluate the Phong tessellation vertex shader on each of them. The Phong normal interpolation already offers a visually smooth shading inside shapes, so there is no need to apply the Phong tessellation operator everywhere on the input mesh. It is needed only on locations where Phong normal interpolation cannot hide the lack of high order continuity of the geometry, namely, on contours and silhouettes. Therefore, we make use of *Adaptive Instanced Tessellation* [Boubekeur and Schlick 07a, Boubekeur and Schlick 08], where the adaptivity factor (i.e., level

```
vec3 p0, p1, p2; // Current coarse triangle vertex positions
vec3 n0, n1, n2; // Current coarse triangle vertex normals
float alpha = 3/4; // Shape factor
vec3 project (vec3 p, vec3 c, vec3 n) {
  return p - dot (p - c, n)*n;
}
// Curved geometry at p={w,u,v} (barycentric coord.).
vec3 PhongGeometry (float u, float v, float w) {
  vec3 p = w*p0 + u*p1 + v*p2;
  vec3 c0 = project (p, p0, n0);
  vec3 c1 = project (p, p1, n1);
  vec3 c2 = project (p, p2, n2);
  vec3 q = w*c0 + u*c1 + v*c2;
  vec3 r = mix (p, q, alpha);
  return r;
}
// Continuous normal field at p
vec3 PhongNormal (float u, float v, float w) {
  vec3 n = normalize (w*n0 + u*n1 + v*n2);
  return n;
}
```

Listing 1.1. Basic GLSL vertex shader for Phong Tessellation. The `PhongGeometry` function defines the displacement to apply on any point of the tessellated polygon. Although watertight (C^0 continuity), we use a different function for interpolating normals (`PhongNormal`) in order to prevent any visual continuity artifact.

of tessellation) is defined on the coarse mesh vertices and edges using a simple *silhouetteness* measure based on vertex normals:

$$d_i = \left(1 - \left\| \mathbf{n}_i^\mathsf{T} \frac{\mathbf{c} - \mathbf{p}_i}{||\mathbf{c} - \mathbf{p}_i||} \right\|\right) m,$$

with $\mathbf{c}$ the position of the camera and m the maximum refinement depth. This value can be mapped on the shape factor to modulate a progressive transition between flat and curved geometry. We refer the reader to our paper [Boubekeur and Alexa 08] for additional analysis on performances and visual quality. Note that other tessellation environments may be used such as NVIDIA SDK's version of Instanced Tessellation [NVIDIA 08] or the ADM Tessellation Library [Tatarchuk et al. 09].

1.2 Extension to Quads

We extend Phong tessellation to quad and tri-quad meshes by redefining the operator over a quadangular polygon $\mathbf{q} = \{\mathbf{v}_\mathrm{i}, \mathbf{v}_\mathrm{j}, \mathbf{v}_\mathrm{k}, \mathbf{v}_\mathrm{l}\}$, a primitive frequently

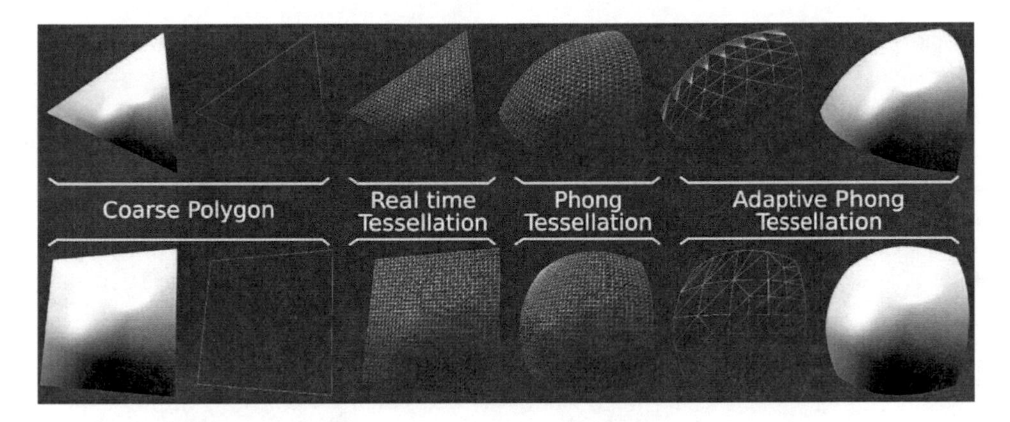

Figure 1.3. Phong tessellation for triangles and quads. Note that, while sampled with triangles, Phong geometry is defined at every point by the four vertices of the input quad.

used in production. We keep the basic idea of interpolation between tangent plane projections and simply replace the barycentric interpolation by a bilinear one (see Figure 1.3). This interpolation involves the four vertices of the quad and is substituted to the barycentric one wherever it appears in the triangle case, including the normal interpolation for shading:

$$\mathbf{p}^*_\alpha(u,v) = (1-\alpha)\mathbf{p}(u,v) + \begin{pmatrix} v(u\pi_i(\mathbf{p}(u,v)) + (1-u)\pi_i(\mathbf{p}(u,v)))+ \\ (1-v)(u\pi_l(\mathbf{p}(u,v)) + (1-u)\pi_k(\mathbf{p}(u,v))) \end{pmatrix}.$$

Triangle and quad Phong tessellation reproduce the same geometry on polygon edges, as only edge vertices are involved on this parameter domain (linear interpolation of orthogonal projections in both cases). Therefore we obtain a

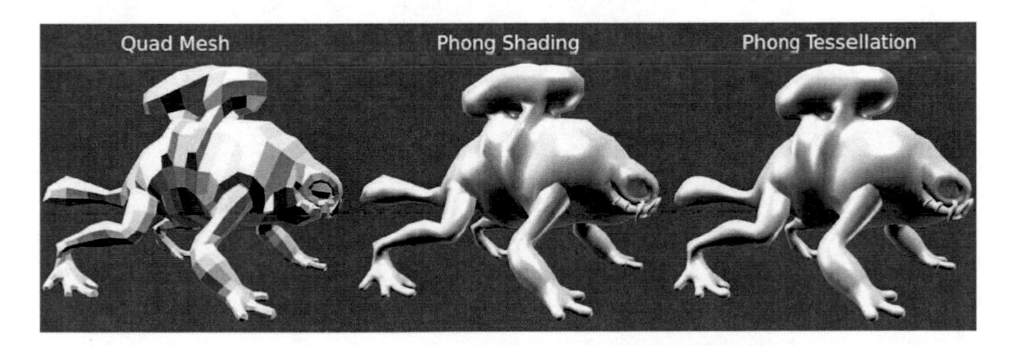

Figure 1.4. Phong tessellation extended to quad and tri-quad meshes.

crack-free tessellation scheme for triangle, tri-quad, and quad meshes. By doing this we avoid the typical trick consisting in triangulating quads, which often results in poor shape quality, even under (triangle) Phong tessellation. Note that while we define a curved geometry over a quad, this imposes no restriction on the way we sample the quad and triangle-based instanced tessellation can still be used safely with two triangular patches for sampling a single quad under quad Phong tessellation projection. Figure 1.4 illustrates the use of this extended operator.

1.3 Results and Discussion

Considering a typical scenario involving a dynamically deforming input coarse mesh (few thousands polygons) tessellated at level 6 (64×64)—a consistent 40% frame rate improvement is measured compared to Curved PN Triangle [Vlachos et al. 01] as well as a lower bandwidth usage for a similarly convincing rendering. Note that the aim of such an operator is to "fix" the strong polygonization artifacts appearing on silhouettes and interior contours, where Phong normal interpolation cannot hide the lack of continuity. Consequently, Phong tessellation differs significantly, for instance, from *subdivision surfaces approximations* [Boubekeur and Schlick 07b, Loop and Schaefer 08], which alterate significantly the perceived shapes in an attempt to generate globally smoother surfaces. Last, Phong tessellation is about 10% slower than flat tessellation when fragment shading is disabled. This number boils down to 2% when lighting and texturing are enabled and used, making it the cheapest way to generate visually smooth and curved geometry out of a dynamic polygonal mesh.

Bibliography

[Boubekeur and Alexa 08] Tamy Boubekeur and Marc Alexa. "Phong Tessellation." *ACM Transaction on Graphics - Special Issue on ACM SIGGRAPH Asia 2008* 27:5 (2008), 1–5.

[Boubekeur and Schlick 05] Tamy Boubekeur and Christophe Schlick. "Generic Mesh Refinement on GPU." In *Proceedings of ACM SIGGRAPH/Eurographics Graphics Hardware*, pp. 99–104, 2005.

[Boubekeur and Schlick 07a] Tamy Boubekeur and Christophe Schlick. *GPU Gems 3*, Chapter Generic Adaptive Mesh Refinement. Addison-Wesley, 2007.

[Boubekeur and Schlick 07b] Tamy Boubekeur and Christophe Schlick. "QAS: Real-time Quadratic Approximation of Subdivision Surfaces." In *Proceedings of Pacific Graphics 2007*, pp. 453–456, 2007.

[Boubekeur and Schlick 08] Tamy Boubekeur and Christophe Schlick. "A Flexible Kernel for Adaptive Mesh Refinement on GPU." *Computer Graphics Forum* 27:1 (2008), 102–114.

[Gee 08] K. Gee. "Direct3D 11 Tessellation." Presentation at Gamefest 2008, 2008.

[Loop and Schaefer 08] Charle Loop and Scott Schaefer. "Approximating Catmull-Clark Subdivision Surfaces with Bicubic Patches." *ACM Transaction on Graphics* 27:1 (2008), 1–8.

[NVIDIA 08] NVIDIA. "NVIDIA OpenGL SDK 10." Technical report, NVIDIA Corp., 2008.

[Phong 75] Bui Tuong Phong. "Illumination for Computer Generated Pictures." *Communications of the ACM* 18:6 (1975), 311–317.

[Shiue et al. 05] Le-Jeng Shiue, Ian Jones, and Jorg Peters. "A Realtime GPU Subdivision Kernel." In *Proceedings of ACM SIGGRAPH*, pp. 1010–1015, 2005.

[Tatarchuk et al. 09] Natalya Tatarchuk, Joshua Barczak, and Bill Bilodeau. "Programming for Real-Time Tessellation on GPU." Technical report, AMD Inc., 2009.

[Vlachos et al. 01] Alex Vlachos, Jorg Peters, Chas Boyd, and Jason Mitchell. "Curved PN Triangles." In *Proceedings of ACM Symposium on Interactive 3D*, pp. 159–166, 2001.

Rule-Based Geometry Synthesis in Real-Time

Milán Magdics and Gergely Klár

2.1 Introduction

Procedural modeling is a very popular modeling technique, since it allows us to create complex models with a small amount of work. Furthermore, algorithmic scene and object representations are often very compact and have low storage costs. Using operations that repeat elements and add random variations, we can represent arbitrarily large and detailed scenes without increasing the storage needs, while the generated objects remain diverse.

Evaluating the algorithmic description on the fly or, in other words, generating only those objects that we see through the camera, results in a runtime storage cost proportional to the complexity of the potentially visible objects, allowing us to store and visualize potentially infinite virtual worlds.

Rule-based modeling, whether using different kinds of grammars like L-systems, or split grammars, is a good trade-off between simplicity, usability, expressiveness, and formalism and has been widely used to generate different kinds of objects such as plants, buildings, road systems, fractals, and subdivision surfaces. This article presents a framework for synthesizing and rendering geometry described by a rule-based representation in real time. The representation is evaluated completely on the GPU, thus, the geometry synthesis can be very fast, and there is no need to copy data between the CPU and the graphics card. By applying frustum culling and rule selection based on the distance from the camera during the synthesis, we can generate only what is required for rendering with dynamic level-of-detail.

Interaction with the procedural scene is also an important problem to solve. The article discusses a technique for discrete collision detection based on a very simple observation: if we use an object's bounding box instead of the viewing frustum for culling, we can get the intersection of the object with the procedural scene the same way as generating objects inside the camera frustum. Thus, the GPU can speed up the calculation of these intersections when the number of objects is large, which is useful for controlling a large number of AI players, having particle systems colliding with the procedural scene, or in physical simulations.

Synthesizing the geometry on the GPU requires shader programs that depend on the current rules. The article shows how to automate shader and client code generation from an arbitrary (but well-defined) grammar description that is easier to create and understand.

2.2 Building Up the Scene with Procedures

The model we use is built on *Procedural Geometric Instancing* (PGI), which was introduced by Hart in [Hart 92,Ebert et al. 02]. PGI is a scene graph model, where nodes are extended with a procedure that executes at the time of instantiation. This procedure can set the parameters of the node based on the parent node or global parameters, and it can create new nodes. The whole scene is built up by instantiating the root node(s).

Since the procedure of a node depends only on the parent node, if two nodes are not descendants of each other, they can be evaluated independently. For example, we can build up the scene graph level by level (which corresponds to the breadth-first traversal of the graph), and evaluate the nodes of a level in parallel, resulting in better performance. We use this kind of geometry generation in our work.

Another simple but important observation is that we get the same result if we start the generation from an interior node instead of the root nodes. This allows us to buffer some parts of the scene graph, and regenerate only a smaller part of the scene in subsequent frames.

2.3 L-systems and the PGI

An *L-system* [Prusinkiewicz and Lindenmayer 91] is a parallel rewriting system defined as a 4-tuple $L = (N, T, \omega, P)$, where T and N is the set of *terminal* and *nonterminal* symbols, ω is the *initial string*, or *axiom* of the system, and P is the set of *production rules*, defining the way nonterminal symbols can be *replaced* with combinations of terminals and other nonterminals. A rule consists of two strings: the *predecessor* and the *successor*. Similarly to formal grammars, the *production* of an L-system iteratively replaces a predecessor with a successor in the current string (this step is often called *applying* of the rule). In contrast to formal grammars, an L-system applies as many rules as possible in an iteration, thus, it is much more capable of describing the growth what we need for fast generation of procedural geometry. The *generated language* of an L-system is a set of symbol sequences (or *strings*) that can be generated by applying zero or more rules to the axiom.

To utilize L-systems in computer graphics we assign geometric meaning to each symbol: for example, a command that changes drawing position or draws a line. Additionally, to make the modeling easier and increase the expressiveness, we can assign parameters to symbols (the concept we get is usually called a *module*). For example, let symbol F denote drawing a line. In this case, the parametrized symbol $F(l)$ may denote drawing a line with length l. This way, any kind of mesh can be described as a sequence of symbols. The graphical objects that can be generated using a specific L-system L are the geometric interpretations of the strings in L's generated language.

The traditional method to evaluate the geometric interpretation of a string is the *turtle graphics model.* In turtle graphics, drawing commands and state modifier commands (i.e. those that change the current global rendering state) are assigned to the symbols. The string is processed sequentially, and every command is evaluated relative to the current rendering state. The main problem of this method is that the evaluation cannot be parallelized, since every command that changes the rendering state has effect on every following commands in the string.

In [Hart 92, Ebert et al. 02], Hart shows how a procedural scene description given in the PGI model can be translated to a context-free L-system (that is, an L-system in which every predecessor consists of exactly one symbol) and vice versa. Thus, in terms of geometric modeling PGI and context-free L-systems are equivalent. Figure 2.1 shows the correspondence of the two models. The most important consequence of the equivalence is that the geometric interpretation of an L-system generated string can be evaluated in parallel, or in other words, a string can be considered as a set of symbols. In our work, we extend Hart's scene-graph-based L-system model to give more flexibility. The details are described in Section 2.4.

Hart also showed how the rules used with the turtle graphics model should be transformed to allow parallel evaluation. Since the turtle graphics model is frequently used in rule-based geometry generation, we will discuss this transformation with respect to our model in Section 2.4.

Modeling objects with PGI requires writing programs, which can be difficult to modelers who are not familiar with programming. Although PGI and rule-based models are very similar, the formalism is different. According to other works [Parish and Müller 01, Wonka et al. 03, Müller et al. 06, Lipp et al. 08], modelers can easily learn and use rule-based modeling methods, which was our main reason to use rules for real-time geometry generation. Please note that when

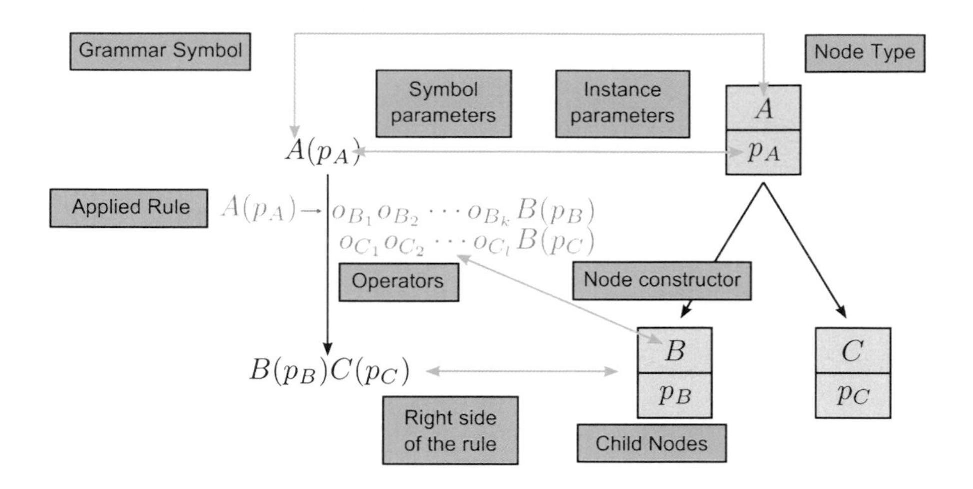

Figure 2.1. Equivalence of the PGI model with L-systems.

using only context-free rules, rules of a sequential grammar G can be evaluated in parallel. As a proof, consider the syntax tree of a string derived in G, the parallel breadth-first traversal of this tree corresponds to the L-system-style rule application. In terms of geometric modeling, when two procedural objects are independent they can always be refined independently, thus, in parallel.

Also note that this does not necessarily hold for context-sensitive rules (i.e., having more than one symbol in the predecessor), as context-sensitive rules are not allowed in our model. Thus, the expressibility of some complex conditions like connectivity and object dependency is very limited in our system.

2.4 Model Details and Implementation

In this section, we will give a detailed overview of the L-system-based model we used and we will describe its GPU implementation. Since our intention is to allow modelers to write rules instead of programs, we will also discuss automatic code generation from a grammar and describe what parts a grammar description should consist of to work with our model. The concrete, XML-based grammar descriptor format we used is included with the demo programs.

2.4.1 Symbols

Representing geometry. In the turtle graphics representation, geometry is created using symbols that represent drawing commands which are relative to the current rendering state. Here, these symbols represent the geometry (an arbitrary mesh) with their parameters and are called *instanced symbols*. Every instanced symbol stores every parameter needed for its drawing (*instancing*), thus, instanced symbols are independent. This enables parallel instancing, and a sequence of symbols becomes a set. In contrast to the turtle graphics model, instanced symbols cannot affect the rendering state in our model. Such symbols given in the turtle graphics model can be treated by splitting them into an instanced symbol and one or more operators (see the definition of operators below). For example, let module $F(l)$ mean drawing a line having length l and then moving the actual drawing position (a global rendering state) by l to the current direction. In the scene-graph-based representation this corresponds to $\overline{F}(l,o)m(\vec{v})$, where $\overline{F}(l,o)$ represents a line with length l and orientation o, and operator $m(\vec{v})$ represents translation, with vector $\vec{v}$ having the same effect as in $F(l)$. In this case, the operator has to follow the instanced symbol to get the same result. After these splitting transformations, the rules can be rewritten as described in Section 2.4.6.

Operators. Symbols representing a command that changes module parameters are called *operators*, we denote them with lower case letters. There is no global rendering state, as every operator modifies only the parameters of the next geometric object. In other words, every operator belongs to one specific instanced symbol in a string and modifies only that symbol's parameters. This way we can easily achieve equivalence with the PGI model, and these operators can hide the programs of the nodes that execute at the time of instantiation. Operators are relative to the left side of the rule, or in terms of PGI the parent node. For example, consider the rule

$$A \rightarrow hBhC$$

where operator h halves the size. The size parameter of both B and C will be the half of A's size.

Representation. In our model, each module is represented by a vertex, which allows us to implement rules and create new modules in the geometry shader. Section 2.4.2 will discuss the benefits of this decision.

Theoretically, different symbols can have different parameters. However, we found this impractical. Having multiple vertex types would result in more passes or additional branching in the shader code. Thus, our approach sacrifices storage space for efficiency and stores generic vertices that are capable of representing any type. The corresponding vertex type (a record in both the CPU and the GPU code) is created automatically as the union of all module data types and a symbol identifier. Thus, symbols appear as identifiers, mapped into integers between 1 and the number of symbols, which is also generated automatically from the grammar description, similarly to [Sowers 08]. To make the implementation of the code generation easier, we required the explicit specification of symbols in the grammar description, although they are implicitly given with the rules.

Since we merge all symbol types into one general record, the vertex type grows bigger. Therefore, it is easier to exceed the limit of the vertex size or the limit of the output size of the geometry shader. To overcome these problems we can use line or triangle primitives (instead of vertices) to represent modules or distribute module data between several primitives (vertex, line or triangle) and perform a single iteration step of the geometry generation in multiple passes. However, the first option alone does not solve the problem of the geometry shader's output limit, and splitting into multiple passes only works if rules do not depend on every attribute value at the same time. Fortunately, this holds in many cases; for example, position, size, and orientation are usually independent. We would like to note here that all our examples used a single vertex.

Modules are stored on the GPU in vertex buffers. In our experience 100 bytes is a reasonable upper bound for the storage requirement of one module (e.g., 25 float attributes can fit into it). Thus, the representation of one million vertices requires about 100 MB of storage, which can fit in a modern GPU's memory several times. Generating much more than one million modules with real-time rates exceeds the capabilities of our implementation, therefore, storage requirements of the algorithm should not cause any problems on high performance graphics cards.

2.4.2 Algorithm Overview

Figure 2.2 shows the overview of the complete algorithm. It consists of three main parts.

The first step is the generation of the procedural scene graph, called *generation step* (see Listing 2.1). This part is similar to the split grammar GPU implementation in [Sowers 08]. Starting with the axiom string, we perform iterations by applying rules of an L-system to generate instanced modules. These are only descriptors of different geometric objects stored as vertices, no concrete geometry is created here. Every operation is performed in the geometry shader, and modules are processed in parallel. The vertex shader simply passes its input without any modifications, and rasterization is disabled. We apply culling first for every module; this ensures that those objects that are not visible in the current frame are not generated. For modules that passed culling and were not terminated, we apply a rule. The left side of the rule is the symbol type of the module, which is stored in a symbol identifier. If there are multiple rules for that symbol, a *rule-selection algorithm* is applied to choose one of them. The chosen rule is applied by creating new modules and initializing them with the proper operators. As a last

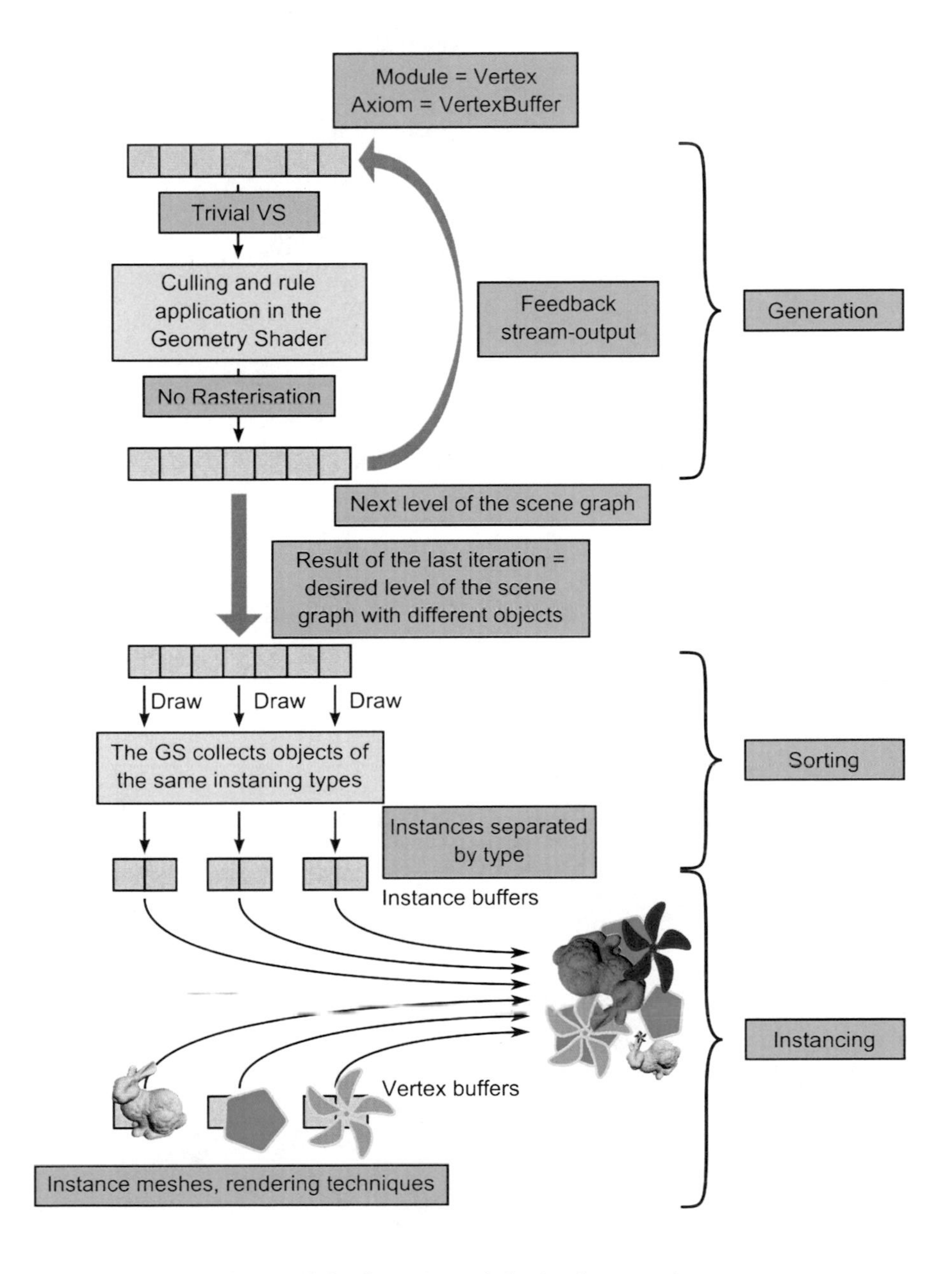

Figure 2.2. Overview of the implementation.

```
// The input of the first iteration is the axiom.
device->IASetVertexBuffers(0, 1, &axiomBuffer,
   &modulStrides, &zeroOffset);
// Apply generation shaders.
generationTechnique->GetPassByIndex(0)->Apply(0);
// Execute an iteration maxDepth times.
for (int depth = 0; depth < maxDepth; ++depth)
{
   // Set dstBuffer as stream output.
   device->SOSetTargets(1, &dstBuffer, &zeroOffset);
   // First draw call has to process the axiom.
   if (0 == depth)
      device->Draw(axiomLength, 0);
   else
      device->DrawAuto();
   // Ping pong
   ID3D10Buffer* nullBuffer = NULL;
   device->SOSetTargets(1, &nullBuffer, &zeroOffset);
   std::swap(dstBuffer, srcBuffer);

   // Set srcBuffer as input
   device->IASetVertexBuffers(0, 1, &srcBuffer,
      &modulStrides, &zeroOffset);
}
```

Listing 2.1. DX10 implementation of the CPU code of the generation step using fixed iteration number.

step, the modules are inserted into the new set of modules (the stream output), which is the input of the next iteration. Since we utilize the geometry shader for the creation of new modules, only the desired number of output is generated. There are no dummy symbols as in Hart's GPU L-system implementation using pixel shaders [Lacz and Hart 04].

After the generation step, modules are sorted by instancing type; that is, what kind of geometry they represent and what technique is used to render them. Sorting is essential for the efficient instancing of the generated descriptors.

Finally, we render the scene using instancing.

2.4.3 Iteration and Termination

We build up the procedural scene graph using breadth-first traversal, generating the subsequent level in every iteration step by performing a draw call and applying one rule for each non-terminated module. In most cases, the depth of the generation can be exactly specified. The concrete depth depends on the grammar as well as other conditions like dynamic level of detail. Thus, the most basic way

```
[maxvertexcount(MAX_SUCCESSOR_LENGTH)]
void gsGeneration(point Module input[1],
                  inout PointStream<Module> stream)
{
    // Perform culling here.

    // Checking termination
    if (input[0].terminated)
    {
        stream.Append(input[0]);
        return;
    }
    // Rule selection and application here.
}
```

Listing 2.2. The role of termination in the geometry shader.

to implement the CPU code is to execute the iteration `maxDepth`-times, where `maxDepth` is the desired level.

Objects usually do not need to be refined to the same level. For example, generating a building may need more iterations then a flower, or we add more details to a building closer to the camera then another building that is far away. To support this, we added a parameter (called `terminated`) to the module type that indicates whether the object is completed or not (see Listing 2.2). The ideal implementation would be to separate the completed objects from those that require further iterations, but the current shader model does not allow inserting different data into multiple stream outputs. We found it too inefficient to insert a sorting step that collects terminated modules and removes them from the working stream. Thus, we keep completed modules in the same buffer with the incomplete modules and simply pass them through the pipeline. Please note that if the shader model would allow dividing the data between multiple buffers, we could simply send the terminated modules into another stream.

Termination can be useful to implement another strategy to determine the generation depth. Sometimes it is better to perform iterations until there has been a change in the current modules. In other words, we can end the iterations if all modules are terminated. Unfortunately we cannot set any kind of global flag in the geometry shader to indicate that there is an incomplete module. However, we can extend the generation with a pixel shader stage, which discards pixels where the `terminated` parameter coming from the geometry shader is true. Thus, if any pixel is refreshed (this can be checked with a single occlusion query), that signals a change in the module set. To avoid writing the same pixels we can extend the Module struct with a module identifier. If M is the maximum number of symbols in successors, d is current generation level, and p denotes the position of

```
[maxvertexcount(MAX_SUCCESSOR_LENGTH)]
void gsGeneration(point Module input[1],
                  inout PointStream<Module> stream)
{
   if ( cullModule( input[0] )
      return;
   // Check for termination. Select and apply rules.
}
```

Listing 2.3. Culling in the geometry shader.

the newly created module in the applied rule, the following formula

$$\texttt{newModule.objectID} = M^d + \texttt{parent.objectID} * M + p$$

gives a unique identifier, which can be easily mapped to the pixels of the render target.

Note that with multiple different stream outputs this strategy is greatly simplified, the iteration has to be continued until the working buffer is not empty.

2.4.4 Culling

The goal of culling is to throw away those modules that do not effect the currently rendered frame. Formally, this corresponds to the application of the rule

$$A \rightarrow \epsilon$$

where ϵ is the empty string. However, adding this rule for every symbol would be too tedious; thus, it is not included in the grammar description. The implementation is simple—the geometry shader simply does not return anything for culled modules (see Listing 2.3).

The culling strategy can be very diverse depending on the procedural model; for example, culling based on size in pixels, alpha value, distance from the camera, or frustum culling.

We implemented spheres and *axis-aligned bounding boxes* (AABBs) as bounding geometries for frustum culling (we use it for interaction in Section 2.5 as well). These are resized to the module's size parameter (possibly three-dimensional) and centered to the module's position (see Listing 2.4). However, any kind of bounding volume can be stored or computed, possibly different for every symbol, and depending on the module parameters.

2.4.5 Rule Selection

Before applying a specific rule, we have to decide which rule to use. First, we have to identify the left side of the rule (the predecessor), which is simply the

```
// Culling a sphere with a plane
bool isecPlaneSphere(float4 plane,float3 center,float radius)
{
   return dot( plane, float4(center,1) ) < -radius;
}

// Frustum culling of a bounding sphere
bool isecFrustumSphere( Module module )
{
   return isecPlaneSphere(p_near, module.pos, module.size)  ||
          isecPlaneSphere(p_far, module.pos, module.size)   ||
          isecPlaneSphere(p_left, module.pos, module.size)  ||
          isecPlaneSphere(p_right, module.pos, module.size) ||
          isecPlaneSphere(p_top, module.pos, module.size)   ||
          isecPlaneSphere(p_bottom, module.pos, module.size);
}

// Main culling function
bool cullModule(Module module)
{
   if (enableModuleCulling)
      // Additional culling conditions can be or-ed
      return isecFrustumSphere(module);
   else
      return false;
}
```

Listing 2.4. An example of the culling function: frustum culling with a bounding sphere.

symbol stored in the symbol ID of the module. These were mapped to numeric constants in compile time, so the selection between rules means a sequence of if-then-else statements as in [Sowers 08], or an equivalent switch-case statement.

We often assign multiple rules to a single symbol; thus, we have to select one. Allowing multiple rules for a symbol is important since this is the way we add variation to the generated geometry or describe different conditions. Traditionally, *stochastic* L-systems are used, where every rule has a probability, and they are chosen randomly. Additionally, we can assign conditions to rules; these are expressions of the module parameters of the predecessor. Thus, the most efficient way to select and apply rules is an if-then-else statement, where the conditions are the conjunction of the equality test with the symbol ID, the condition of the rule, and the test of the generated random value for handling probability, and the body of the if statement is the rule application, as in [Sowers 08] (see Listing 2.5).

In our model (see Listing 2.6), we use rule selection strategies to select between rules of a given predecessor (this is mostly a modification only in the terminology). The idea comes from [Parish and Müller 01], where the authors used the

```
[maxvertexcount(MAX_SUCCESSOR_LENGTH)]
void gsGeneration(point Module input[1],
                  inout PointStream<Module> stream)
{
   // Culling and termination check.

   // Any kind of one-dimensional GPU noise can be used here.
   float random = getRandomBetween_0_1();

   Module parent = input[0];
   if(parent.symbolID == ID_of_the_predecessor &&
      conditions_of_the_rule &&
      // Intervals of the allowed random values,
      // an_upper_bound-a_lower_bound is the
      // probability of the rule.
      a_lower_bound <= random &&
      random < an_upper_bound)
   {
      // Create and initialize successor modules here.
   }
   else if (/*...*/) {/*...*/}
   // Similar if statements for the remaining rules
}
```

Listing 2.5. Rule selection with if-then-else statements.

concepts *global goals* and *local constraints* to select rules and set module parameters. Identifiers are generated for rules in compile time (of the shader), and we obtain the length of each rule and store these as compile time constants. Rule selection strategies are functions that return a rule identifier and a rule length. The returned rule lengths are not necessarily constants, thus, we can simulate variable length rules like split or repeat in [Müller et al. 06]:

```
void selectRule_ModuleName(in Module parent,
                           out int rule_length,
                           out int ruleID);
```

Rule selection methods are discussed in more detail in Section 2.4.7.

Instead of the if-then-else statements described above, we used a different structure. It is only an insignificant technical detail, but we found it very comfortable to use. We made these changes to improve the quality and readability of the generated code by separating its main parts, however, it adds a small overhead to the generation. Code quality is not the most important goal in real-time computer graphics (unlike performance), but when the number of rules becomes larger and we would like to modify the code, it can help a lot. Although code generation is automated, nearly the same way as Sowers, we may want to modify the

```
void selectRule(in Module module, out int rule_length,
                out int ruleID)
{
   switch (module.symbolID)
   {
      case a_symbol_ID:
         // Invokes the proper strategy assigned to the ↩
            symbol
         // Such function is created for every symbol
         selectRule_symbolsName(module, rule_length, ruleID);
         break;
      // A case for other symbols
   };
}

[maxvertexcount(MAX_SUCCESSOR_LENGTH)]
void gsGeneration(point Module input[1],
                  inout PointStream<Module> stream)
{
   // Culling and termination check

   // Selecting a rule and getting its successor length
   // (number of modules in it)
   int rule_length, rule_id;
   selectRule( input[0], rule_length, rule_id );

   // Getting the successor modules and streaming them out
   for ( int i = 1; i < rule_length; ++i )
   {
      stream.Append(getNextModule(input[0], rule_id, i));
   }
}
```

Listing 2.6. A less efficient, but more readable implementation of the rule selection and application.

generated parts for debugging, optimize, or experiment with new rule selection strategies or operators.

To support automatic code generation, different rule selection strategies have to be implemented. These can be assigned to symbols by the modelers in the grammar description. Thus, the task of automatic code generation is simply the invocation of the proper pre-implemented function.

2.4.6 Rule Application

In our model a rule successor is a set of symbols, where operators can be assigned to every symbol. For example, the rules of the Sierpinski-triangle formally would

```
// Main module creation function
Module getNextModule(Module parent, // predecessor
                     int ruleID,    // ID of the selected rule
                     int index)     // position in the successor
{
   Module output;
   switch (ruleID)
   {
      // A case is created for every rule.
      case a_specific_ruleID:
         // A similar function is created for every ruleID.
         return getNextModule_a_RuleID( parent, index );
      // ...
   };
   return output;
}
// Module creation function for a rule
Module getNextModule_a_RuleID(Module parent, int index )
{
   Module output = parent;
   switch(index)
   {
      // The number of cases equals the number of
      // symbols in the successor.
      case i:
         output.symbolID = id_of_the_ith_symbol;
         // Invocation of the proper operators
      // ...
   };
}
```

Listing 2.7. Rule application functions, that return one module.

look like

$$T \to \{s(0.5)m(-0.5, -h/4)T, s(0.5)m(0.5, -h/4)T, s(0.5)m(0, h/4)T\},$$

where s multiplies the inherited size by the given parameter, $m(x, y)$ adds the vector $(x, y) * P.size$ to the parameter, and h denotes here the height of the unit triangle. In our XML-based grammar descriptor, this rule can be expressed as

```
<rule>
   <predecessor module="T" />
   <successor module="T">
      <operator name="resize" value="0.5" />
      <operator name="move_xy" x="-0.5" y="0.2165" />
   </successor>
```

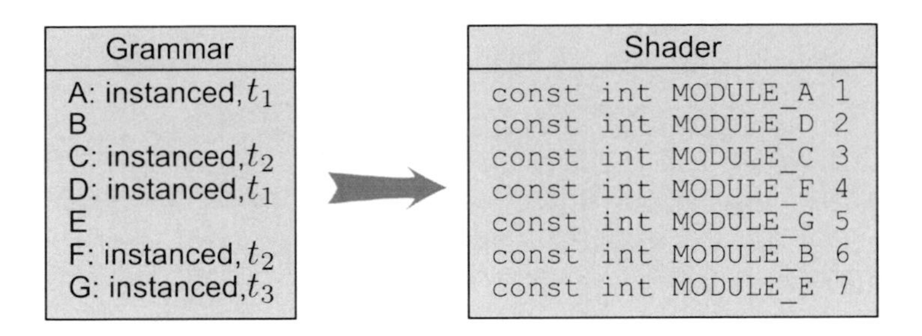

Figure 2.3. An example of the generated symbol IDs. Instancing types are denoted with t_i for simplicity.

```
    ...the remaining two modules are given similarly...
</rule>
```

If we have rules given in the turtle graphics style, we can easily rewrite them to get the same result. First, we have to split symbols representing drawing commands that change the rendering state as described in Section 2.4.1 (see Listing 2.7). Then, for each drawing command, we have to calculate the current rendering state at the point of its invocation, which is simply the concatenation of all operators before the symbol. For example, in the following rule

$$A \to \{o_1 B o_2 C o_3 o_4 D\},$$

the rendering state at C is the result of $o_1 o_2$. Thus, the rule is transformed to

$$A \to \{o_1 B, o_1 o_2 C, o_1 o_2 o_3 o_4 D\}.$$

Automatic code generation goes the same way as for rule selections. We have to pre-implement operators; the generated code will invoke these.

The performance of our system strongly depends on the number of different branches. Sorting the vertex buffer by symbol ID could greatly improve performance when there are many rules. However, adding a sorting step after each iteration would result in a significant performance loss (see Figure 2.3). Thus, performance issues of dynamic branching should be taken into account when designing the rules. Since the successor is a set (although theoretically there is no order between the symbols), it affects the order in which modules are inserted into the stream. For a very basic example, let us consider an L-system with the following three rules:

$$A \to BCBC$$
$$B \to BB$$
$$C \to CC$$

The first rule is theoretically equivalent to

$$A \to BBCC$$

If we compare a simple derivation using these rules, we can see that in the second case, the vertex buffer is sorted, therefore parallel threads will follow the same path except in one case. In contrast, using the unsorted vertex buffer, it is more likely that parallel threads always have to follow the path for two rules.

$$A \to BCBC \to BBCCBBCC \to BBBBCCCCBBBBCCCC$$
$$A \to BBCC \to BBBBCCCC \to BBBBBBBBCCCCCCCC$$

2.4.7 Expressiveness of the Model: Selection Strategies and Operators

As mentioned in the previous sections, automatic code generation becomes simple if we use predefined and implemented rule selection functions and operators. From the viewpoint of geometric modeling, these are the soul of the whole system, they determine usability and expressiveness, therefore, they should be well-designed.

The role of selection functions and operators is to hide programs from modelers. There is no restriction on the complexity of these programs, they can contain branching or even loops. Since in our knowledge, there are no standards of rule-based modeling, it is only a matter of taste what complex functions we allow. We would like to note here that branching in operators or selection functions can be replaced by using more rules and vice versa. Therefore, by allowing more complex operators, modeling of a specific scene can be represented with less rules in most cases, but makes it harder to learn and understand the operators. It is also important to note that the complexity of the shaders is about the same in both cases.

Many examples for both rule selection and operators can be found in the demo programs. To name a few, rule selection based on level-of-detail, stochastic selection, perturbation of a module attribute with a random value, changing its size or position or time dependent operators.

2.4.8 Sorting

Different types of objects may be rendered differently. For example, a window object and a trunk of a bonsai tree probably have different meshes and shaders. Thus, after the generation step we sort the created modules by instancing type. Any kind of sorting algorithm can be used here, we used bucket sort.

The symbol identifiers are sorted in compile time of the shader code such that the identifiers of the instanced symbols have smaller values and form a continuous interval. There is no reason to forbid assigning the same instancing type to two or more symbols. To handle this case, we sort the identifiers of the instanced

```
[maxvertexcount(1)]
void gsSort(point Module input[1],
            inout PointStream<SortedModule> stream)
{
   // In every step minID and maxID are set by the CPU code.
   if (minID <= input[0].symbolID && input[0].symbolID <= maxID)
      // Unused attributes are removed; the rest are simply
      // copied.
      stream.Append(convertToSorted(input[0]));
}
```

Listing 2.8. Geometry shader code of the sorting.

symbols such that symbols with the same instancing type also form a continuous interval. Thus, modules with a specific instancing type can be collected using two arithmetic comparisons in the geometry shader. Modules with the current instancing type are emitted by the geometry shader, while others are pruned. As in the generation step, we use a trivial vertex shader and no rasterization. As an output of the geometry shader we use a different structure. To save space, those attributes that are not used in the instancing step (e.g., the symbol ID) are removed (see Listing 2.8).

The naive implementation of the bucket sort fills one bucket in every pass. If we denote the number of modules generated by the production step by N, this approach requires I passes, and N modules go through the pipeline in each pass (see Listing 2.9).

Since N is usually much larger than I, we should optimize the naive approach to reduce the size of the input of each pass. We can achieve this by using the same structure as a binary tree sorting: we recursively divide the modules into two parts, until each part corresponds to only one instancing type. Since only one stream output is allowed, we have to process each level of the binary tree sequentially. The leaves of this tree are the original buckets, thus the total number of passes needed is $2I - 1$, but the average size of the input of a pass will be approximately

$$N\frac{\log I}{(2I - 1)}$$

(assuming that instancing types have equal probability). Note that since all information we need for implementing the shader code that performs these passes is the value of I and the ID of the instanced symbols, this code can be generated automatically. Collecting more than one instancing type in a pass does not require modifications in the shader code, since the IDs are connected, continuous intervals.

```
const int moduleIDStart = grammarDescriptor->getSymbolIDStart();
// Query the number of emitted modules.
D3D10_QUERY_DESC queryDesc;
queryDesc.Query = D3D10_QUERY_SO_STATISTICS;
queryDesc.MiscFlags = 0;
ID3D10Query * pQuery;
device->CreateQuery(&queryDesc, &pQuery);
int offset = 0;
for(int i = 0;
    i < grammarDescriptor->getInstancingTypeNumber(); ++i)
{
   pQuery->Begin();

   // Set interval borders.
   int minID = grammarDescriptor->getInstancedIDInterval(i).min;
   int maxID = grammarDescriptor->getInstancedIDInterval(i).max;
   minIDEffectVar->AsScalar()->SetInt(minID+moduleIDStart);
   maxIDEffectVar->AsScalar()->SetInt(maxID+moduleIDStart);
   sortingShader->GetPassByIndex(0)->Apply(0);

   device->SOSetTargets(1, &instancedModuleBuffer, &offset);
   device->DrawAuto();

   // Get the result of the query.
   pQuery->End();
   D3D10_QUERY_DATA_SO_STATISTICS queryData;
   while(S_OK != pQuery->GetData(&queryData,
      sizeof(D3D10_QUERY_DATA_SO_STATISTICS), 0))
   {}
   instancedModuleNumbers[i] = queryData.NumPrimitivesWritten;
   offset += instancedModuleNumbers[i];
   instancedModuleOffsets[i] = offset;
   pQuery->Release();
}
```

Listing 2.9. CPU code of the naive sorting.

Note that both sorting methods would gain a significant speedup in performance from the possibility to have multiple different stream outputs.

The buckets can be represented as a part of a single vertex buffer, or as separate vertex buffers. Separate vertex buffers can be wasteful if we do not assume anything about the distribution of the generated symbols, since we have to allocate a buffer having the same size as the result of the buffer used in the generation step to ensure that all modules fit into the buckets. Using a single buffer requires us to play with offsets and it needs only the same amount of memory as the buffer used in the generation step.

```
const unsigned int moduleStride = sizeof(SortedModule);
for(int i = 0;
    i < grammarDescriptor->getInstancingTypeNumber(); ++i)
{
   // Set instance buffers.
   device->IASetVertexBuffers(1, 1, &instancedModuleBuffer,
                              &moduleStride,
                              &instancedModuleOffsets[i]);
   // Set the proper rendering technique.
   setRenderingTechnique(i);
   // Rendering the mesh. This sets the vertex buffer as well.
   renderMesh(i,                        // instancing type ID
              instancedModuleNumbers[i] // number of instances
}
```

Listing 2.10. Our CPU implementation of the instancing step in DX10.

2.4.9 Instancing

Now we have the generated object descriptors separated into buckets. Our last task is to render them. An *instancing type*, which is an ordered pair consisting of a mesh and a rendering technique, is assigned to every instanced symbol in the grammar description.

The rendering is performed using instancing; we bind the buckets as instance buffers and the corresponding meshes as vertex buffers (see Listing 2.10). Thus, the number of draw calls needed equals to the number of different instancing types used in the grammar description. Note that the geometry is fully synthesized on the GPU, no readback to the CPU is needed.

Using the instance information in the rendering code is straightforward. For example, assuming that every mesh is centered at the origin and normalized to have unit size, the world position of a vertex in an instance mesh can be calculated in the vertex shader as

```
output.pos.xyz = input.basic_pos;       // mesh coordinates
output.pos.xyz *= input.size;           // instance size
output.pos.xyz += input.inst_pos.xyz;   // instance position
output.pos.w = 1;
```

2.5 Interaction with the Procedural Scene

Interaction with the environment is an important task in many applications (see Figure 2.4). To address this in our procedurally created scenes, we have to evaluate the intersection of an object with the procedural scene. We can use this information later to generate any kind of response. In the following, we refer the

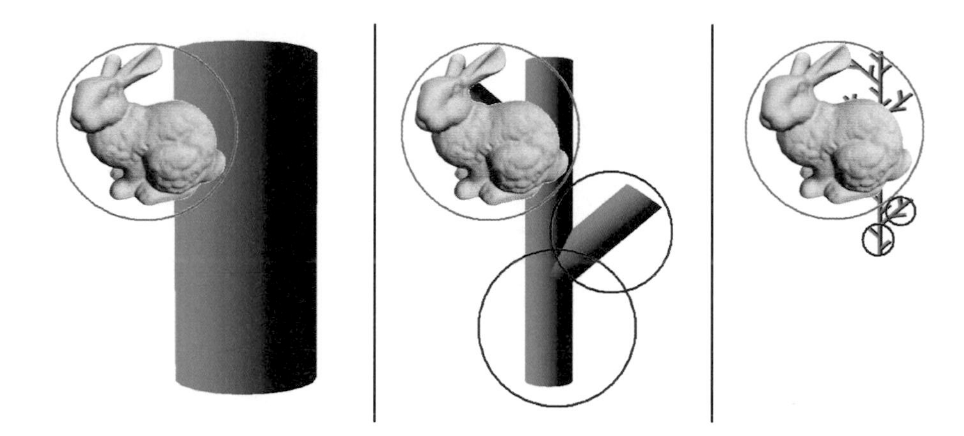

Figure 2.4. Intersection of an active object (the bunny) with a procedural fern. The orange circle shows the bounding sphere of the bunny, the red circles show the culled modules.

nodes of the procedural scene graph as *passive objects* and the objects that the procedural scene interacts with as *active objects*.

By assigning a bounding volume to every node in the procedural scene graph we can obtain a bounding volume hierarchy. This hierarchy is used as usual. During intersection tests, a node in the graph is thrown away, if it has no intersection with the current active object. Implementing this in our model can be achieved by creating a new module type and a new culling strategy. Naturally, it is designed for a large number of active objects (which is common in physical simulations, in games having AI-players, or particle systems), so we can utilize the parallel computing capabilities of the GPU. For only a few active objects, a CPU implementation would be sufficiently fast.

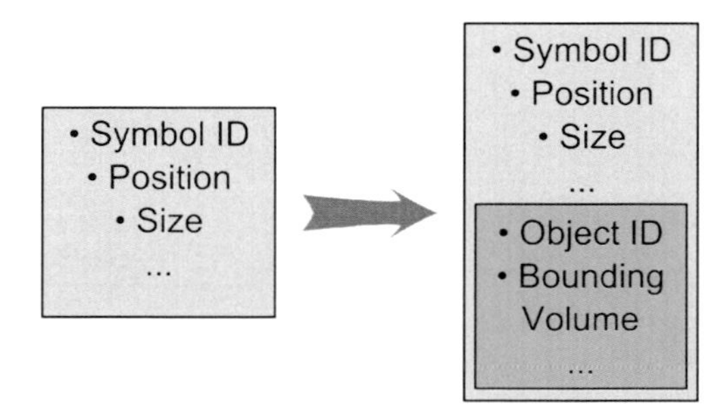

Figure 2.5. Extending the module type with information on an active object.

```
bool cullIntersect( Module module )
{
    float dist = length(module.pos-module.object_pos);
    return dist > module.size+module.object_radius;
}
```

Listing 2.11. HLSL implementation of the new culling strategy in the case of bounding spheres.

First, the original module type is extended with the parameters of the active object. Thus, instead of representing a single node of the procedural scene graph, now a module represents both the node (a passive object) and its intersection with an active object. Figure 2.5 shows the new module type.

If the represented node of the procedural scene does not intersect with the represented active object, the new culling strategy throws away a module. Listing 2.11 shows the HLSL implementation of the culling strategy in the case of bounding spheres.

The rest of the algorithm remains the same. Since we are not interested in rendering the intersections, we perform only the generation step. Every iteration generates the intersection of the active objects with the corresponding level of the procedural scene graph. Therefore, with the last iteration, we obtain the intersection with the leaf nodes.

Figure 2.6. A procedurally created labyrinth using recursive subdivisions and its intersection with balls.

After the GPU has finished the production, we can read back the results to the CPU and process it. The result is a vertex buffer in which, for each active object, there is a vertex for every procedural leaf node it intersects with. Each vertex stores the active object's ID (and its position and bounding volume information which are now irrelevant), and all information about the intersected procedural node. The appropriate collision response can be performed by processing this data sequentially on the CPU.

Figure 2.6 shows an example. The labyrinth is created procedurally using recursive subdivision rules (the exact rules are included with the demo programs). The balls are the (nonprocedural) active objects. The lower right picture shows the intersection of the balls with the scene, which is the result of the generation step.

2.6 Results

We implemented our model using C++, DirectX10, and HLSL. Tests were performed on a system with an ATI Radeon 4850 graphics card, a 3 GHz Intel Core 2 Duo CPU, and 4 GByte RAM.

As a performance test, we measured the times required to perform one iteration of the generation step and the perform the complete sorting steps. The input consisted of 100, 000 vertices in both cases. In our experience, the running times scaled approximately linearly with the size of the input for larger numbers, thus, we can estimate the values for different input size.

2.6.1 Performance of the Generation Step

In most cases, the successor of the rules consists of multiple symbols. This means that the size of the input is at least doubled in an iteration step. Therefore the performance of the whole generation is mostly affected by the last iteration. In other words, we can approximate the running time of the whole generation independently of the grammar and generation depth by measuring only one iteration.

The performance is mostly affected by dynamic branching. Thus we performed an iteration for up to 20 rules for a fixed input size of 100, 000 modules. For every symbol we had one rule, as the number of nonterminal symbols was equal to the number of rules. Rules created two new modules, and set one attribute. Culling and termination check was disabled.

We also tested the effect of the sortedness of the input buffer on the performance. Let us denote the nonterminal symbols with $A_1 ... A_n$. We initialized the vertex buffer as

$$A_1^k A_2^k ... A_n^k A_1^k A_2^k ... A_n^k ...$$

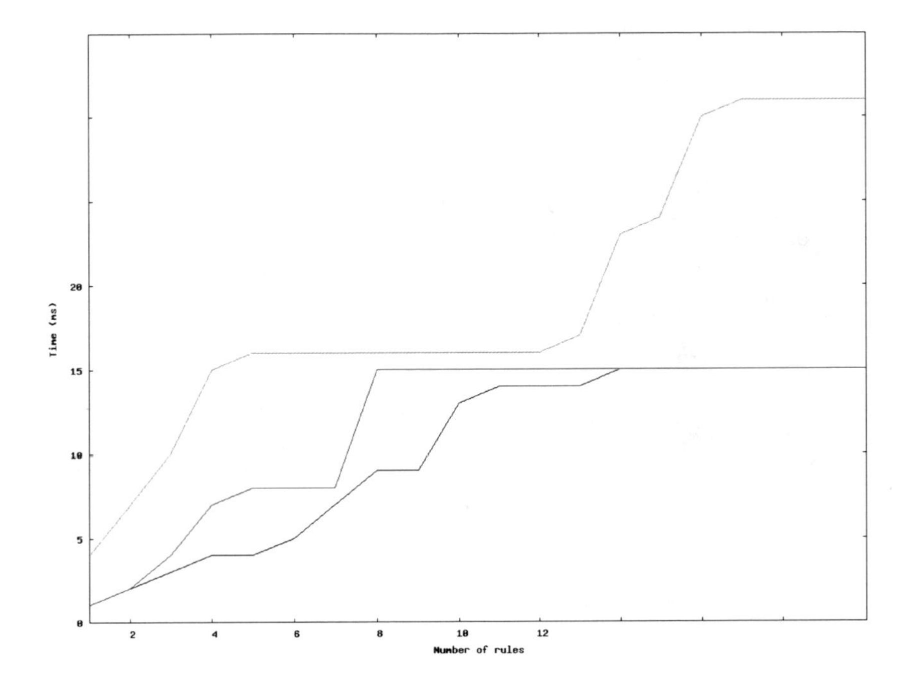

Figure 2.7. Time needed to perform an iteration for 100,000 modules.

where A_i^k denotes repetition of A_i k-times. Thus, the sortedness of the buffer is increased with the value of k.

Figure 2.7 shows our results.

2.6.2 Performance of the Sorting Step

To test the performance of the sorting, we measured the running time of one pass (which collects modules with a specific instancing type or types), which was between 1 and 14 ms, depending on the sortedness. The times for a specific grammar can be approximated using the formulas described in Section 2.4.8.

2.6.3 Examples

To test the complete algorithm including rendering, we designed several simple examples: see Figures 2.8 and 2.9.

Collision detection was measured on the labyrinth example shown in Section 2.4.9. The average FPS rate was 25 for 1000 balls and labyrinth size of 128×128, including readback to the CPU and rendering the scene.

Figure 2.8. Rough model of a city generated from one vertex, using 22 rules, rendered at 22 FPS.

Figure 2.9. Sierpinski pyramids extended with random coloring, rendered at 55 FPS.

2.7 Conclusion

In this article, we proposed a framework for real-time generation of procedural geometry described by context-free L-systems or equivalent models. Our results showed that the system is capable of synthesizing complex geometry with real-time rates.

Additionally, we showed a simple modification to allow discrete collision detection with the procedural scene by calculating the intersection of any given objects with the procedurally generated environment. Our implementation worked well for a large number of objects.

Assuming the rule selection strategies and operators are fixed and previously implemented, every part of both the shader and the CPU code can be generated automatically. Thus, the modelers do not need to learn any programming.

Bibliography

[Ebert et al. 02] David S. Ebert, Kenton F. Musgrave, Darwyn Peachey, Ken Perlin, and Steven Worley. *Texturing and Modeling: A Procedural Approach.* Morgan Kaufmann, 2002.

[Hart 92] John C. Hart. "The Object Instancing Paradigm for Linear Fractal Modeling." In *Proceedings of the Conference on Graphics Interface '92*, pp. 224–231. San Francisco: Morgan Kaufmann Publishers, 1992.

[Lacz and Hart 04] P. Lacz and JC Hart. "Procedural Geometric Synthesis on the GPU." In *Proceedings of the ACM Workshop on General Purpose Computing on Graphics Processors*, 2, 2, pp. 23–23. New York: ACM, 2004. Available online (http://graphics.cs.uiuc.edu/~jch/papers/).

[Lipp et al. 08] Markus Lipp, Peter Wonka, and Michael Wimmer. "Interactive Visual Editing of Grammars for Procedural Architecture." 2008. Article no. 102. Available online (http://www.cg.tuwien.ac.at/research/publications/2008/LIPP-2008-IEV/).

[Müller et al. 06] Pascal Müller, Peter Wonka, Simon Haegler, Andreas Ulmer, and Luc Van Gool. "Procedural Modeling of Buildings." In *SIGGRAPH '06: ACM SIGGRAPH 2006 Papers*, pp. 614–623. New York: ACM, 2006.

[Parish and Müller 01] Yoav I. H. Parish and Pascal Müller. "Procedural Modeling of Cities." In *SIGGRAPH '01: Proceedings of the 28th Annual Conference on Computer Graphics and Interactive Techniques*, pp. 301–308. New York: ACM, 2001.

[Prusinkiewicz and Lindenmayer 91] Przemyslaw Prusinkiewicz and Aristid Lindenmayer. *The Algorithmic Beauty of Plants (The Virtual Laboratory).* Springer, 1991. Available online (http://www.amazon.ca/exec/obidos/redirect?tag=citeulike09-20&path=ASIN/0387972978).

[Sowers 08] B. Sowers. "Increasing the Performance and Realism of Procedurally Generated Buildings." Master's thesis, College of Engineering and Mineral Resources, West Virginia University, 2008.

[Wonka et al. 03] Peter Wonka, Michael Wimmer, François Sillion, and William Ribarsky. "Instant Architecture." *ACM Transactions on Graphics* 22:4 (2003), 669–677. Proceeding. Available online (http://artis.inrialpes.fr/Publications/2003/WWSR03).

GPU-Based NURBS Geometry Evaluation and Rendering

Graham Hemingway

Non-uniform rational B-spline (NURBS) geometry is used in a wide range of three-dimensional applications. NURBS can represent nearly any curve or surface and require only a small set of values to precisely define every point within their domain. Their utility comes from their precise mathematical definition and their relative ease of use compared to other forms of mathematically-defined geometry, such as Bezier or Hermite curves. Increased precision comes at a cost though. NURBS geometry imposes a relatively heavy computational burden. Unlike simple geometry meshes, every NURBS vertex must be individually calculated using a parametric equation and a set of characteristic values. This computational cost has limited NURBS use to applications where the need for precision outweighs the increased computational overhead.

In this chapter we will demonstrate a method for using the GPU to calculate NURBS geometry and make the vertex data readily available for either rendering or further processing. Compared to evaluation on the CPU, this method yields significant performance improvements without drawbacks in precision or flexibility. We will also discuss an extension to the general NURBS evaluation method that allows for calculating trimmed NURBS surfaces. All of the methods discussed are applicable to a wide range of GPUs. Throughout our discussion we will use OpenGL and GLSL to illustrate our examples, but DirectX could just as easily be used.

3.1 A Bit of NURBS Background

Before discussing how to calculate NURBS on the GPU, it is important to have a good working understanding of the mathematics and terminology behind NURBS.

There are numerous approaches to representing curves and surface on computers. The most common in graphics is to use a free-form mesh of vertices. In a mesh, individual points are defined, but the geometry between two points is not fully defined. While this works well in many situations, some applications need to have geometry accurately specified throughout a curve or surface. For example, the process of designing mechanical parts demands extremely precise definition of all geometric entities since inaccuracies from the model can result in unexpected physical phenomenon, errors in manufacturing, and increased design cost.

The two methods most commonly used to mathematically express curves and surfaces are implicit and parametric functions. They provide a rigorous mathematical basis from which to construct geometry. An implicit function for a two-dimensional curve takes the form of $f(x, y) = 0$. The variables x and y vary in a coordinated way and yield a curve on the x, y plane. Similarly, a three-dimensional surface can be represented by an implicit function in the form $f(x, y, z) = 0$.

Parametric functions for two-dimensional curves take the following form:

$$C(u) = (x(u), y(u)) \qquad a \leq u \leq b.$$

$C(u)$ is a vector-valued function of the independent variable u. a and b can be arbitrary, but in general it is assumed that the parametric value varies over the interval $[0, 1]$. The prototypical example of a parametric curve is the one for a unit circle centered at the origin:

$$\begin{aligned} x(t) &= \cos(t), \\ y(t) &= \sin(t). \end{aligned}$$

Parametric equations for surfaces are a simple extension from curves and have the form $S(u, v) = (x(u, v), y(u, v), z(u, v))$. Note that two independent variables, u and v, are now present. The use of two variables gives each surface a rectangular parametric space. This is important to keep in mind throughout the discussion of both NURBS surfaces and later, trimmed surfaces.

NURBS are a class of parametric equations that can represent either curves or surfaces depending on their formulation. As mentioned above, a NURBS curve has one parametric value, while a NURBS surface has two. Several pieces of information are needed to fully define NURBS geometry. Let us first consider what is needed to define a curve through three-dimensional space.

First, we require a set of characteristic points called the *control points*. These points define the general geometric shape of the curve, though the curve generally does not intersect them. In Figure 3.1 the blue dots represent the control points for the purple curve, which is the NURBS curve we are trying to define. The second piece of information is the *knot vector* and will be discussed further below. Finally, we need to specify the set of *control point weights*. These are used to

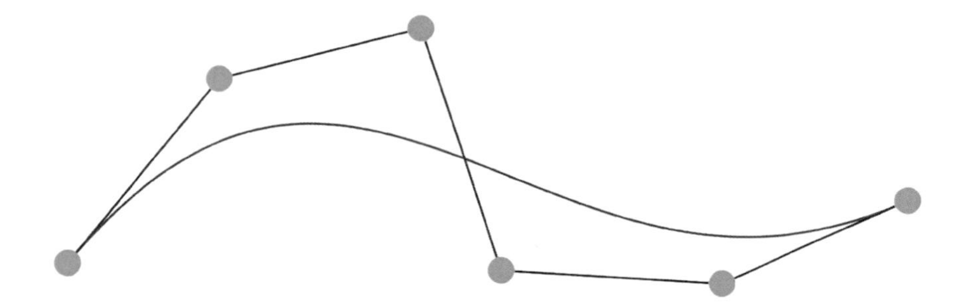

Figure 3.1. A simple NURBS curve

"rationalize" the curve and allow for certain shapes, such as circles or conic sections, to be modeled. With these pieces of information in hand, we can proceed to define the general equation for NURBS curves:

$$C(u) = \frac{\sum_{i=0}^{n-1} N_{i,p}(u) w_i \mathbf{P}_i}{\sum_{i=0}^{n-1} N_{i,p}(u) w_i} \qquad 0 \leq u \leq 1. \tag{3.1}$$

Given some parametric input, u, Equation (3.1) says that the result, $C(u)$, is a point in space with the same number of dimensions as the control points, $\mathbf{P}_i$. If three-dimensional control points are used, then the resulting point is also in three-dimensional space. The degree of the curve, p, determines the number of control points that are taken into account when calculating the result. The control point weights are w_i. The summation is done from 0 to $n-1$, which is the number of control points minus one. The final component of Equation (3.1) is the B-spline basis function, $N_{i,p}$. It is this basis function that gives NURBS geometry its special properties.

Before discussing the basis function we need to get a better understanding of the knot vector. The knot vector dictates how the control points relate to the NURBS curve. There are many different ways to construct a knot vector, but we are mostly concerned with non-periodic and non-uniform knot vectors. Knot vectors frequently take the following form:

$$U = \{\underbrace{a, \ldots, a}_{p}, u_p, \ldots, u_{m-p-1}, \underbrace{b, \ldots, b}_{p}\}. \tag{3.2}$$

The number of knot values, m, in a knot vector is determined by the degree of the curve, p, and the number of control points, n, with the relation $m = n+p+1$. A curve with a knot vector in the form of Equation (3.2) will start and end exactly at the first and last control points. This is achieved through repeating a and b, typically 0 and 1, respectively, p times. The remainder of the knot values are non-decreasing and lie between a and b. An example knot vector for a third degree curve with five control points is $\{0, 0, 0, 0.25, 0.5, 0.75, 1, 1, 1\}$.

Now we are ready to consider the B-spline basis function itself:

$$N_{i,0}(u) = \{ \begin{array}{ll} 1 & \text{if } u_i \leq u < u_{i+1} \\ 0 & \text{otherwise} \end{array}$$

$$N_{i,p} = \frac{u - u_i}{u_{i+p} - u_i} N_{i,p-1}(u) + \frac{u_{i+p+1} - u}{u_{i+p+1} - u_{i+1}} N_{i+1,p-1}(u). \tag{3.3}$$

Equation (3.3) defines the basis function for all NURBS geometry. The top portion of the equation determines if a particular control point will be included in calculating the output value based on the parametric value, u and its relationship to the values in the knot vector, called the *span*. The bottom portion of the equation calculates the actual basis function value, and is quite recursive. This recursiveness is, in large part, the source of the computational complexity involved in evaluating NURBS geometry. The top portion of the equation requires finding which knot values are closest to the parametric value, both above and below. This also causes a good deal of computation to search through the knot vector.

The construction of the B-spline basis function imparts many unique and valuable properties upon NURBS geometry. One of the most important is that the basis function provides local support. This means that if one control point is repositioned, the NURBS curve will only change shape in the area close to that control point. This property makes designing complex geometry with NURBS much easier than other representations that alter the entire curve if any of the control points are altered. There are numerous other interesting properties of NURBS but we leave the reader to explore those.

Equation (3.4) shows that the equation for a NURBS surface is just an extension of that for a NURBS curve. A second parametric value, v, is added which necessitates a second basis function and some additional summation. Once you have become familiar with calculating NURBS curves, moving to surfaces is straightforward:

$$S(u, v) = \frac{\sum_{i=0}^{n-1} \sum_{j=0}^{m-1} N_{i,p}(u) N_{j,q}(v) w_{i,j} \mathbf{P}_{i,j}}{\sum_{i=0}^{n-1} \sum_{j=0}^{m-1} N_{i,p}(u) N_{j,q}(v) w_{i,j}} \qquad 0 \leq u, v \leq 1. \tag{3.4}$$

Calculating trimmed NURBS surfaces is the next step beyond standard NURBS surfaces. Imagine projecting a circle onto a surface and then removing, or simply not rendering, the surface within the circle. This type of complex geometry is very difficult to realize with a single NURBS surface but can easily be created using a trimming profile created from NURBS curves and projected onto an underlying NURBS surface. We will discuss a method for evaluating and rendering trimmed NURBS surfaces later in this chapter.

This has been a very high-level overview of NURBS geometry. Hopefully you now posses the basic concepts and terminology necessary to understand the remainder of this chapter. NURBS are incredibly powerful and can do many things beyond simple curves and surfaces. For further information about NURBS we highly recommend reading [Piegl and Tiller 97] as it is an excellent reference for NURBS background and computation.

3.2 Related Work

There has been a great deal of work on computerized methods for evaluating and displaying NURBS geometry. All methods tend to build on the fundamental NURBS algorithms [Piegl and Tiller 97] to some extent. Algorithms for efficiently rendering NURBS [Abi-Ezzi and Subramanian 94] and for tessellating and rendering trimmed surfaces [Kumar and Manocha 95, Guthe et al. 02, Rockwood et al. 89] have been developed. Prior to the advent of suitable GPUs, most effort was focused on rapid evaluation of the geometry on a traditional CPU.

Once GPUs became sufficiently programmable, methods to exploit their processing power to evaluate NURBS were developed. Starting with [Guthe et al. 05], and expanded upon by [Krishnamurthy et al. 07] and [Kanai 07], there are now a range of approaches for evaluating NURBS geometry. These methods allow for generalized NURBS geometry to be evaluated and rendered using the GPU, thus greatly reducing the computation cost of using precise geometry. In this chapter we demonstrate a slightly modified version of these approaches.

Trimmed NURBS surfaces have specific additional requirements beyond basic NURBS. In [Guthe et al. 05] and [Guthe et al. 06], a fundamental insight was made by proposing to separate the evaluation of the underlying NURBS surface from the generation of a trim texture, which is blended onto the surface at render time. Due to a lack of general programming support in the available GPU APIs at that time, their proposed approach was limited in what could be accomplished on the GPU and significant portions of the algorithm executed on the CPU. In their approach, NURBS geometry is replaced with bi-cubic approximations calculated on the CPU. Since its original publication, additional research has extended this approach. [Krishnamurthy et al. 07] greatly reduces dependencies on CPU precalculation and uses an alpha blending approach for trim-texture generation that allows for non-view port parallel trimming profiles.

3.3 NURBS Surface and Curve Evaluation

Now that you have a working understanding of the mathematics behind NURBS we can move on to how we use the GPU to evaluate and render curves and surfaces. Our method closely follows the approaches in [Krishnamurthy et al. 07] and [Kanai 07]. At a high level, all necessary data is fed into one fragment shader via textures and uniform variables. The shader outputs into a framebuffer object (FBO) with an associated textures in GL_RGBA32F_ARB internal format. The 32-

```
void GenerateCurve(GLfloat start, GLfloat stop, int lod) {
    //Determine step size.
    GLfloat step = (stop - start) / (lod - 1);

    // Set up shader and uniform values.
    glUseProgram(this->_curveProgram);
    glUniform4i(this->_curveParam1, degree, numCP, lod, 1);
    glUniform2f(this->_curveParam2, start, step);

    // Generate the control points texture.
    this->GenerateControlPointsTexture();
    // Create a knotpoint texture.
    this->GenerateKnotPointsTexture();

    // Bind to framebuffer object.
    glBindFramebufferEXT(GL_FRAMEBUFFER_EXT, this->_framebuffer);
    // Set up the viewport and polygon mode.
    glPolygonMode(GL_FRONT_AND_BACK, GL_FILL);
    glViewport(0, 0, lod, 1);
    // Ready the input quad.
    GLfloat quad[] = { -1.0,-1.0,-1.0,1.0,1.0,1.0,1.0,-1.0 };
    glVertexPointer(2, GL_FLOAT, 0, quad);

    // Render the quad.
    glDrawArrays(GL_QUADS, 0, 4);
    // Bind to the PBO.
    glBindBuffer(GL_PIXEL_PACK_BUFFER, buffer);
    glBufferData(GL_PIXEL_PACK_BUFFER, sizeOfBuffer,
         NULL, GL_DYNAMIC_DRAW);
    // Read the pixel data.
    glReadPixels(0, 0, lod, 1, GL_RGBA, GL_FLOAT, NULL);
}
```

Listing 3.1. NURBS curve evaluation.

```
void GenerateKnotPointsTexture(void) {
        // Set up Texture Parameters.
        glPixelStorei(GL_UNPACK_ALIGNMENT, 1);
        // Set up Texture.
        glActiveTexture(GL_TEXTURE2);
        glBindTexture(GL_TEXTURE_RECTANGLE_ARB, this->_knotTex);
        glTexSubImage2D(GL_TEXTURE_RECTANGLE_ARB, 0, 0, 0,
                        this->_numKnots, 1, GL_RGBA,
                        GL_FLOAT, this->_knotVector);
}
```

Listing 3.2. NURBS knot vector texture creation.

bit float texture provides an appropriate level of precision for mathematically defined geometry.

Our curve evaluation function is designed to allow all or part of a curve to be generated at any desired level of detail. We assume the parametric step between vertices is fixed. The primary C++ function, shown in Listing 3.1, takes a starting parametric value, an ending parametric value, and a level of detail value. From these we calculate the step size and what the dimensions of the rendering viewport must be. In order to limit CPU-bound precalculation we rely upon the normal rasterization process to calculate a portion of the parametric values for each vertex. The compiled shader is enabled, and several parameters and textures are passed to the shader via uniform variables. We will cover these variables shortly. Control points and knot vectors are loaded into textures (see Listing 3.2) for use in the shader. A centered unit quad is then rendered into a viewport that is sized to exactly the same dimensions as the desired output texture. This results in each rasterized pixel having a fragment coordinate corresponding to its relative parametric value. Listing 3.6 and its discussion will go into more detail on this point. The remainder of Listing 3.1 is the extraction of the evaluated NURBS curve data from the framebuffer via a pixel buffer object (PBO) that copies the vertex data from the output texture into a VBO buffer thus avoiding any client-side copies. The VBO is then ready to be rendered or used elsewhere.

Calculating NURBS geometry requires several inputs to the shader. As mentioned above, these are passed into the shader using uniform variables and textures. Each control point needs three float values to define it in space and also weight, making a total of four floats per control point that are packed into a 32-bit float texture. Listing 3.3 shows the NURBS curve evaluation fragment shader header which defines all of the global variables. Control points are passed them into a `sampler2DRect`. The *cp(index)* macro makes reading through the remainder of the code a bit easier. The knot vector is just a vector of floats so it too is passed into a `sampler2DRect` and has an access macro.

```
//extension GL_ARB_texture_rectangle : require

//Uniform textures
uniform sampler2DRect knotPoints;
#define kp(index) (texture2DRect(knotPoints, vec2(index)).r)
uniform sampler2DRect controlPoints;
#define cp(index) (texture2DRect(controlPoints, vec2(index)))

//Uniform inputs
uniform ivec4 numParams; // { degree, cp, texWidth, texHeight }
uniform vec2 fltParams;  // { start, step }
float bv[8];
```

Listing 3.3. NURBS curve shader globals.

The remaining variables are necessary to define the many parameters involved in calculating the curve. There is an integer vector of four values: curve degree, number of control points, width of the output texture and height of the output texture—which for curves is always 1. Additionally, there is a float vector of two values: parametric value start and parametric step. These are used to determine what the parametric value actually is and will become more evident in a moment.

In the NURBS background section we mentioned that a key step in calculating NURBS curves is to determine where a given parametric value falls in the knot vector. This value is called the *span* value. This value must be calculated for every parametric value, and therefore for each vertex being evaluated. Listing 3.4

```
int FindSpan(float u) {
   // Check the special case.
   if (u >= kp(numParams.y)) return numParams.y-1;

   // Set up search.
   ivec3 lmh = ivec3(numParams.x, 0, numParams.y);
   lmh.y = (lmh.x + lmh.z) / 2;
   // Start binary search.
   while ((u < kp(lmh.y)) || (u >= kp(lmh.y+1))) {
      if (u < kp(lmh.y)) lmh.z = lmh.y;
      else lmh.x = lmh.y;
      lmh.y = (lmh.x + lmh.z) / 2;
   }
   // Return the span value.
   return lmh.y;
}
```

Listing 3.4. NURBS curve shader span function.

```
void BasisValues(float u, int span) {
    ivec2 jr;
    vec2 tmp;
    vec4 left;
    vec4 right;

    //Basis[0] is always 1.0.
    bv[0] = 1.0;
    //Calculate basis values.
    for (jr.x=1; jr.x<=numParams.x; jr.x++) {
        left[jr.x] = u - kp(span+1-jr.x);
        right[jr.x] = kp(span+jr.x) - u;
        tmp.x = 0.0;
        for (jr.y=0; jr.y<jr.x; jr.y++) {
            tmp.y = bv[jr.y] / (right[jr.y+1] + left[jr.x-jr.y]);
            bv[jr.y] = tmp.x + right[jr.y+1] * tmp.y;
            tmp.x = left[jr.x-jr.y] * tmp.y;
        }
        bv[jr.x] = tmp.x;
    }
}
```

Listing 3.5. NURBS curve shader basis value function.

show the shader code that implements an efficient binary search through the knot vector to find the span value.

The function in Listing 3.5 is for calculating B-spline basis values. The parametric value and the span are passed and the array of basis values are calculated. It is important to note that in this implementation a global float vector, *bv* (defined in Listing 3.3), is used to store the basis values. This vector has eight components; therefore, this implementation is limited to seventh degree curves or lower. This was done because within our application we did not have a need for higher than seventh degree curves, but the implementation could be set arbitrarily high or low.

Finally, we come to the `main()` function of the curve evaluation shader in Listing 3.6. The first task of the shader is to determine this vertex's parametric value. Because the output texture is a rectangle, the `gl_FragCoord` values will be mostly whole numbers. We say *mostly* because some graphics card vendors will use 1.5 instead of 1.0, for example. The `floor()` function compensates for this fact. Using `fltParam.x` (initial parametric value) and `fltParam.y` (parametric step size) and `inPos.x` (number of vertices into the curve) we can calculate u, the actual parametric value for this vertex.

The remainder of the `main` function finds the span value, using the parametric value, then calculates the basis values and the final vertex x-, y-, and

```
void main(void) {
  int j, span, index;

  //Calculate u value.
  vec2 inPos = floor(gl_FragCoord.xy);
  float u = fltParams.x + inPos.x * fltParams.y;

  //Find the span for the vertex.
  span = FindSpan(u);
  BasisValues(u, span);

  //Make sure to zero the results.
  vec3 pos = vec3(0.0);
  float w = 0.0;
  //Loop through each basis value.
  for (j=0; j<=numParams.x; j++) {
    index = span - numParams.x + j;
    w = w + cp(index).a * bv[j];
    pos = pos + cp(index).rgb * cp(index).a * bv[j];
  }
  //Set the position (do the w divide) and write out.
  pos = pos / w;
  gl_FragColor = vec4(pos, 1.0);
}
```

Listing 3.6. NURBS curve shader main function.

z-coordinates implementing Equation (3.1). The final `vec4` value is written out to `gl_FragColor` which will be a pixel within the output texture. This value will then be converted from a pixel to a vertex using the PBO readback as seen in Listing 3.1. Figure 3.2 shows a sketch using a number of NURBS curves, including lines, circles, and arcs.

Evaluating NURBS surfaces is only slightly more complex. Knot vector textures for both u and v must be created and passed into the shader, as do more uniform variables for parametric start and step in the v direction. The only other major change is the need to calculate normals for each vertex. It turns out that the basis values already being calculated can be used to calculate the normals, so little additional computation is needed. Likewise the texture coordinates are the parametric values, u and v, so no additional computation is necessary. The NURBS surface fragment shader renders everything into an FBO with three associated textures, all in `GL_RGBA32F_ARB` internal format. Each piece of output data (vertex, normal, and texture coordinate) is rendered into a separate texture. As with curve evaluation, the textures are copied into separate VBOs using a PBO and `glReadPixels()`.

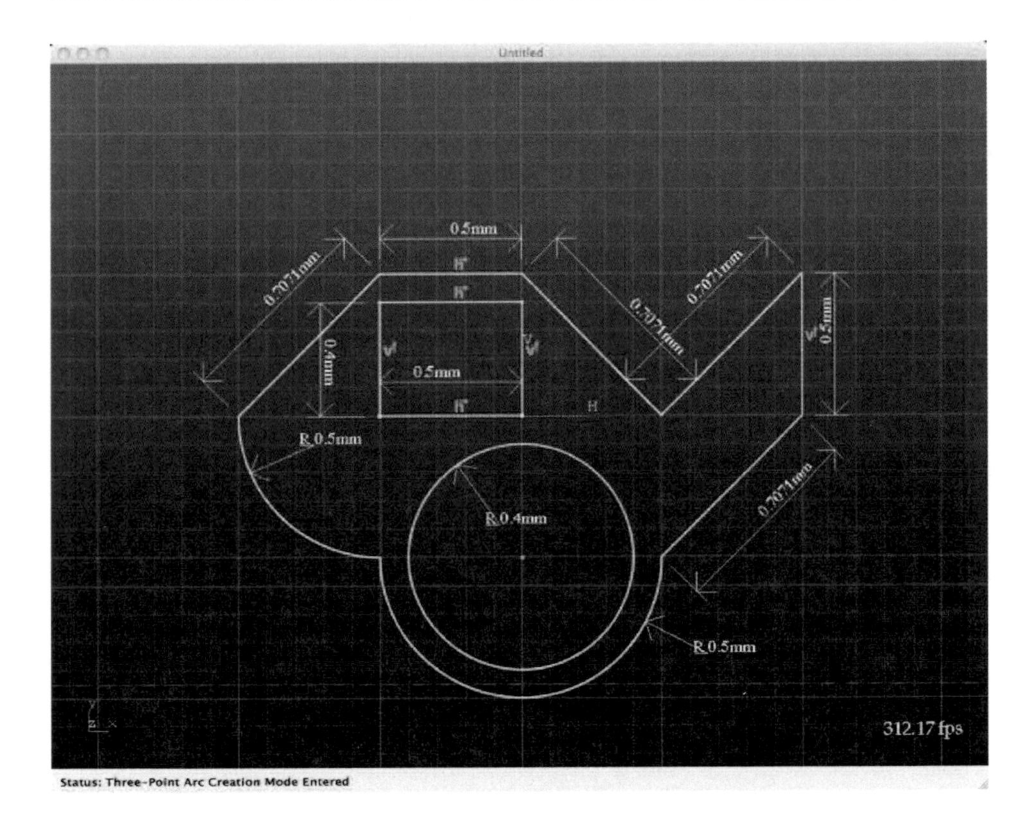

Figure 3.2. A sketch composed of NURBS curves.

3.4 Trimmed NURBS Surface Evaluation

Our approach for evaluating and rendering trimmed NURBS surfaces is comprised of five primary steps. As mentioned above, the separation of the underlying surface from that of the trimming texture is a key insight and our workflow for generating trimmed surfaces reflects this separation.

Figure 3.3 illustrates the steps and products of our approach. The five primary steps involved in our approach all execute fully on the GPU and the resulting outputs are stored as textures or are converted to VBOs. The steps are: generation of NURBS surface vertices, normals, and texture coordinates, generation of trimming profile vertices, inverse evaluation of profile vertices, rendering inverse profile vertices into the trim texture, and visually rendering the final trimmed surface. Each of these steps is covered in the following sections.

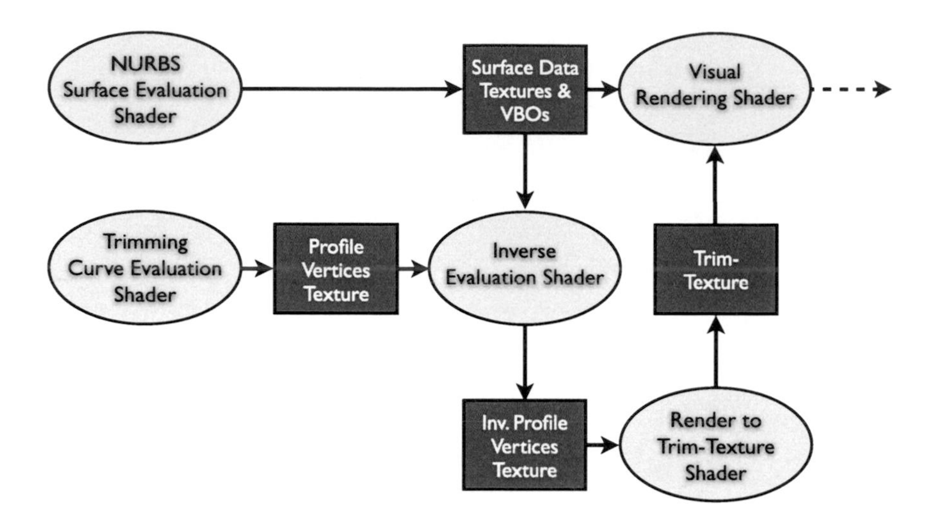

Figure 3.3. Trimmed NURBS surface workflow.

3.4.1 NURBS Surface and Curve Evaluation

The first step in evaluating a trimmed surface is to evaluate the underlying NURBS surface and all of the curves in the trimming profile. The evaluation process is exactly the same as discussed above, except that the final project is not read from the FBO into a VBO via a PBO, but is instead kept as textures. The reason for this is that the evaluated vertices are needed as input into the trim-texture generation process. Our implementation of this process requires that the data be formatted as textures and fed into a shader, so there is no need to copy the data into VBOs.

An additional important point regards the trim profiles. A profile can be composed of numerous different lines or curves and may contain one or more holes. Figure 3.2 is an excellent example as it demonstrates a reasonably complex trimming profile with several lines, curves, and two holes. Each curve in the profile must be individually evaluated at the appropriate level of detail, and all of this vertex information is then assembled into a single data texture. It is important to strictly order the trim profile vertices clockwise around the profile. The need for this will be discussed later.

3.4.2 Trimming Profile Projection

Once both the surface and the trimming profile have been evaluated, the next step is to project the profile vertices onto the surface. Profile projection serves two purposes: to map the trimming profile vertices into the parametric domain of the underlying surface, and to ensure that the trimming profile lies exactly on the surface. Without projecting the profile onto the surface the resulting trim texture can suffer from a inaccuracies and visual artifacts.

Many methods exist for point projection onto parametric geometry, also called *inverse evaluation*. Many of these algorithms use some form of iteration to search the parametric space for the correct value. [Piegl and Tiller 97] proposes a Newton interation-based method based on Equation (3.5) for inverse evaluation of a NURBS curve. In this equation $C(u_i)$ is the value of the curve evaluated at parametric value u_i, and P is the point to be projected:

$$u_{i+1} = u_i - \frac{f(u_i)}{f'(u_i)} = u_i - \frac{C'(u_i) \cdot (C(u) - P)}{C'' \cdot (C(u_i) - P) + |C'(u_i)^2|}. \tag{3.5}$$

Iteration continues until $|C - P| < tol$, $|u_i - u_{i+1}| < tol$, or another stopping tolerance is met. It is straightforward to extend this method to NURBS surfaces. [Krishnamurthy et al. 08] demonstrates a GPU-based approach for inverse evaluation. It uses an axis-aligned bounding box (AABB) algorithm instead of Newton Iteration.

Our approach is relatively simplistic in an attempt to yield an easier to implement, though still high performance, solution. Only one pass through the GPU is necessary to project all profile vertices onto the surface. The pass takes as input the outputs from the previous evaluation step, namely one texture filled with surface vertex data, and one texture filled with profile vertex data. The fragment shader is set up so that each fragment processed corresponds to a single profile vertex resulting in the primary input to the inversion fragment shader being a single profile vertex. While it would be possible to use the iterative Equation (3.5) to find the parametric value, it would require evaluating the surface several times instead of making use of the large set of evaluated surface vertex data already available. Instead of iteration, we use a simple brute force search through the existing surface vertex data.

The first step in the fragment shader is to find the surface vertex closest to the *profile* input vertex, called `pt`. A nested for-loop checks each vertex in the surface and reserves the index of the closest. We name this vertex the *prime* vertex, and its record its location with `markU` and `markV`.

The second step determines the two vertices closest to the *profile* and *prime* vertices, one each in the u and v directions of the surface, named `uPt` and `vPt`. This effectively determines in which quadrant from the *prime* vertex the profile vertex lies. A triangle is formed by *prime*, `uPt` and `vPt`.

The final step projects the *profile* vertex into this triangle and determines its barycentric coordinates. These coordinates are then easily translated into the parametric (u, v) values for the underlying surface which are written to the output texture.

The for-loop search through the entire texture of surface data is by far the most expensive portion of this algorithm. Obviously, the performance impact of inverse evaluation grows in direct proportion to the surface LOD, as a higher LOD demands a larger set of surface vertex data, which corresponds to slower inverse evaluation. On the positive side, through experimentation we have found the visual improvement from elevating surface LOD tends to plateau before performance is noticeably impacted by the inversion process.

One potential drawback of our approach relates to highly non-planar surfaces. In these situations the estimation approach employed by the inversion shader proves to be inadequate. Highly non-planar surfaces tend to have a higher occurrence of multiple surface points being equally close to the *profile* vertex. Imagine a point at the center of a sphere—all locations on the sphere are valid inversions. Our approach looks only for the single closest vertex on the surface, not for a set of such vertices. Accurately generating trimming profiles such that they lie as close to the surface as possible is a practical, though not perfect, remediation for this problem.

```
// Uniform Inputs
uniform sampler2DRect    verts;
uniform sampler2DRect    surfData;
#define sd(i,j)    (texture2DRect(surfData, vec2(i,j)))
uniform ivec2   params;    // { surfWidth, surfHeight }

void main(void) {
   // Variable setup etc.
   vec4 pt = texture2DRect(verts, floor(gl_FragCoord.xy));
   ...
   // Loop through all points on surface.
   for (v=0; v<params.y; v++) {
      for (u=0; u<params.x; u++) {
         // Determine distance from point to surface point.
         surfPt = sd(u,v);
         dist = distance(pt, surfPt);
         // Is this smaller than current smallest?
         if (dist <= minDist) {
            // Capture the location
            markU = u;
            markV = v;
            minDist = dist;
         }
      }
   }
   ...
   // Determine neighborhood points.
   if (markU > 0) {
      left = vec4( sd(markU-1, markV) );
      leftDist = distance(left, pt);
   }
```

```
    else leftDist = 10000.0;
    if (markU < params.x-1) {
        right = vec4( sd(markU+1, markV) );
        rightDist = distance(right, pt);
    }
    else rightDist = 10000.0;
    if (markV > 0) {
        bottom = vec4( sd(markU, markV-1) );
        bottomDist = distance(bottom, pt);
    }
    else bottomDist = 10000.0;
    if (markV < params.y-1) {
        top = vec4( sd(markU, markV+1) );
        topDist = distance(top, pt);
    }
    else topDist = 10000.0;
    ...
    // Find the quadrant and hPt and vPt.
    if (leftDist < rightDist) { hSign = -1.0; hPt = left; }
    else { hSign = 1.0; hPt = right; }
    if (bottomDist < topDist) { vSign = -1.0; vPt = bottom; }
    else { vSign = 1.0; vPt = top; }
    // Convert triangle values to [u,v] using hPt, vPt, and signs.
    ...
    // Record [u,v] value into output texture.
    gl_FragColor = vec4(uValue, vValue, 0.0, 1.0);
}
```

Listing 3.7. Trimmed surface point inversion fragment shader.

3.4.3 Trim Texture Generation

Finally, the trim texture itself can be generated. [Guthe et al. 05] and [Shreiner et al. 05] offer similar methods for rendering concave polygons properly, both using a stencil buffer-based approach. When rendering the trimming profile as a triangle fan, hence the strict clockwise ordering, pixels within the trimming profile will be covered by an odd number of triangles and pixels outside the profile will have an even coverage. Instead of counting coverage, the stencil buffer can flip single bits to track odd versus even coverage. The end result is a properly rendered texture.

After inverse evaluation, all of the profile vertices lie in the $[0,1]$ range in both u and v, and the strictly clockwise ordering of the vertices around the profile has been maintained. A square viewport is set and an FBO is created with a stencil buffer and an attached texture sized according to the desired LOD for the trim texture. A single rendering pass, leveraging the stencil buffer but with no specialized shaders, generates the final trim texture.

```
// Trim texture
uniform sampler2DRect trimTexture;
// Texture size parameters
uniform vec2 texSize;

void main(void) {
   // Calculate trim texture lookup location.
   vec2 inPos = vec2(gl_TexCoord[0].s * texSize.x,
           gl_TexCoord[0].t * texSize.y);
   // Fetch trim texture value.
   vec4 texColor = texture2DRect(trimTexture, inPos);
   // Discard fragment if not present.
   if (texColor.x == 0.0) discard;
   gl_FragColor = gl_Color;
}
```

Listing 3.8. Trimmed surface visualization fragment shader.

3.4.4 Rendering Trimmed Surfaces

Once the vertices of the surface have been evaluated and the trim texture generated, rendering the trimmed surface may occur. For non-trimmed NURBS surfaces, rendering is simple. References to the surface's vertex—normal, index, and texture coordinate VBOs—are set in the GL state machine and a single call, with no specialized shaders, to `glDrawElements()` is made.

Only a slight adjustment to the pipeline is required in order to render a trimmed surface. A small fragment shader is inserted that checks the trim texture to see if the surface is present or not. Listing 3.8 shows the fragment shader necessary to properly display a trimmed surface using its trim texture.

In the shader, after the trim texture color is retrieved from the texture, a quick check is made to see if it is colored or not. If not, the *discard* command is used due to a lower computational cost as compared to setting the alpha value of the fragment to zero. Otherwise the fragment's color is set to the appropriate display color.

3.5 Results and Conclusion

The approach for evaluating and rendering NURBS geometry presented in this paper results in signification performance improvements compared to on the CPU. To evaluate our approach we performed all benchmarks on a MacBook Pro with a 2.53 GHz Intel Core 2 Duo processor, 4GB of RAM, and an NVIDIA GeForce 9600M GT graphics processor with 256MB of video memory at a 1440×900

Figure 3.4. A three-dimensional object composed of NURBS geometry.

screen resolution. Figure 3.4 shows our benchmark case. The model is composed of twelve surfaces, some trimmed and some not, and every edge is outlined with a NURBS curve. Results from our experiments are summarized in Table 3.1.

In order to test regeneration speed the models were initially set to completely regenerate all data every frame. Using a strictly CPU-based implementation we achieved a frame rate of 8.2fps. Using GPU-based generation our approach resulted in a sustained throughput of roughly ten million vertices per second or a frame rate of 107.8fps. These results show that GPU-based geometry and trim-texture generation provide a significant performance increase. More interestingly, if we allowed the models to generate once and then render purely from stored data—being free to rotate and reorient randomly each frame as would be seen

Algorithm	Benchmark Model
CPU-based with constant regen.	8.2fps
GPU w/ constant regen.	107.8fps
GPU w/ minimal regen.	752.4fps

Table 3.1. Performance benchmark results.

in real-world usage—the frame rate jumped to 752.4fps. In other words, not having to regenerate each frame resulted in a further seven fold improvement in performance.

In this chapter we demonstrated how to evaluate and render mathematically well-defined NURBS geometry. Our approach is both flexible and provides much higher performance compared to evaluating the NURBS geometry on the CPU. The code for the work described in this paper is available for download at http://wildcat-cad.googlecode.com.

Bibliography

[Abi-Ezzi and Subramanian 94] S. S. Abi-Ezzi and S. Subramanian. "Fast Dynamic Tessellation of Trimmed NURBS Surfaces." *Computer Graphics Forum* 13:3 (1994), 107–126.

[Guthe et al. 02] Michael Guthe, J. Meseth, and Reinhard Klein. "Fast and Memory Efficient View-Dependent Trimmed NURBS Rendering." In *Pacific Graphis 2002*, 2002.

[Guthe et al. 05] Michael Guthe, Ákos Balázs, and Reinhard Klein. "GPU-based Trimming and Tessellation of NURBS and T-spline Surfaces." *ACM Transactions on Graphics* 24:3 (2005), 1016–1023.

[Guthe et al. 06] Michael Guthe, Ákos Balázs, and Reinhard Klein. "GPU-based Appearance Preserving Trimmed NURBS Rendering." *Journal of WSCG* 14 (2006), 1–8.

[Kanai 07] Takashi Kanai. "Fragment-based Evaluation of Non-Uniform B-spline Surfaces on GPUs." *Computer-Aided Design and Applications* 4:3 (2007), 287–294.

[Krishnamurthy et al. 07] Adarsh Krishnamurthy, Rahul Khardekar, and Sara McMains. "Direct Evaluation of NURBS Curves and Surfaces on the GPU." In *SPM '07: Proceedings of the 2007 ACM symposium on Solid and Physical Modeling*, pp. 329–334, 2007.

[Krishnamurthy et al. 08] Adarsh Krishnamurthy, Rahul Khardekar, Sara McMains, Kirk Haller, and Gershon Elber. "Performing Efficient NURBS Modeling Operations on the GPU." In *SPM '08: Proceedings of the 2008 ACM Symposium on Solid and Physical Modeling*, pp. 257–268. ACM, 2008.

[Kumar and Manocha 95] S. Kumar and Dinesh Manocha. "Efficient Rendering of Trimmed NURBS Surfaces." *Computer-Aided Design* 27:7 (1995), 509–521.

[Piegl and Tiller 97] Les Piegl and Wayne Tiller. *The NURBS Book*, Second edition. Springer-Verlag, 1997.

[Rockwood et al. 89] A. Rockwood, K. Heaton, and T. Davis. "Real-Time Rendering of Trimmed Surfaces." *ACM Transactions on Graphics* 23 (1989), 107–116.

[Shreiner et al. 05] Dave Shreiner, Mason Woo, Jackie Neider, and Tom Davis. *OpenGL Programming Guide*, Fifth edition. Addison Wesley, 2005.

Polygonal-Functional Hybrids for Computer Animation and Games

D. Kravtsov, O. Fryazinov, V. Adzhiev,
A. Pasko, and P. Comninos

4.1 Introduction

The modern world of computer graphics is mostly dominated by polygonal models. Due to their scalability and ease of rendering such models have various applications in a wide range of fields. Unfortunately some shape modeling and

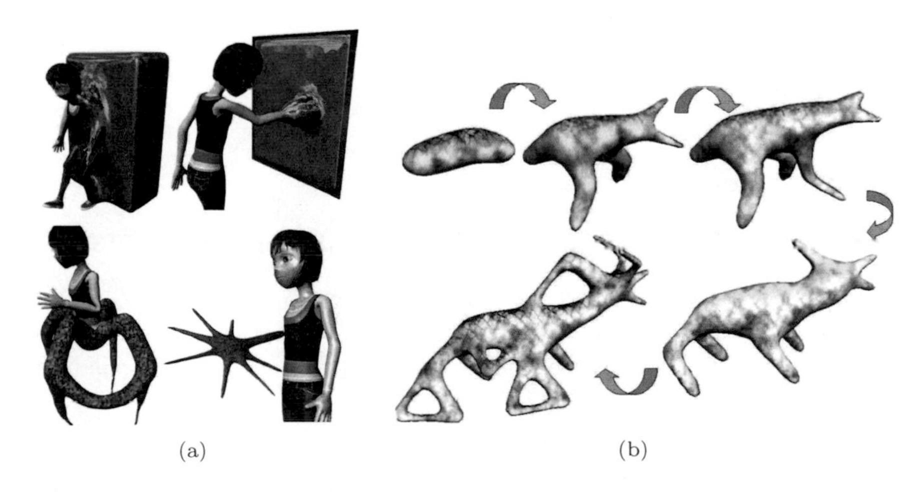

(a) (b)

Figure 4.1. (a) Mimicked viscoelastic behavior and hybrid characters (Model "Andy" is courtesy of John Doublestein); (b) Iterations of character growth controlled by the user.

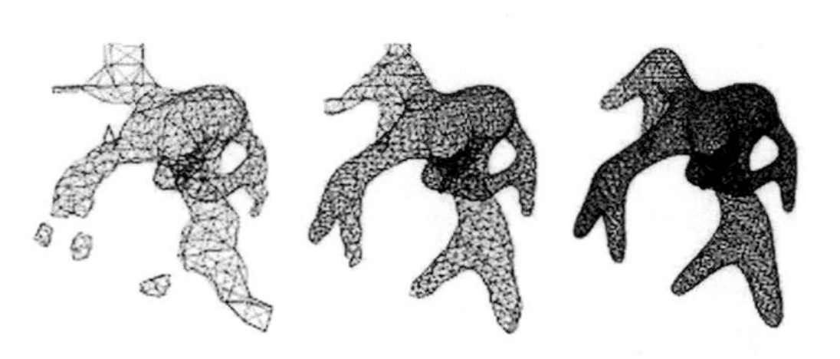

Figure 4.2. Variation of polygonization resolution.

animation problems can hardly be overcome using polygonal models alone. For example, dramatic changes of the shape (involving changes of topology) or metamorphosis between different shapes can not be performed easily. On the other hand, *function representation* (FRep) [Pasko et al. 95] allows us to overcome some of the problems and simplify the process of the major model modification. We propose to use a hybrid model, where we combine together both polygonal and FRep models. Hence we can take advantages of different model representations performing model evaluation entirely on the GPU. Our approach allows us to

- Produce animations involving dramatic changes of the shape (e.g., metamorphosis, viscoelastic behavior, character modifications, etc.) in short times (Figure 4.1(a)).
- Interactively create complex shapes with changing topology (Figure 4.1(b)) and specified LOD (Figure 4.2).
- Integrate existing animated polygonal models and FRep models within a single model.

4.2 Background

4.2.1 Implicit Surfaces and Function Representation (FRep)

FRep [Pasko et al. 95] incorporates implicit surfaces and more generic types of procedural objects. Any point in space can be classified to find out if it belongs to FRep object. FRep object can be defined by a scalar function and an inequality:

$$f(\mathbf{p}) : R^3 \to R \begin{cases} f(\mathbf{p}) > T, \mathbf{p} \text{ is inside the object,} \\ f(\mathbf{p}) = T, \mathbf{p} \text{ is on the object's boundary,} \\ f(\mathbf{p}) < T, \mathbf{p} \text{ is outside the object,} \end{cases}$$

where $\mathbf{p}$ is an arbitrary point in three-dimensional space and T is a threshold value (or isovalue). The subset $\{\mathbf{p} \in R^3 : f(\mathbf{p}) \geq T\}$ is called a solid object and

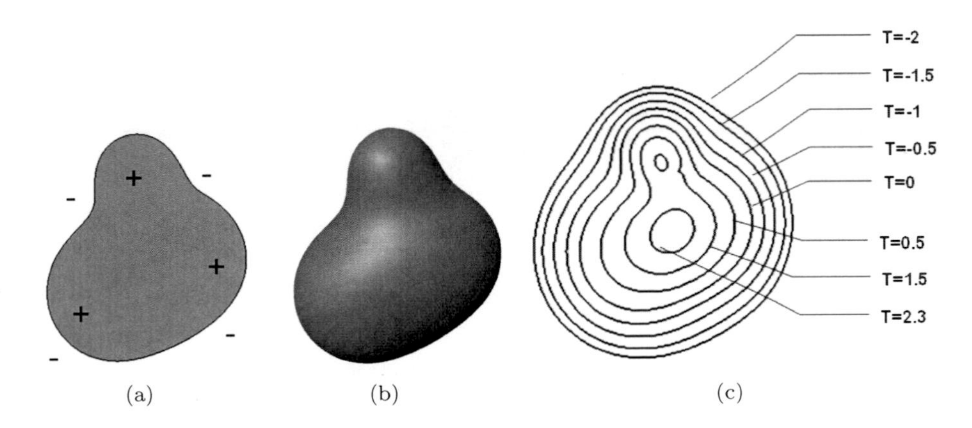

Figure 4.3. Scalar field (defining function): (a) the sign of a scalar field; (b) the extracted implicit surface (T=*0*); (c) different iso-surfaces for different values of T.

the subset $\{\mathbf{p} \in R^3 : f(\mathbf{p}) = T\}$ is called an iso-surface (see Figure 4.3). Function $f(\mathbf{p})$ can be a signed distance field or an arbitrary scalar field.

The first derivative of the function can be used to compute the gradient and the normal on the object's surface:

$$\nabla f(\mathbf{p}) = (\frac{\partial f(\mathbf{p})}{\partial x}, \frac{\partial f(\mathbf{p})}{\partial y}, \frac{\partial f(\mathbf{p})}{\partial z}); \nabla f, \mathbf{p} \in R^3,$$

$$\vec{n}(\mathbf{p}) = -\frac{\nabla f(\mathbf{p})}{\|\nabla f(\mathbf{p})\|}; \vec{n}(\mathbf{p}), \mathbf{p} \in R^3. \tag{4.1}$$

Unfortunately, only a small subset of these models is well known to a wider audience. One of the most popular types of implicit surfaces are blobs [Blinn 82], also known as metaballs or soft objects. Individual blobby objects defined by positions of their centers and radii can be smoothly blended with each other. Superposition of these simple primitives provides an opportunity to build more complex shapes with changing topology, which is usually hard to achieve with purely polygonal models.

Implicit objects are also known for easy definition of metamorphosis sequences (also known as morphing). One only needs to interpolate between values of two signed distance fields to retrieve an intermediate object (Figure 4.4):

$$f(\mathbf{p}, t) = f_\alpha(t) \cdot f_1(\mathbf{p}) + f_\beta(t) \cdot f_2(\mathbf{p}); \mathbf{p} \in R^3, t \in R,$$

where $f_\alpha(t)$ and $f_\beta(t)$ are continuous scalar functions. Parameter t is usually defined on the $[0; 1]$ interval and interpolating functions $f_\alpha(t)$, $f_\beta(t)$ are chosen

Figure 4.4. Metamorphosis sequence.

to satisfy the following constraints:

$$f(\mathbf{p}, 0) = f_1(\mathbf{p}); f(\mathbf{p}, 1) = f_2(\mathbf{p}).$$

Another known advantage of implicit surfaces is an easy implementation of constructive solid geometry (CSG) operations [Ricci 73]. Arbitrary objects can be combined together to produce shapes of high complexity. R-functions-based CSG operations preserving C^1–continuity were applied in FReps [Pasko et al. 95]:

$$S_1 \cup S_2 = f_1(\mathbf{p}) + f_2(\mathbf{p}) + \sqrt{f_1^2(\mathbf{p}) + f_2^2(\mathbf{p})},$$
$$S_1 \cap S_2 = f_1(\mathbf{p}) + f_2(\mathbf{p}) - \sqrt{f_1^2(\mathbf{p}) + f_2^2(\mathbf{p})}.$$

Preserving C^1 continuity of the resulting function may be especially important to overcome rendering artifacts when estimating gradient of the scalar field produced by the object. Another important operation available in FRep is *blending union.* This operation allows to perform smooth blending between two objects controlling the shape of the resulting object:

$$f_b(f_1, f_2) = f_1 + f_2 + \sqrt{f_1^2 + f_2^2} + \frac{a_0}{1 + \left(\frac{f_1}{a_1}\right)^2 + \left(\frac{f_2}{a_2}\right)^2},$$

where a_1 controls the contribution of the first object, a_2 controls the contribution of the second object and a_0 controls the overall "shift" from the resulting object. Blending set-theoretic operations allow us to dramatically change the resulting shape controlling the influence of each of the initial shapes being blended, as well as controlling the overall offset from the resulting shape (see Figure 4.5).

This is a small subset of FRep features that we will use to show a number of interesting applications later.

4.2.2 Convolution Surfaces

Aforementioned metaballs can be considered a subset of so called *convolution surfaces* [Bloomenthal and Shoemake 91]. These surfaces are defined by a lower-dimensional skeleton and a function-defining surface profile:

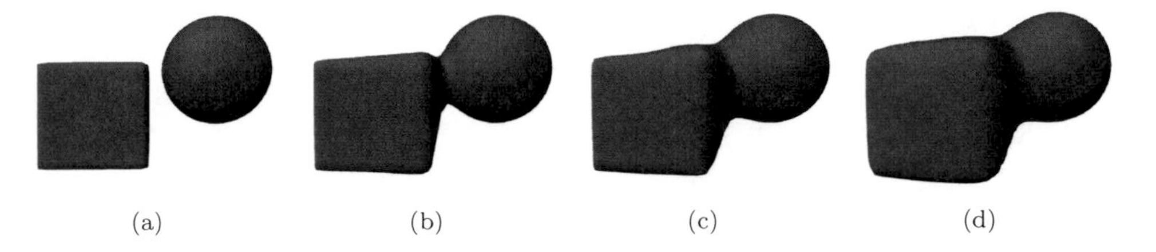

(a) (b) (c) (d)

Figure 4.5. Changing blending parameters.

$$f(\mathbf{p}) = \int_{R^3} g(\mathbf{r})h(\mathbf{p}-\mathbf{r})d\mathbf{r} = g \otimes h,$$

where $g(\mathbf{r})$ defines the geometry of the primitive (i.e., the skeleton function), $h(\mathbf{p})$ is a kernel function (similar to various potential functions used for metaballs), and $g(\mathbf{r})$ equals to "1," if point r belongs to the skeleton and equals to "0" everywhere else. Resulting convolution surface $f(\mathbf{p}) = T$ also depends on the threshold value T.

Convolution surfaces exhibit an important superposition property:

$$(g_1 + g_2) \otimes h = (g_1 \otimes h) + (g_2 \otimes h). \tag{4.2}$$

This means that the field produced by two independent skeletons is the same as the field produced by the combination of these skeletons, i.e., fields produced

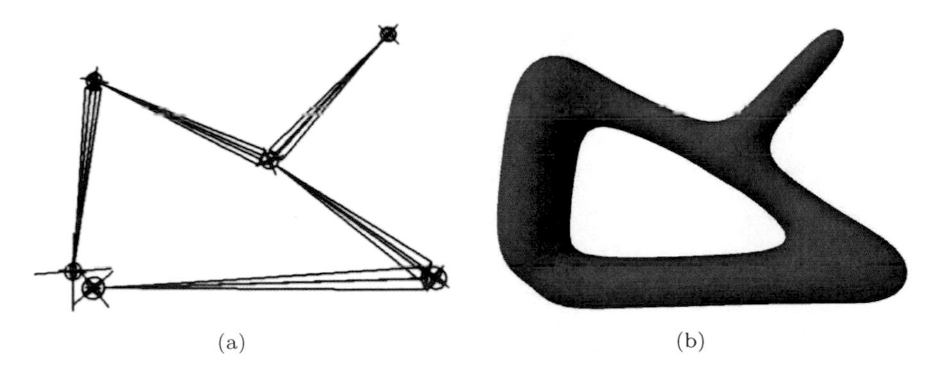

(a) (b)

Figure 4.6. Convolution surfaces: (a) underlying skeleton; (b) produced convolution surface.

by different skeletal elements blend together resulting in a smooth surface. Convolution surfaces can be defined by points, line segments, arcs, and triangles. Analytical solutions were obtained for a number of kernel functions [McCormack and Sherstyuk 98]. We will refer to Cauchy kernel:

$$h(d) = \frac{1}{(1+s^2d^2)^2}; d > 0,$$

where d specifies the Euclidean distance from a point of interest in space and s is a scalar value controlling the radius of the convolution surface. Let us write an equation for a convolution surface produced by a line segment. Given a line segment

$$\mathbf{r}(t) = \mathbf{b} + t\mathbf{a}; 0 \leq t \leq l,$$

where $\mathbf{b}$ is the segment base (position vector), $\mathbf{a}$ is the segment axis (direction vector), and l is the segment length. For an arbitrary point, $\mathbf{p} \in R^3$, the squared distance between $\mathbf{r}(t)$ and $\mathbf{p}$ is defined as

$$d^2(t) = |\mathbf{p} - \mathbf{b}|^2 + t^2 - 2t((\mathbf{p} - \mathbf{b}) \cdot \mathbf{a}).$$

A field function for an arbitrary point $\mathbf{p}$ is then defined as

$$\begin{aligned} f(\mathbf{p}) &= \int_0^l \frac{dt}{(1+s^2d^2(t))^2} \\ &= \frac{x}{2m^2\left(m^2+s^2x^2\right)} + \frac{l-x}{2m^2n^2} + \frac{1}{2sm^3}\left(\arctan\left[\frac{sx}{m}\right] + \arctan\left[\frac{s(l-x)}{m}\right]\right), \end{aligned}$$

where x is the coordinate on the segment's axis,

$$\begin{aligned} x &= (\mathbf{p} - \mathbf{b}) \cdot \mathbf{a}, \\ m^2 &= 1 + s^2(q^2 - x^2), \\ n^2 &= 1 + s^2(q^2 + l^2 - 2lx). \end{aligned}$$

According to Equation (4.2) the field produced by N line segments is defined as follows

$$F(\mathbf{p}) = \sum_{i=1}^{N} f_i(\mathbf{p})$$

where $f_i(\mathbf{p})$ is the field produced by the i-th line segment. An improved version of Cauchy kernel can be used to vary the radius along the line segment [Jin et al. 01].

A few other kernels can be used as well. But polynomial kernels require windowing (i.e., limiting function values within particular intervals), resulting in less smooth convolution surfaces. Besides, evaluation procedure requires more branching instructions, which is often undesirable when performing computations on the GPU. Some other kernels with infinite support are either more computationally expensive or provide less control over the resulting shape. In this article we will only refer to convolution surfaces produced by the line segments using Cauchy kernel.

4.2.3 Rendering FRep Models

Even though FReps have a lot of advantages, visualizing them is not as straightforward as visualizing polygonal models. It is often desirable to convert an FRep object to a polygonal mesh for efficient rendering. One of the well-known methods used for the extraction of isosurfaces from a scalar field is called Marching Cubes [Lorensen and Cline 87]. There are a number of other methods solving the same problem, but Marching Cubes is still popular due to its high speed and ease of implementation.[1] Texturing of isosurfaces requires additional attention as well. Traditional UV-unwrapping is not suitable for complex dynamic models. Known parameterization methods can be applied to calculate UV-coordinates of the extracted surface in real time (for instance, spherical or cubic projection often used for rendering of liquid substances; see Figure 4.7). Triplanar texturing [Geiss 07] provides a better way of texturing of complex functional objects. Another option could be the usage of procedural solid textures implemented in a shader [Ebert et al. 02]. Though pure procedural textures are not always well suited for rendering of arbitrary objects.

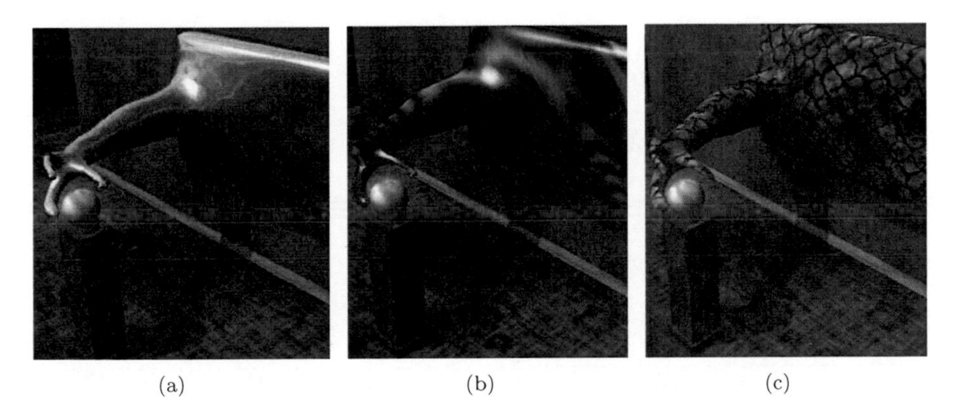

(a) (b) (c)

Figure 4.7. Texturing: (a) cubemap; (b) procedural shader; (c) triplanar texturing.

[1]Original Marching Cubes is known to have a number of ambiguous cases, though.

4.3 Working with FRep Models Using the GPU

Modern GPUs allow evaluation and rendering of certain types of FRep models entirely on the GPU in real time. Three main steps need to be performed:

1. Sample distance function values in the volume and save the results to a temporary buffer.
2. Extract isosurface and its normals from discretized data set.
3. Render extracted isosurface.

Sampling of distance functions can be performed in a vertex shader [Uralsky 06] or a fragment shader. In the latter case, the volume can be sliced with a set of two-dimensional planes or directly rendered to a volume texture. Isosurface extraction from discretized data set can be performed with the help of geometry shaders [Geiss 07,Tatarchuk et al. 07]. All of the above steps can be performed on any DirectX 10 compatible hardware. The code accompanying this article, available on the CRC Press website, is based on NVIDIA CUDA SDK [NVIDIA 09], which allows performing generic computations on the GPU without the necessity to overcome limitations of the existing graphics APIs. Moreover, CUDA SDK already includes an illustrative example of Marching Cubes running on the GPU. We used this code as a starting point for the implementation of our approach running on the GPU.

In the following sections we will describe each of the aforementioned steps in detail.

4.3.1 Function Evaluation

First of all, model parameters need to be updated and uploaded to the GPU. These parameters are stored in the constant memory and need to be modified

```
// Parameters of the segments defining convolution surface
__constant__
CONVOLUTION_SEGMENT segmentsOnDevice[segmentsNumMax];
// Other parameters of the model
...
// Copy segments from CPU to constant GPU memory.
cudaMemcpyToSymbol( segmentsOnDevice, segmentsOnHost,
                    sizeof(CONVOLUTION_SEGMENT));
// Copy other parameters.
...
```

Listing 4.1. Update parameters.

before model evaluation. The `cudaMemcpyToSymbol` function can be used for this purpose, as shown in Listing 4.1.

The volume where the defining function will be evaluated needs to be defined. This volume is uniformly divided in a number of cells according to required resolution. Values of the function at the corners of each cell are evaluated in parallel threads, one thread per value.

We need to save the sampled function values to a temporary buffer in order to avoid function re-evaluations in the future. Writing to global memory is a relatively slow operation and should only be performed if computationally expensive functions are evaluated. Otherwise, the time required to save and load the results may appear to be significantly higher than the time needed for function evaluation (see Section 4.3.2). Depending on the computational complexity of the function being evaluated it may be beneficial to perform more than one evaluation in a kernel and store temporary results in a shared memory. After this, temporary results need to be copied from shared memory to global memory in one instruction, thus achieving coalescing.

In many circumstances high-precision function evaluation is not required, and faster math intrinsics available in CUDA can be used. They can either be called directly or automatically enabled via `use_fast_math` CUDA compiler option (refer to CUDA documentation for more details on the topic).

4.3.2 Isosurface Extraction

We have already mentioned that the Marching Cubes (MC) algorithm is commonly used to extract an isosurface from the scalar field. The MC algorithm works with individual cells uniformly distributed in the volume. The size and the number of the cells is determined by the required quality of the resulting isosurface. The algorithm allows us to find a set of polygons representing a surface patch of the functional object enclosed in each individual cell. Each cell is handled by an independent thread. Here are a number of steps required to efficiently extract an isosurface on the GPU.

1. For each cell,
    (a) Write out the number of vertices it contains.
    (b) Write out the flag indicating whether it contains any geometry.
2. Find the number of nonempty cells.
3. Create a group of all nonempty cells using the flags information from Step 1(b).

4. Generate the table of vertex buffer offsets for nonempty cells.
5. For each nonempty cell,
   (a) Find the number of vertices it outputs.
   (b) Generate vertices of the triangles being output from the cell.
   (c) Generate normal for each vertex being output.
   (d) Save vertices and normals using offset generated at Step 4.

This may look complicated at first, because a number of additional issues arise when performing polygonization on parallel computing devices. First of all, we want to find a set of cells that actually contain geometry in them. Usually the majority of the cells do not contain any geometry, as they are situated completely inside or outside the object, thus having no intersections with the surface of the object. It is important to discard such cells early in order to avoid redundant computations (Step 3). Secondly, each nonempty cell outputs from one to five triangles. For each cell we need to know the offset in the vertex buffer where the vertices will be output. But this offset depends on the number of vertices output from preceding cells. In the case of sequential MC this offset can be iteratively increased, while processing each cell one after another. But it gets more complicated when the cells are processed in parallel. This problem is solved at Step 4 using CUDA Data Parallel Primitives Library [Sengupta et al. 08]:

Step 1. In order to find out whether a cell is empty or not, its MC case index needs to be determined. To do so, we need to retrieve function values at eight corners of each cell and determine its MC case index. At this point we use the data sampled before (see Section 4.3.1), as shown in Listing 4.2.

```
// Based on original source code provided by NVIDIA Corporation

// Get MC case index depending on function values.
__device__ uint getMCIndex(const float* field,float threshold)
{
   uint indexMC;
   indexMC =  uint(field[0] < threshold);
   indexMC |= uint(field[1] < threshold) << 1;
   indexMC |= uint(field[2] < threshold) << 2;
   indexMC |= uint(field[3] < threshold) << 3;
   indexMC |= uint(field[4] < threshold) << 4;
   indexMC |= uint(field[5] < threshold) << 5;
   indexMC |= uint(field[6] < threshold) << 6;
   indexMC |= uint(field[7] < threshold) << 7;

   return cubeindex;
}
```

```
// Sample volume data set at the specified point.
__device__ float sampleVolume(uint3 point, uint3 gridSize)
{
   p.x = min(point.x, gridSize.x - 1);
   p.y = min(point.y, gridSize.y - 1);
   p.z = min(point.z, gridSize.z - 1);
   uint i = (point.z * gridSize.x * gridSize.y) +
            (point.y * gridSize.x) + point.x;
   return tex1Dfetch(volumeTex, i);
}

// Output number of vertices that need to be generated for
// current cell and output flag indicating whether current
// cell contains any triangles at all.
__global__ void
preprocessCell(...)
{
   ...
   float field[8];

   // Retrieve function values at eight corners of a cube.
   field[0]= sampleVolume(gridPos, gridSize);
   field[1]= sampleVolume(gridPos + make_uint3(1,0,0),gridSize);
   field[2]= sampleVolume(gridPos + make_uint3(1,1,0),gridSize);
   field[3]= sampleVolume(gridPos + make_uint3(0,1,0),gridSize);
   field[4]= sampleVolume(gridPos + make_uint3(0,0,1),gridSize);
   field[5]= sampleVolume(gridPos + make_uint3(1,0,1),gridSize);
   field[6]= sampleVolume(gridPos + make_uint3(1,1,1),gridSize);
   field[7]= sampleVolume(gridPos + make_uint3(0,1,1),gridSize);

   // Find out case index in the MC table.
   uint indexMC =  getMCIndex(field, threshold);

   // Read number of vertices produced by this case.
   uint numVerts = tex1Dfetch(numVertsTex, indexMC);
   if (cellIndex < numCells) {
      // Save the number of vertices for later usage.
      cellVerts[cellIndex] = numVerts;
      // Flag indicating whether this cell outputs any triangles
      cells[cellIndex] = (numVerts > 0);
   }
}
```

Listing 4.2. Step 1 of isosurface extraction.

In this case, `numVertsTex` is a table containing the number of triangle vertices contained in a cell corresponding to a certain MC case and `volumeTex` is the sampled volume data that was earlier bound to a one-dimensional texture, as shown in Listing 4.3. It is preferable to fetch data from textures rather than from global device memory as in this case texture cache can be utilized to reduce memory access times.

Step 2. Once output vertex information for each cell has been retrieved, the scan algorithm (also known as "Parallel Prefix Sum," see Listing 4.4) can be used.

```
struct cudaChannelFormatDesc channelDesc =
   cudaCreateChannelDesc( 32,0,0,0,cudaChannelFormatKindFloat );

   cudaBindTexture(0, volumeTex, d_volume, channelDesc);
```

Listing 4.3. Fetching data from textures to utilize texture cache.

This operation allows to generate an array, in which each element contains the sum of all preceding values of the input array (see Figure 4.8) [Sengupta et al. 07].

In our case input array `d_cells` contains either "0" (empty cell) or "1" (nonempty cell). Hence each element of the array generated by an exclusive scan operation (`d_cellsScan`) applied to the input array contains the number of nonempty cells preceding it. The values of such elements can be interpreted as sequential indices of nonempty cell. The last element of the generated array is equal to the total number of all nonempty cells except the last one.

```
// Scan array of nonempty cells.
cudppScan( scanPlanExclusive, d_cellsScan,
           d_cells, numCells);
// Copy the value of the last element from the GPU.
uint CellNumber, lastCellIsEmpty;
cudaMemcpy((void *) &CellNumber,
           (void *) (d_cellsScan + numCells - 1),
           sizeof(uint), cudaMemcpyDeviceToHost);
// Add the value from the last cell as it may be nonempty too.
cudaMemcpy((void *) &lastCellIsEmpty,
           (void *) (d_cells + numCells - 1),
           sizeof(uint), cudaMemcpyDeviceToHost);

// Final number of nonempty cells
CellNumber += lastCellIsEmpty;
```

Listing 4.4. Parallel Prefix Sum.

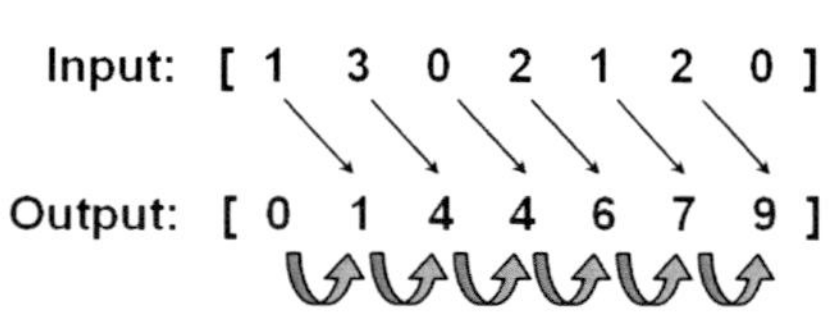

Input: [1 3 0 2 1 2 0]

Output: [0 1 4 4 6 7 9]

Figure 4.8. Exclusive scan.

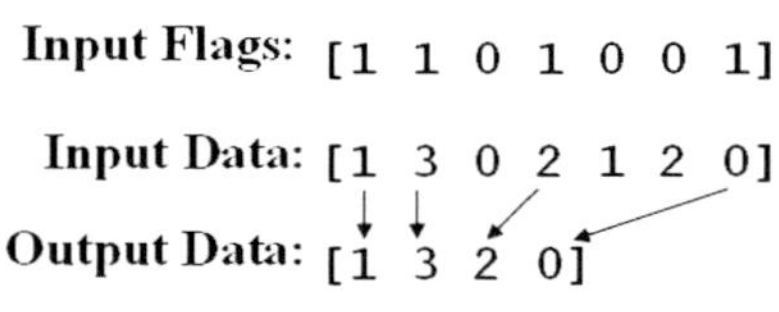

Input Flags: [1 1 0 1 0 0 1]

Input Data: [1 3 0 2 1 2 0]

Output Data: [1 3 2 0]

Figure 4.9. Compact algorithm.

Step 3. Stream compaction (a.k.a. *enumerate operation*) is used to generate an array containing indices of only nonempty cells. "Stream compaction" operation requires two input arrays. First array contains boolean values indicating whether respective elements from the second array need to be copied to the output array. Example input data sets and output generated by this operation are shown in Figure 4.9. In our case we provide an array of flags `d_cells` and a "scanned" array of nonempty cell indices `d_cellsScan`:

```
cudppCompact(compactPlan, d_compactedCells,d_CellNumber,
             d_cellsScan, d_cells, numCells);
```

After this step, `d_compactedCells` contains the set of indices of all nonempty cells.

Step 4. As was mentioned earlier, the generation of vertex buffer offsets (see Listing 4.5) for each cell is performed using scan operation (similar to Step 2). We apply an exclusive scan again, as we need each element in the array to contain the sum of previous elements excluding the current element. The first element in the array of offsets should be equal to "0." After the application of the scan operation each element of `d_cellVBOffsets` contains the total number of vertices contained in preceding cells (i.e. offset in the vertex buffer that can be used to output the vertices from the cell):

```
cudppScan( scanPlanExclusive, d_cellVBOffsets,
           d_cellVerts, numCells);
```

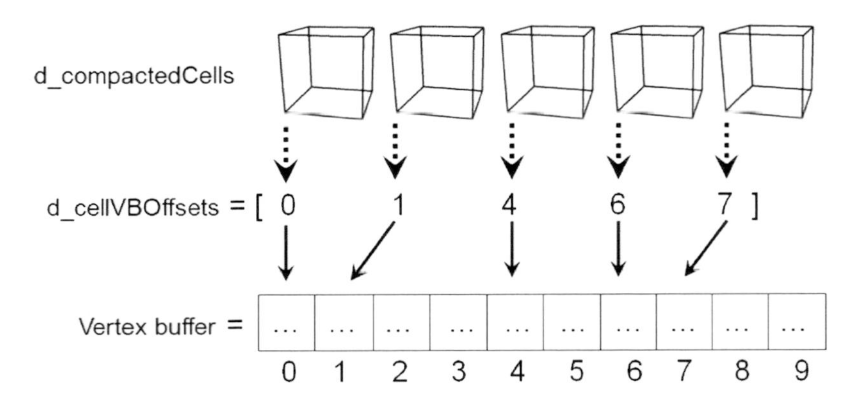

Figure 4.10. Offsets in the vertex buffer for each cell.

```
__device__
float3 interpolatePosition(float threshold,
                           float3 cellVertex1,float3 cellVertex2,
                           float funcValue1, float funcValue2)
{
   float t = (threshold - funcValue1) /
             (funcValue2 - funcValue1);
   return lerp(cellVertex1, cellVertex2, t);
}

// Write out vertices and normals.
__global__ void generateTriangles(...)
{
   ...
   // Vertices placed at the corners of current cell
   float3 cellVertices[8];
   ...
   // Field values at the eight corners of current cell
   float field[8];
   field[0] = sampleVolume(gridPos, gridSize);
   ...
   // Array of vertices placed along 12 edges of the cell
   // shared between different threads
   __shared__ float3 vertices[12 * NTHREADS];

   // Find positions of 12 vertices along all edges:
   vertices[threadIdx.x] = interpolatePosition(threshold,
                              cellVertices[0], cellVertices[1],
                              field[0], field[1]);

   vertices[NTHREADS+threadIdx.x] =
           interpolatePosition(threshold,
                              cellVertices[1], cellVertices[2],
                              field[1], field[2]);
   ...
   // Last vertex
   vertlist[(NTHREADS*11)+threadIdx.x] =
           interpolatePosition(threshold,
                              cellVertices[3], cellVertices[7],
                              field[3], field[7]);
   // Wait while threads are filling ''vertices'' buffer.
   __syncthreads();
   ...
```

Listing 4.5. Generation of vertex buffer offsets.

Step 5. Finally, we can start generating triangles and writing them to the specified vertex buffer. Each vertex being generated is placed along one of the 12 edges of the cell. The function value at each generated vertex is expected to be equal

```
...
// Get the number of triangles that need to be output
// in this MC case.
uint indexMC = getMCIndex(field, threshold);
uint numVerts = tex1Dfetch(numVertsTex, indexMC);

for(int i=0; i < numVerts; i++) {
   // Find the offset of this vertex in the vertex buffer.
   uint vertexOffset = cellVBOffsets[cellIndex] + i;
   if (vertexOffset >= maxVerts) {
      continue;
   }

   // Will get the vertex from the appropriate
   // edge of the cell
   uint edge = tex1Dfetch(triTex, (indexMC << 4) + i);

   // Write out vertex position to VB.
   float3 p = vertlist[(edge*NTHREADS)+threadIdx.x];
   positions[vertexOffset] = make_float4(p, 1.0f);
   // Evaluate normal at this point.
   normals[vertexOffset] = calcNormal(p);
}
```

Listing 4.6. Generating triangles.

to the **threshold** value (in this case vertex is placed on the extracted isosurface) Thus we can linearly interpolate function values along each edge in order to find locations where vertices have to be placed.[2]

You can see that the **vertices** array is placed in shared memory. This is done to decrease the amount of local storage required to run the kernel. Additionally, this memory is accessed in a special way. Each vertex of the cell is placed with a stride of NTHREADS elements, thus helping to avoid bank conflicts between the threads; i.e., consecutive threads access consecutive memory addresses and such memory requests can be serialized.[3] These are well-known optimization techniques often used in CUDA applications.

Once all output vertex positions have been retrieved, they need to be connected to form a set of triangles. We need to find the MC case index again, which is used to read the set of vertex indices from the MC triangles table. After this step a set of vertices and normals can be output to a vertex buffer as shown in Listing 4.6. In this case, the MC triangles table was earlier mapped to a one-dimensional texture **triTex** (in a fashion similar to the mapping of volume texture performed in Step 1).

---

[2]Non-linear interpolation schemes can be used to improve the quality of the resulting mesh. [Tatarchuk et al. 07].

[3]This need not be done for devices with CUDA Compute Capability 1.2 and above.

```
__device__
float4 calcNormal(float3 p)
{
    float f = fieldFunc(p.x, p.y, p.z);
    const float delta = 0.01f;
    // Approximate derivative:
    float dx = fieldFunc(p.x + delta, p.y, p.z) - f;
    float dy = fieldFunc(p.x, p.y + delta, p.z) - f;
    float dz = fieldFunc(p.x, p.y, p.z + delta) - f;
    return make_float4(dx, dy, dz, 0.f);
}
```

Listing 4.7. Generate normals.

Per-vertex analytic normals (see Equation (4.1)) are retrieved using forward differences approximation as shown in Listing 4.7.

It is important to note that polygonization does not have to be performed for each frame. Depending on the available processing power, mesh extraction can be performed only once for a number of frames. Alpha blending between the extracted meshes can be applied to perform smoother transition between them. *Note*: An issue of loading from and saving to global memory was mentioned in Section 4.3.1. From the code provided in this section it can be seen, that in case function values are not stored in the memory, eight function evaluations need to be performed for each cell (see `sampleVolume()` function). For instance, on a $64 \times 64 \times 64$ grid one would need to perform at least two million function evaluations (that is eight evaluations for each of 262,144 cells) only to find out the MC case index for each cell. Add about 10–20% of this number to get the total number of all required function evaluations. This includes actual vertex position and normal calculations for nonempty cells. For a number of example applications that will follow, saving to global memory results in better performance as memory latency is hidden by the expensive function evaluations.

4.3.3 Rendering

Once the vertex buffer has been filled with the geometry information it can be rendered as any conventional polygonal model. We only need to enable a shader making the extracted isosurface look more visually interesting (see Section 4.2.3).

It is worth noting that FRep models can be rendered using ray-casting [Fryazinov et al. 08], thus avoiding the necessity to perform the complex isosurface extraction procedure. At the moment, however, only relatively simple models can be rendered at high resolution in real time [Kravtsov et al. 08].

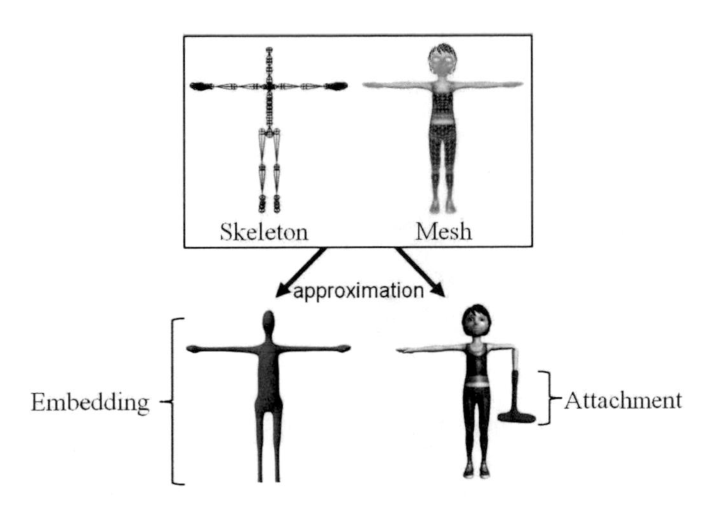

Figure 4.11. Possible approximations.

4.4 Applications

In Section 4.2 we have briefly described a small subset of FReps. In this section we will demonstrate a number of applications (see Figure 4.1) that can be implemented using the combination of FRep objects and polygonal meshes.

4.4.1 Approach Outline

A rigging skeleton is commonly used to animate polygonal meshes. Convolution surfaces described in Section 4.2.2, as well as a number of other implicit surfaces, can also be animated using similar skeletons.[4] We will use a skeleton as a base for the integration of FReps and polygonal models. The cases we will refer to can be generally classified as follows:

1. Embedding an FRep object inside a mesh object or coating mesh objects with FRep objects (Figure 4.11).

2. Attaching an FRep object to the mesh (Figure 4.11).

3. Attaching a polygonal object to the FRep object (Figure 4.19).

We will provide detailed descriptions of each case.

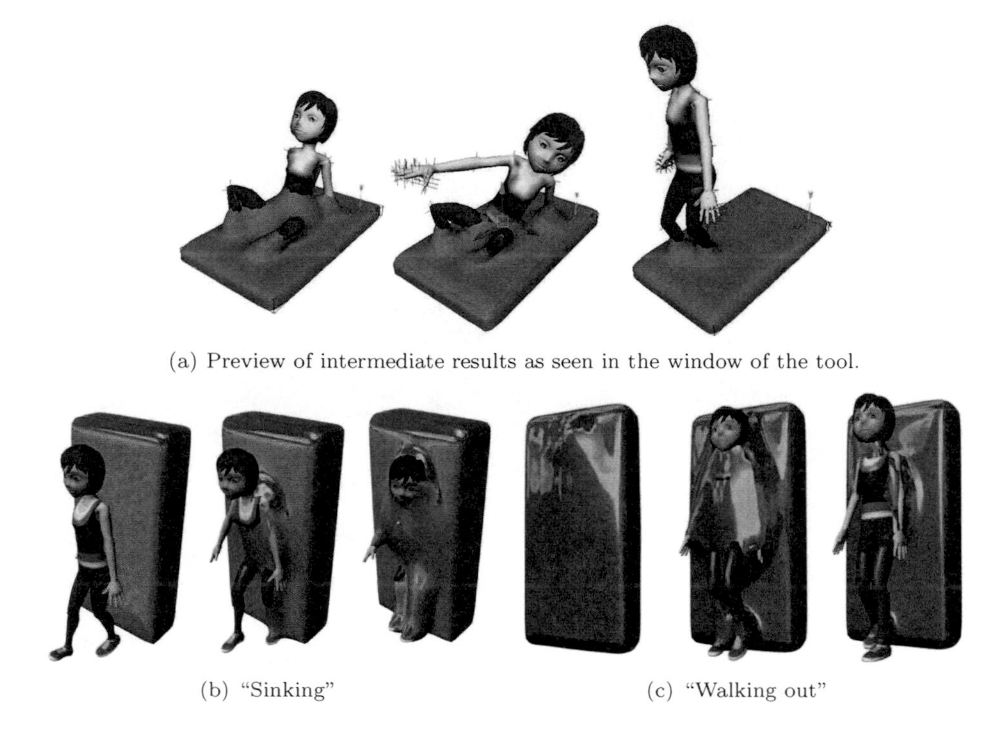

(a) Preview of intermediate results as seen in the window of the tool.

(b) "Sinking" (c) "Walking out"

Figure 4.12. The interaction of an animated object with viscous liquid.

4.4.2 Embedding an FRep Object inside a Mesh Object

The embedding of an FRep object allows us to mimic the interaction of a viscous object with an animated character (Figure 4.12) as well as "supra-natural" behavior of the liquid material (Figure 4.16).

In this case we approximate an animated mesh with a convolution surface using the rigging skeleton. The resulting convolution surface is expected to be completely hidden inside the mesh. In most cases the approximation needs to be performed only once for the character's bind pose. We can estimate the parameters of the embedded convolution surface using the available information. For the initial approximation we use the rigging skeleton. Given the set of bones of the rigging skeleton, where each bone is a line segment in three-dimensional space, we use the set of these segments as the basis for an initial convolution

[4]We have chosen convolution surfaces mainly because of their relatively simple defining function and an absence of bulges and other unwanted artifacts.

Figure 4.13. Synchronization of polygonal and functional objects.

skeleton.[5] To calculate the radius of the convolution surface for each segment, we calculate the minimal distance between each line segment. For the set of rigging skeletal bones $s_i \in \mathbf{S}$ (where $\mathbf{S}$ is a set of skeletal bones) the radius of the ith convolution surface associated with the ith bone is

$$r_i = \min_{p_j \in \mathbf{P}} (\text{dist}(s_i, p_j)),$$

where p_j is the jth face of the polygonal mesh, $\mathbf{P}$ is a connected set of faces and $\text{dist}(s_i, p_j)$ is the distance between the bone s_i and the face p_j. Thus each individual convolution surface is fitted inside the mesh in its initial position.

After the initial approximation, a global optimization is usually required to achieve a better approximation of the given polygonal mesh using the embedded convolution surface (more details are provided in [Kravtsov et al. 08]). An additional embedding optimization step is usually necessary because only individual convolution surfaces are considered at the initial approximation step. In fact, the fields produced by all convolution surfaces sum up, which is equivalent to the increase of the radius of the individual convolution surfaces. This is especially noticeable in the locations near the skeleton branches. Alternatively, instead of global non-linear optimization, an artist can manipulate radius values of individual convolutions to achieve better embedding. Note that, if we wish to apply the blending union operation described in Section 4.2.1, the quality of the initial approximation does not play a significant part in this process.

After the approximation step, the segments of the convolution skeleton are transformed relative to the transformation of the rigging skeleton. Hence, the motion of the convolution surface is synchronized with the motion of the animated mesh (see Figure 4.13). Once the approximation of the animated mesh has been retrieved, we can apply FRep operations to achieve a number of effects (see Figure 4.14).

As the first application of our technique, we mimic the interaction of a viscous object with an animated object using the blending union of two implicit surfaces.

[5] A convolution skeleton does not need to have the same configuration as the original rigging skeleton. For instance, it can have different number of bones or positions of the bones can differ. The only requirement is that the convolution skeleton be defined relative to the rigging skeleton.

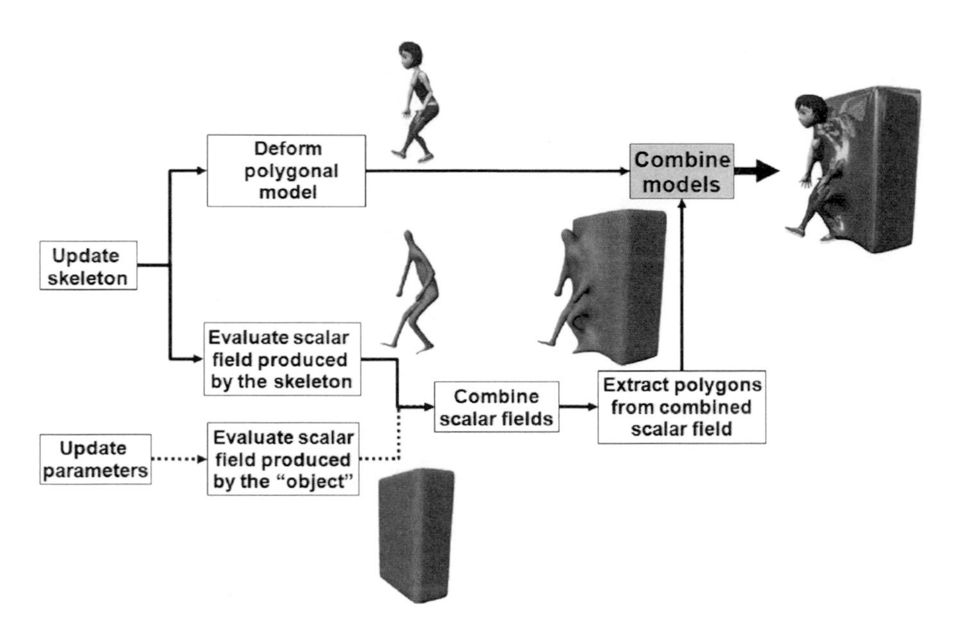

Figure 4.14. Approach outline.

As we mentioned above, the FRep object corresponding to the initial mesh is an embedded convolution surface. The second FRep object, representing the viscous object, can be modeled using a set of implicit primitives. If the defining functions of both objects have distance properties, the shape of the surface resulting from the blending operation depends on the distance between the original objects. The further the objects are from each other the less they are deformed (see Figure 4.15). The behavior of the blended shape visually resembles adhesion, stretching, and breach of the viscous material. If the blended shape is rendered together

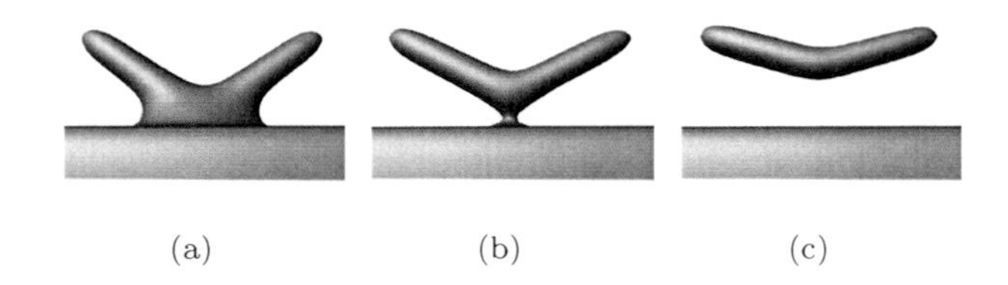

Figure 4.15. Phases of interaction between animated blended objects: (a) two objects and a single blend shape during blending; (b) the boundary case before two shapes separate; (c) two separate shapes with some deformation showing the objects' reciprocal attraction.

with the polygonal mesh, a part of the convolution surface embedded within the mesh becomes visible due to the deformation, thus contributing to the material interacting with the mesh (see Figure 4.12). Therefore, the quality of the initial approximation of the mesh by the convolution surface does not play a significant part in this application. It is more important just to fully embed the convolution surface into the mesh when no deformation is applied. Surely, such an approach is aimed at achieving verisimilitude rather than physically correct results.

The aforementioned approach can be used to model "supra-natural" behavior of the liquid material. In such an animation effect the convolution surface radii are increased over time, which creates the effect of the liquid flowing up the mesh and gradually engulfing it. It is possible to automatically generate this sort of

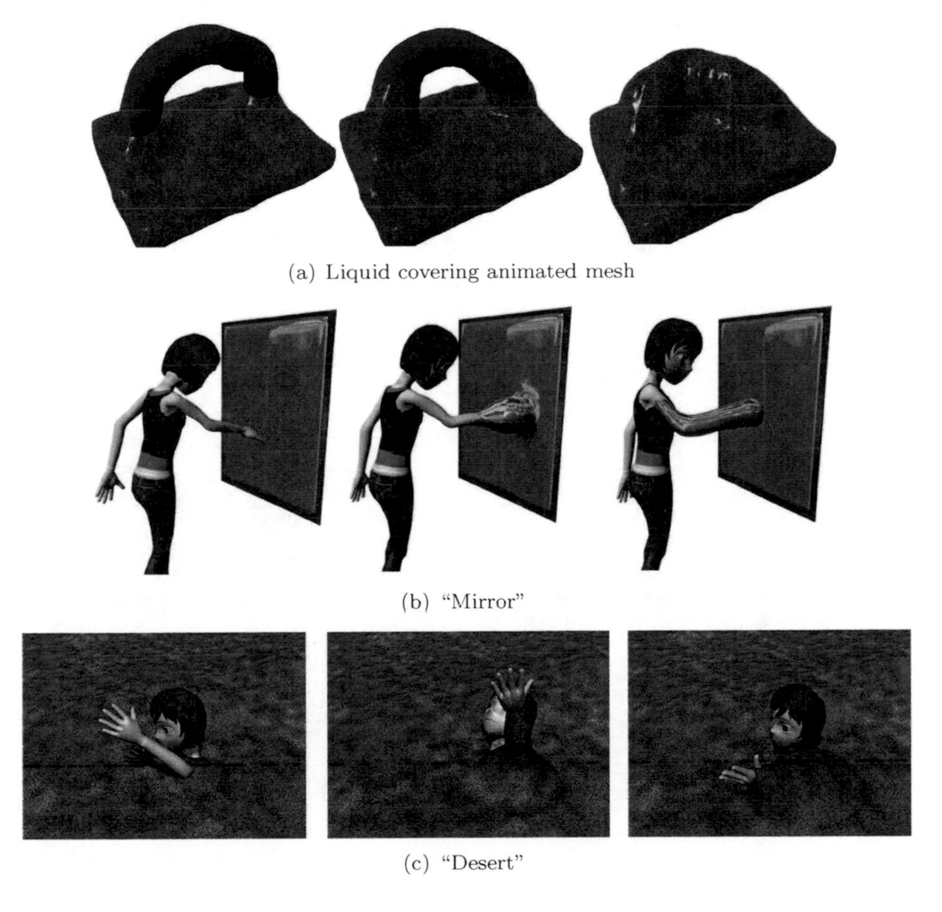

(a) Liquid covering animated mesh

(b) "Mirror"

(c) "Desert"

Figure 4.16. The interaction of an animated object with viscous liquid.

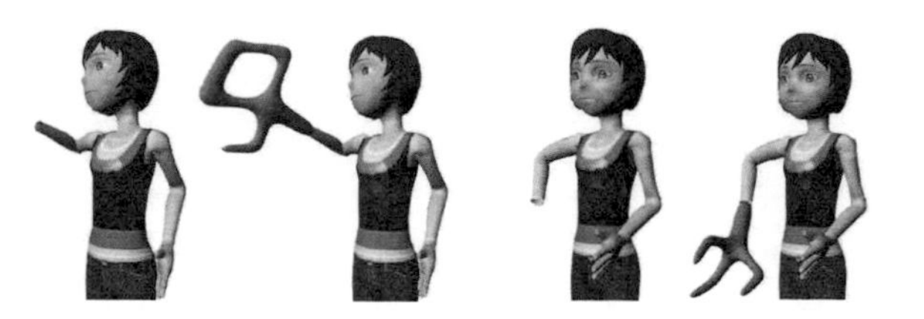

Figure 4.17. Attachment of functional object.

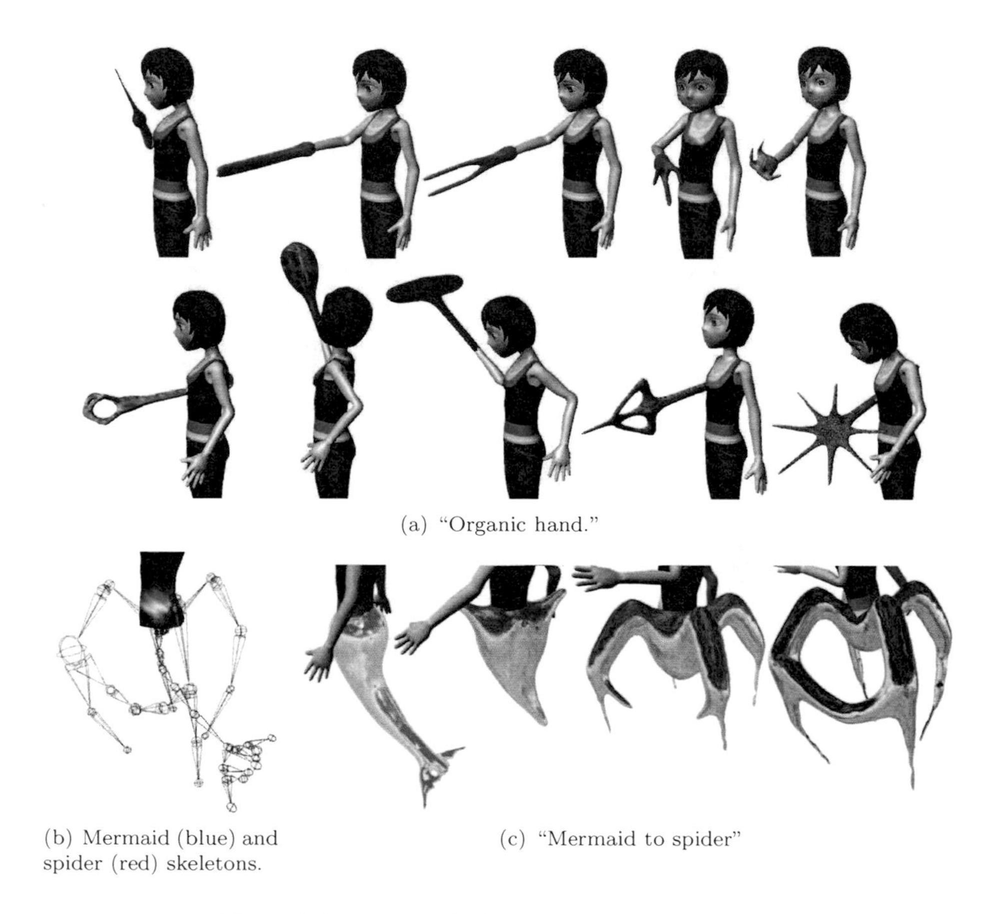

(a) "Organic hand."

(b) Mermaid (blue) and spider (red) skeletons.

(c) "Mermaid to spider"

Figure 4.18. Controlled metamorphosis sequences.

animation. The artist only needs to specify the first and last joint of the skeletal chain as well as the final "thickness" of the liquid flowing over the mesh (see Figure 4.16).

4.4.3 Attaching an FRep Object to the Mesh

In this case we attach an implicit surface to the mesh. To do so, we attach a skeleton defining convolution surface to the rigging skeleton that is animated in a usual way (Figure 4.17). Optionally the implicit surface can be fitted at its boundary attachment to the polygonal mesh. Animation of the skeleton defining the convolution surface leads to the automatic changes of the attached functional object, which means the resulting shape can be dramatically changed. No additional blending is required as convolution surfaces are automatically blended with each other. This approach can be used for the creation of easily metamorphosing parts of animated characters (Figure 4.18(a)). It is also possible to perform metamorphosis between implicit limbs with quite different geometry and topology. The user just needs to specify two skeletons (Figure 4.18(b)) and the time needed to morph from one to another. The intermediate meshes are generated automatically (Figure 4.18(c)).

4.4.4 Attaching a Polygonal Object to the FRep Object

In this case, the skeleton controlling the functional object is defined independently. The FRep object can be placed in the virtual environment as a self-contained entity. Various special effects can be implemented for this entity. The interaction

Figure 4.19. Attachment of polygonal object to the functional object.

Grid resolution for polygonization	"Supra-natural" (11 segments)	Andy (45 segments)	Hybrid Andy (10 segments)
32x32x32	3 ms	9 ms	2 ms
64x64x64	7 ms	22 ms	3 ms
128x128x128	30 ms	95 ms	14 ms

Table 4.1. Average times for mesh generation (milliseconds/frame) on an NVIDIA GeForce 8800 Ultra, 768 MB of RAM. Andy is a mesh model with an embedded convolution surface (Figure 1.12). Hybrid Andy is shown in Figure 1.18(a).

with polygonal objects can be performed using the "implicit skeleton." The polygonal objects can be attached to this skeleton and follow its motion (see Figure 1.19). Collision detection with an FRep object can also be implemented in a relatively simple way, as the scalar field produced by such an object has distance properties.

One can notice that implicit surfaces are a great tool in defining complex dynamic shapes with arbitrary topology. They can also be used for the creation and modification of the user-generated content (similar to EA's "Spore"). The user can define the skeleton and tweak its parameters seeing the resulting shape in real time (see Figure 1.1(b)). After the extraction of the convolution surface, it can be assigned skinning weights and later used in the virtual environment. LODs for such a mesh can be generated automatically via variation of the polygonization grid resolution (Figure 1.2).

We show the times required to evaluate the field and extract the mesh in Table 1.1.

4.5 Tools

Any technique loses its value if no appropriate tools are available for people who are actually producing the content. We wanted to demonstrate that the proposed approach can be employed in a conventional animation pipeline with

Grid resolution for polygonization	"Supra-natural" (11 segments)	Andy (45 segments)	Hybrid Andy (10 segments)
20x20x20	25 ms	80 ms	30 ms
30x30x30	80 ms	220 ms	60 ms
50x50x50	310 ms	930 ms	260 ms
70x70x70	810 ms	2580 ms	670 ms

Table 4.2. Average times for mesh generation (milliseconds/frame) on a PC with a Dual Core Intel Xeon (2.66 GHz), 2 GB of RAM.

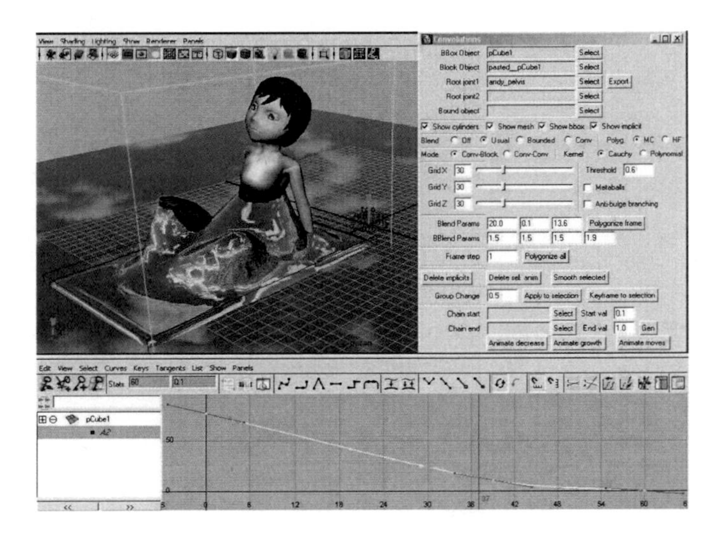

Figure 4.20. A screenshot of the working environment.

near real-time preview without a significant effort. Thus we have implemented our approach as a plug-in for Maya. We have chosen Maya as it is a popular tool for modeling and animation used by a lot of professional artists. Our plug-in performs polygonization on the CPU and feeds the extracted mesh back to the modeling package. All the scenes illustrating the aforementioned ases were defined using a set of developed plug-ins. Even though all calculations are performed on the CPU, intermediate results can be seen in the editor in near real time (Figure 1.20). The actual times for a number of models are shown in Table 1.2. Alternatively, the mesh extracted on the GPU could be copied to RAM and provided to Maya for further manipulation and rendering.

Integration of our technique into an existing animation package can decrease the learning curve for the user. Users are free to produce an animation sequence in a way that they are accustomed to within the familiar software environment, while having an opportunity to see the results of his actions in near real time. Thus, the incorporation of the plug-in in a general-purpose animation software package allows the user to easily integrate the produced animation into complex scenes developed using this package.

4.6 Limitations

The applied blending operation is based on the distance properties of the functions defining the initial geometric objects being blended. The scalar fields produced by known convolution surface kernels significantly decrease as the distance from the line segment increases. At a particular distance from the line segment, the values of such a field are almost equal to zero, and no blend shape is generated at these distances by the blending operation. Thus it is hard to model the interaction between the mesh and the viscous object at large distances. In such cases, an approximation of the mesh with a set of blended quadric surfaces could provide better results. Additionally as the distance between the two blended objects increases, the deformation of the convolution surfaces decreases until these surfaces are again embedded into the polygonal mesh and are no longer visible. The proposed method does not allow us to easily model the separation of droplets of the viscous liquid from the mesh. If this effect is desired, some additional particles modeling this effect could be attached to the mesh. It is also possible to introduce particles to the viscous object. These can improve the visual quality and dynamism of the resulting animation sequence. Simplified particle-based physical models can be applied to the implicit model to improve the default behavior of the viscous object. A metaball representation of the particles is frequently used to integrate these particles into the implicit model. Particle positions retrieved after the physical simulation could be used to add metaballs to the final model. This would allow for partial modeling of physically correct behavior within the existing geometric model.

Also, the proposed approximation for polygonal meshes can only provide good results for typical skeletal characters with axial symmetry. Other types of meshes may require additional efforts in order to achieve better approximation.

4.7 Conclusion

We have outlined a number of advantages of function representation (FRep) and demonstrated a number of applications suitable for computer animation and games. This representation has low memory requirements. Natural resolution independence of the original model allows us to adjust rendering quality according to available hardware specs. The discretization of the model can be performed in parallel, so that it is well suited for modern GPUs and CPUs with an ever increasing number of internal cores.

We believe that FReps have many more useful applications in the fields of computer animation and games.

4.8 Source Code

In the web material, you can find source code implementing the proposed approach based on NVIDIA CUDA SDK release 2.1.

Bibliography

[Blinn 82] James F. Blinn. "A Generalization of Algebraic Surface Drawing." *ACM Transactions on Graphics* 1:3 (1982), 235–256.

[Bloomenthal and Shoemake 91] Jules Bloomenthal and Ken Shoemake. "Convolution Surfaces." *SIGGRAPH Computer Graphics* 25:4 (1991), 251–256.

[Ebert et al. 02] David S. Ebert, Kenton F. Musgrave, Darwyn Peachey, Ken Perlin, and Steven Worley. *Texturing & Modeling: A Procedural Approach*, Third edition. The Morgan Kaufmann Series in Computer Graphics, Morgan Kaufmann, 2002.

[Fryazinov et al. 08] O. Fryazinov, A. Pasko, and Adzhiev V. "An Exact Representation of Polygonal Objects by C^1-continuous Scalar Fields Based on Binary Space Partitioning." Technical Report TR-NCCA-2008-03, The National Centre for Computer Animation, Bournemouth University, UK, 2008.

[Geiss 07] Ryan Geiss. *GPU GEMS 3*, Chapter Generating Complex Procedural Terrains Using the GPU, pp. 7–38. Addison-Wesley Professional, 2007.

[Jin et al. 01] Xiaogang Jin, Chiew-Lan Tai, Jieging Feng, and Qunsheng Peng. "Convolution Surfaces for Line Skeletons with Polynomial Weight Distributions." *Journal of Graphics Tools* 6:3 (2001), 17–28.

[Kravtsov et al. 08] D. Kravtsov, O. Fryazinov, V. Adzhiev, A. Pasko, and P. Comninos. "Embedded Implicit Stand-ins for Animated Meshes: A Case of Hybrid Modelling." Technical Report "TR-NCCA-2008-01", The National Centre for Computer Animation, Bournemouth University, UK, 2008.

[Lorensen and Cline 87] William E. Lorensen and Harvey E. Cline. "Marching Cubes: A High Resolution 3D Surface Construction Algorithm." In *SIGGRAPH '87: Proceedings of the 14th Annual Conference on Computer Graphics and Interactive Techniques*, 21, 21, pp. 163–169. ACM Press, 1987.

[McCormack and Sherstyuk 98] Jon McCormack and Andrei Sherstyuk. "Creating and Rendering Convolution Surfaces." *Computer Graphics Forum* 17:2 (1998), 113–120.

[NVIDIA 09] NVIDIA. Compute Unified Device Architecture (CUDA). 2009.

[Pasko et al. 95] A. Pasko, V. Adzhiev, A. Sourin, and V. Savchenko. "Function Representation in Geometric Modeling: Concepts, Implementation and Applications." *The Visual Computer* 11:8 (1995), 429–446.

[Ricci 73] A. Ricci. "A Constructive Geometry for Computer Graphics." *The Computer Journal* 16 (1973), 157–160.

[Sengupta et al. 07] Shubhabrata Sengupta, Mark Harris, Yao Zhang, and John D. Owens. "Scan Primitives for GPU Computing." In *Graphics Hardware 2007*, pp. 97–106, 2007.

[Sengupta et al. 08] Shubhabrata Sengupta, Mark Harris, and Michael Garland. "Efficient Parallel Scan Algorithms for GPUs." Technical Report NVR-2008-003, NVIDIA Corporation, 2008.

[Tatarchuk et al. 07] Natalya Tatarchuk, Jeremy Shopf, and Christopher DeCoro. "Real-Time Isosurface Extraction Using the GPU Programmable Geometry Pipeline." In *SIGGRAPH '07: ACM SIGGRAPH 2007 courses*, pp. 122–137. New York: ACM, 2007.

[Uralsky 06] Y. Uralsky. "DX10: Practical Metaballs and Implicit Surfaces." Technical report, NVIDIA Corporation, 2006.

Terrain and Ocean Rendering with Hardware Tessellation

Xavier Bonaventura

Currently, one of the biggest challenges in computer graphics is the reproduction of detail in scenes. To get more realistic scenes you need high-detail models, which slow the computer. To increase the number of frames per second, you can use low-detail models; however, that doesn't seems realistic. The solution is to combine high-detail models near the camera and low-detail models away from the camera, but this is not easy.

One of the most popular techniques uses a set of models of different levels of detail and, in the runtime, changes them depending on their distance from the camera. This process is done in the CPU and is a problem because the CPU is not intended for this type of work—a lot of time is wasted sending meshes from the CPU to the GPU.

In DirectX 10, you could change the detail of meshes into the GPU by performing tessellation into the geometry shader, but it's not really the best solution. The output of the geometry shader is limited and is not intended for this type of serial work.

The best solution for tessellation is the recently developed tessellator stage in DirectX 11. This stage, together with the hull and the domain shader, allows the programmer to tessellate very quickly into the GPU. With this method you can send low-level detail meshes to the GPU and generate the missing geometry to the GPU depending on the camera distance, angle, or whatever you want.

In this article we will take a look at the new stages in DirectX 11 and how they work. To do this we will explain a simple implementation of terrain rendering and an implementation of water rendering as it appeared in *ShaderX*^{6} [Szécsi and Arman 08], but using these tools.

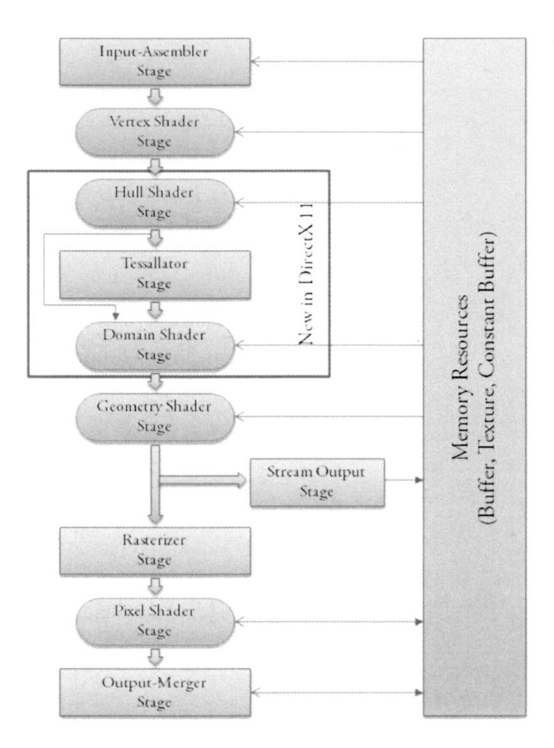

Figure 5.1. DirectX 11 pipeline.

5.1 DirectX 11 Graphics Pipeline

The DirectX 11 graphics pipeline [Microsoft 09] adds three new stages to the DirectX 10: the hull shader stage, tessellator stage, and domain shader stage (see Figure 5.1). The first and third are programmable and the second is configurable. These come after the vertex shader and before the geometry shader, and they are intended to do tessellation into the graphic card.

5.1.1 Hull Shader Stage

The hull shader stage is the first part of the tessellation block. The data used in it is new in DirectX 11, and it uses a new primitive topology called *control point patch list*. As its name suggests, it represents a collection of control points wherein the number in every patch can go from 1 to 32. These control points are required to define the mesh.

The output data in this stage is composed of two parts—one is the input control points that can be modified, and the other is some constant data that will be used in the tessellator and domain shader stages.

To calculate the output data there are two functions. The first is executed for every patch, and there you can calculate the tessellation factor for every edge of the patch and inside it. It is defined in the high-level shader language (HLSL) code and the attribute `[patchconstantfunc(''func_name'')]` must be specified.

The other function, the main one, is executed for every control point in the patch, and there you can manipulate this control point. In both functions, you have the information of all the control points in the patch; in addition, in the second function, you have the ID of the control point that you are processing.

An example of a hull shader header is as follows:

```
HS_CONSTANT_DATA_OUTPUT TerrainConstantHS(
  InputPatch<VS_CONTROL_POINT_OUTPUT, INPUT_PATCH_SIZE> ip,
  uint PatchID : SV_PrimitiveID )

[domain(''quad'')]
[partitioning(''integer'')]
[outputtopology(''triangle_cw'')]
[outputcontrolpoints(OUTPUT_PATCH_SIZE)]
[patchconstantfunc(''TerrainConstantHS'')]
HS_OUTPUT hsTerrain(
  InputPatch<VS_CONTROL_POINT_OUTPUT, INPUT_PATCH_SIZE> p,
  uint i : SV_OutputControlPointID,
  uint PatchID : SV_PrimitiveID )
```

For this example, we clarify some of the elements.

- `HS_CONSTANT_DATA_OUTPUT` is a struct and it must contain `SV_TessFactor` and `SV_InsideTessFactor`. Their types are dependent on `[domain(type_str)]`.
- `INPUT_PATCH_SIZE` is an integer and it must match with the `control point primitive`.
- `[domain(''quad'')]` can be either `''quad''`, `''tri''`, or `''isoline''`.
- `[partitioning(''integer'')]` can be either `''fractional_odd''`, `''fractional_even''`, `''integer''`, or `''pow2''`.
- `[outputtopology(''triangle_cw'')]` can be either `''line''`, `''triangle_cw''`, or `''triangle_ccw''`.
- `OUTPUT_PATCH_SIZE` will affect the number of times the main function will be executed.

5.1.2 Tessellator Stage

The tessellator stage is a part of the new pipeline wherein the programmer, setting some values, can change its behavior although he cannot program it. It is executed once per patch; the input data are the control points and the tessellation factors, which are the output from the hull shader. This stage is necessary to divide a quad, triangle, or line among many of them, depending on the tessellation factor and the type of partitioning defined (Figure 5.2). The tessellation factor on each edge defines how many divisions you want. The output is UV or UVW coordinates which go from 0 to 1 and define the position of new vertices relative to the patch. If the patch is a triangle, then UVW represents the position in barycentric coordinates. If the patch is a quad, then the UV coordinates are the position within the quad.

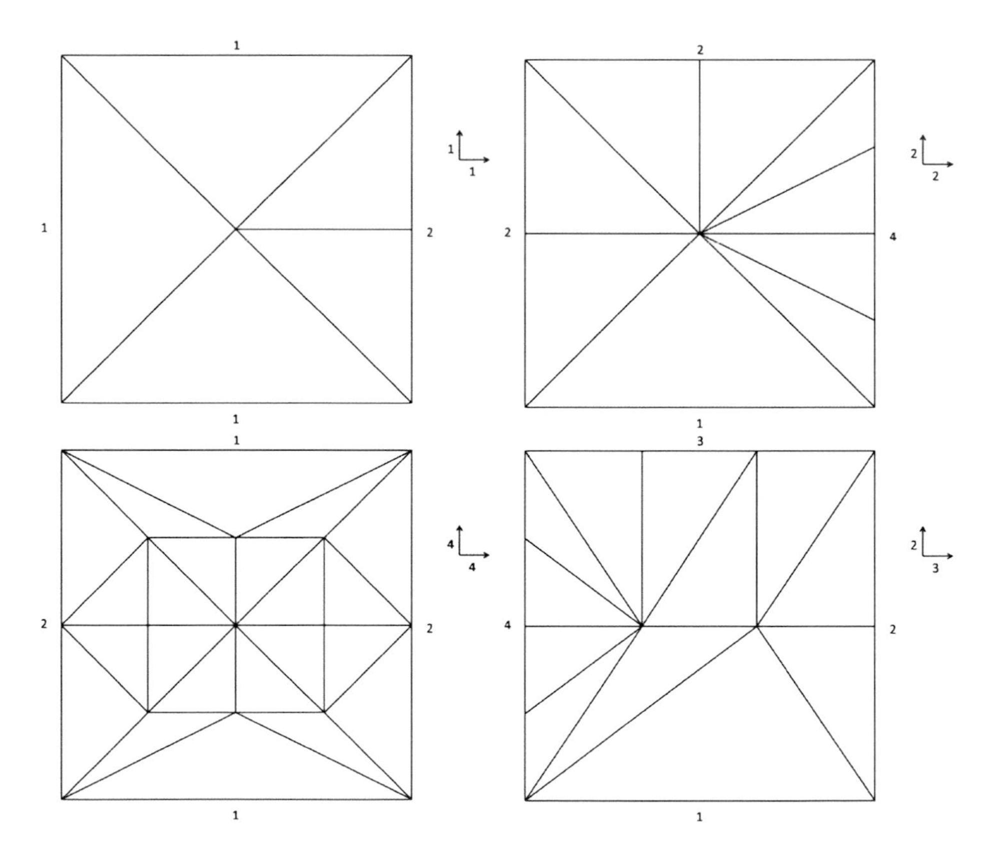

Figure 5.2. Tessellation in a quad domain: to define the tessellation on the edges, we start on the left edge and then we conitnue counterclockwise. To define the tessellation inside, first we input the horizontal and then the vertical tessellation factor.

5.1.3 Domain Shader Stage

The domain shader stage is the last stage of the tessellation. This part is executed once per UV coordinate generated in the tessellator stage, and it accesses the information from the hull shader stage output (control points and constant data) and UV coordinates. In this final stage, the aim is to calculate the final position of every vertex generated and all the associated information as normal, color, texture coordinate, etc.

Below is an example of a domain shader header:

```
[domain(‘‘quad’’)]
DS_OUTPUT dsTerrain(
  HS_CONSTANT_DATA_OUTPUT input,
  float2 UV : SV_DomainLocation,
  const OutputPatch<HS_OUTPUT, OUTPUT_PATCH_SIZE> patch )
```

The type of `SV_DomainLocation` can be different in other domain shaders. If the `[domain(_)]` is `‘‘quad’’` or `‘‘isoline’’`, its type is `float2`, but if the domain is `‘‘tri’’`, its type is `float3`.

5.2 Definition of Geometry

To tessellate the terrain and water, we need to define the initial geometry—to do this, we will divide a big quad into different patches. The tessellation can be

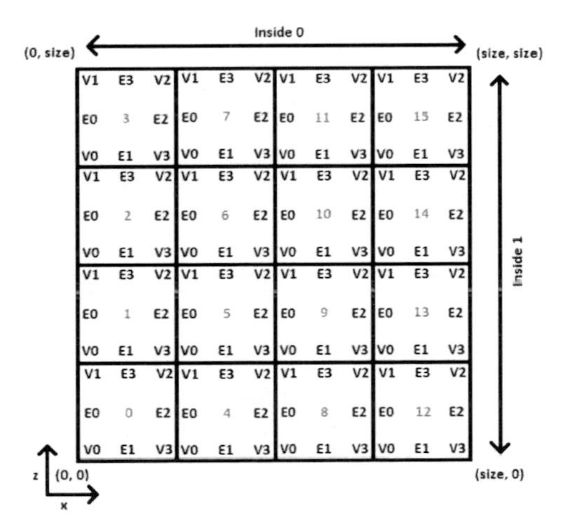

Figure 5.3. Division of the terrain into a grid: Vx and Ex represent vertices and edges, respectively, in every patch where x is the index to access. Inside 0 and Inside 1 represent the directions of the tessellation inside a patch.

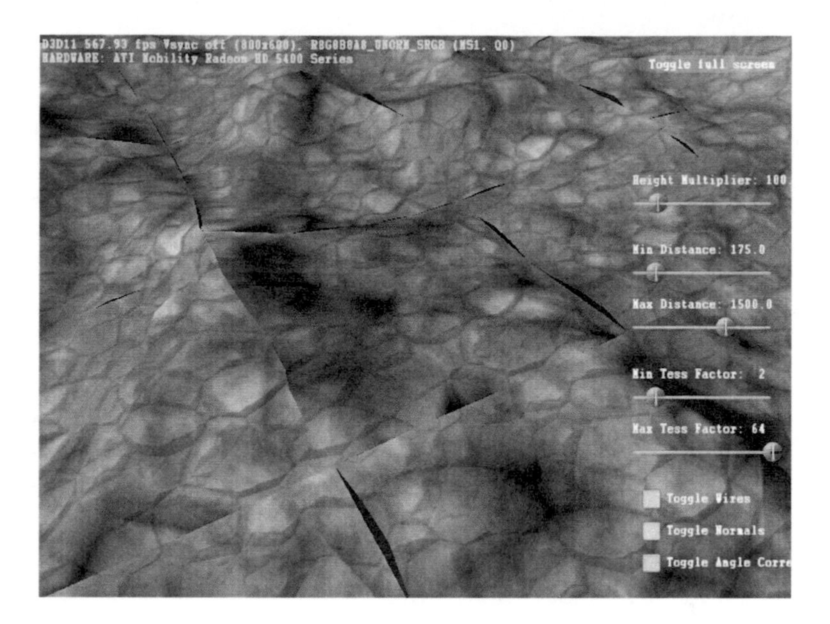

Figure 5.4. Lines between patches when tessellation factors are wrong.

applied to a lot of shapes, but we will use the most intuitive shape: a patch with four points. We will divide the terrain into patches of the same size, like a grid (Figure 5.3). For every patch, we will have to decide the tessellation factor on every edge and inside. It's very important that two patches that share the same edge have the same tessellation factor, otherwise you will see some lines between patches (Figure 5.4).

5.2.1 Tessellation Factor

In the tessellator stage you can define different kinds of tessellations (`fractional_even`, `fractional_odd`, `integer`, or `pow2`). In the terrain rendering we will use the integer, but in addition we will impose one more restriction: this value must be a power of two, to avoid a wave effect. This way, when a new vertex appears it will remain until the tessellation factor decreases again and its x- and z-values will not change. The only value that will change will be the y-coordinate, to avoid popping. In the ocean rendering we will not impose the power-of-two restriction because the ocean is dynamic and the wave effect goes unnoticed.

The tessellation factor ranges from 1 to 64 when the type of partitioning is integer, but the user has to define the minimum and the maximum values. We do not want the minimum of the tessellation factor to be 1 if there are few patches, or the maximum to be 64 if the computer is slow. In the terrain rendering, the minimum and the maximum value must be powers of two.

We also have to define two distances between the camera and the point where we want to calculate the tessellation factor: the minimum and the maximum. When the distance is minimum the tessellation factor is maximum and when the distance is maximum the tessellation factor is minimum.

The tessellation factor will be 2^x where x is a number whose range will be from $\log_2$ MaxTessellation to $\log_2$ MinTessellation linearly interpolated between the minimum and the maximum distance. In terrain rendering, this x will be rounded to the nearest integer to get a power-of-two tessellation factor (see Equation (5.1)). In the ocean rendering, x will not be rounded (see Equation (5.2)).

$$te(d) = \begin{cases} 2^{\max(te_{\log 2})}, & \text{for } d \leq \min(d), \\ 2^{\text{round}(\text{diff}(te_{\log 2})(1-\frac{d-\min(d)}{\text{diff}(d)})+\min(te_{\log 2}))}, & \text{for } \min(d) < d < \max(d), \\ 2^{\min(te_{\log 2})}, & \text{for } d \geq \max(d). \end{cases} \tag{5.1}$$

$$te(d) = \begin{cases} 2^{\max(te_{\log 2})}, & \text{for } d \leq \min(d), \\ 2^{\text{diff}(te_{\log 2})(1-\frac{d-\min(d)}{\text{diff}(d)})+\min(te_{\log 2})}, & \text{for } \min(d) < d < \max(d), \\ 2^{\min(te_{\log 2})}, & \text{for } d \geq \max(d), \end{cases} \tag{5.2}$$

where $\text{diff}(x) = \max(x) - \min(x)$ and d is the distance from the point where we want to calculate the tessellation factor to the camera. The distances defined by the user are $\min(d)$ and $\max(d)$ and $\min(te_{\log_2})$ and $\max(te_{\log_2})$ are the tessellation factors defined by the user. For the tessellation factors, we use the $\log_2$ in order to get a range from 0 to 6 instead of from 1 to 64. The final value $te(d)$ is calculated five times for every patch, using different distances—four for the edges and one for inside.

As we said before, when an edge is shared by two patches the tessellation factor must be the same. To do this we will calculate five different distances in every patch, one for each edge and one inside. To calculate the tessellation factor for each edge, we calculate the distance between the camera and the central point of the edge. This way, in two adjacent patches with the same edge, the distance at the middle point of this edge will be the same because they share the two vertices that we use to calculate it. To calculate the tessellation factor inside the patch in U and V directions, we calculate the distance between the camera position and the middle point of the patch.

5.3 Vertex Position, Vertex Normal, and Texture Coordinates

In the domain shader we have to reconstruct every final vertex and we have to calculate the position, normal, and texture coordinates. This is the part where the difference between terrain and ocean rendering is more important.

5.3.1 Terrain

In terrain rendering (see Figure 5.5) we can easy calculate the x- and z-coordinates with a single interpolation between the position of the vertices of the patch, but we also need the y-coordinate that represents the height of the terrain at every point and the texture coordinates. Since we have defined the terrain, to calculate the texture coordinates, we have only to take the final x- and z-positions and divide by the size of the terrain. This is because the positions of the terrain range from 0 to the size of the terrain, and we want values from 0 to 1 to match all the texture over it.

Once we have the texture coordinates, to get the height and the normal of the terrain in a vertex, we read the information from a heightmap and a normal map in world coordinates combined in one texture. To apply this information, we have to use mipmap levels or we will see some popping when new vertices appear. To reduce this popping we get the value from a texture in which the concentration of points is the same compared to the concentration in the area where the vertex is located. To do this, we linearly interpolate between the minimum and the maximum mipmap levels depending on the distance (see Equation (5.3)).

Four patches that share a vertex have to use the same mipmap level in that vertex to be coherent; for this reason, we calculate one mipmap level for each vertex in a patch. Then, to calculate the mipmap level for the other vertices, we have only to interpolate between the mipmap levels of the vertices of the patch, where $\text{diff}(x) = \max(x) - \min(x)$, $M = \text{MipmapLevel}$, and d is the distance from the point to the camera:

$$\text{Mipmap}(d) = \begin{cases} \min(M), & \text{for } d \leq \min(d), \\ \text{diff}(M)\frac{d-\min(d)}{\text{diff}(d)} + \min(M), & \text{for } \min(d) < d < \max(d), \\ \max(M), & \text{for } d \geq \max(d). \end{cases} \quad (5.3)$$

To calculate the minimum and the maximum values for the mipmap level variables, we use the following equations, where textSize is the size of the texture that we use for the terrain:

$$\min(\text{MipmapLevel}) = \log_2(\text{textSize}) - \log_2(\text{sqrtNumPatch} \cdot 2^{\max(te)})$$
$$\max(\text{MipmapLevel}) = \log_2(\text{textSize}) - \log_2(\text{sqrtNumPatch} \cdot 2^{\min(te)})$$

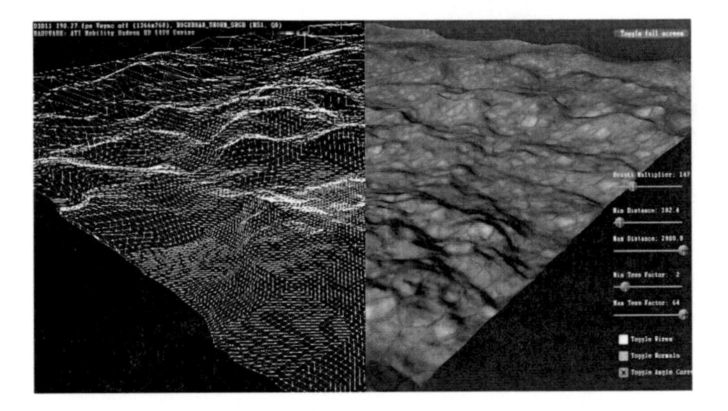

Figure 5.5. Terrain rendering.

We have to keep in mind that we use only squared textures with a power-of-two size. If the minimum value is less than 0, we use 0.

5.3.2 Ocean

Calculating the final vertex position in ocean rendering (see Figure 5.6) is more difficult than calculating terrain rendering. For this, we do not have a heightmap, and we have to calculate the final position depending on the waves and the position in the world coordinate space. To get real motion, we will use the technique explained in *ShaderX*^{6} developed by Szécsi and Arman [Szécsi and Arman 08].

First, we have to imagine a single wave with a wavelength (λ), an amplitude (a), and a direction (k). Its velocity (v) can be represented by

$$v = \sqrt{\frac{g\lambda}{2\pi}}.$$

Then the phase (φ) at time (t) in a point (p) is

$$\varphi = \frac{2\pi}{\lambda}(p \cdot k + vt)$$

Finally, the displacement (s) to apply at that point is

$$s = a[-\cos\varphi, \sin\varphi].$$

An ocean is not a simple wave, and we have to combine all the waves to get a realistic motion:

$$p_\Sigma = p + \sum_{i=0}^{n} s(p, a_i, \lambda_i, k_i).$$

All a_i, λ_i, and k_i values are stored in a file.

Figure 5.6. Ocean rendering.

At every vertex, we need to calculate the normal—it's not necessary to access a normal texture because it can be calculated by a formula. We have a parametric function where the parameters are the x- and z-coordinates that we have used to calculate the position; if we calculate the cross product of the two partial derivatives, we get the normal vector at that point:

$$N = \frac{\partial p_\Sigma}{\partial z} \times \frac{\partial p_\Sigma}{\partial x}.$$

5.4 Tessellation Correction Depending on the Camera Angle

So far we have assumed that the tessellation factor depends only on the distance from the camera; nevertheless, it's not the same if you see a patch from one direction or from another. For this reason, we apply a correction to the tessellation factor that we have calculated previously. This correction will not be applied to the final tessellation factor because we want the final one to be a power-of-two value; this correction will be applied to the unrounded value x that we use in 2^x to calculate the final tessellation factor.

The angle to calculate this correction will be the angle between the unit vector over an edge ($\widehat{e}$) and the unit vector that goes from the middle of this edge to the camera ($\widehat{c}$).

To calculate the correction, we will use this formula (see Figure 5.7):

$$\frac{\frac{\pi}{2} - \arccos\left(\left|\widehat{c} \cdot \widehat{e}\right|\right)}{\frac{\pi}{2}} \text{rank} + 1 - \frac{\text{rank}}{2}.$$

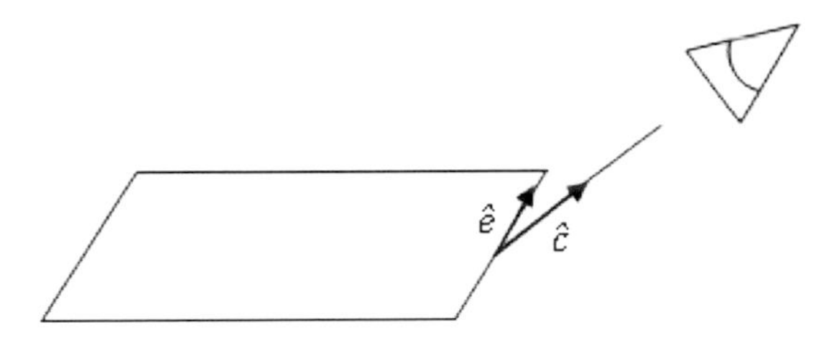

Figure 5.7. Angle correction.

The rank value is used to decide how important the angle is compared with the distance. The programmer can decide the value, but is advised to use values from 0 to 1 to reduce or increase the tessellation factor by 50%. If you decide to use a rank of 0.4 then the tessellation factor will be multiplied by a value between 0.8 and 1.2, depending on the angle.

To be consistent with this modification, we have to apply the correction to the value that we use to access the mipmap level in a texture. It is very important to understand that four patches that share the same vertex have the same mipmap value at that vertex. To calculate the angle of the camera at this point, we calculate the mean of the angles between the camera ($\widehat{c}$) and every vector over the edges ($\widehat{v0}$, $\widehat{v1}$, $\widehat{v2}$, $\widehat{v3}$) that share the point (see Figure 5.8):

$$\frac{\frac{\pi}{2} - \frac{\arccos\left(\left|\widehat{c}\cdot\widehat{v0}\right|\right)+\arccos\left(\left|\widehat{c}\cdot\widehat{v1}\right|\right)+\arccos\left(\left|\widehat{c}\cdot\widehat{v2}\right|\right)+\arccos\left(\left|\widehat{c}\cdot\widehat{v3}\right|\right)}{4}}{\frac{\pi}{2}}\text{rank} + 1 - \frac{\text{rank}}{2}.$$

We don't know the information about the vertex of the other patches for the hull shader, but these vectors can be calculated in the vertex shader because we know the size of the terrain and the number of patches.

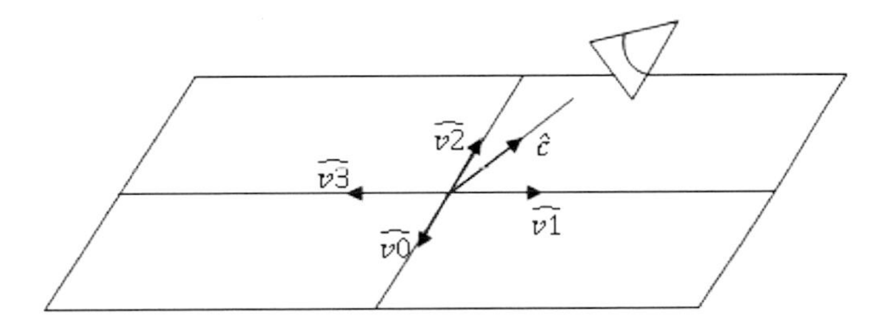

Figure 5.8. Mipmap angle correction.

5.5 Conclusions

As we have shown in this article, hardware tessellation is a powerful tool that reduces the information transfer from the CPU to the GPU. Three new stages added to the graphic pipeline allow great flexibility to use hardware tessellation advantageously. We have seen the application in two fields—terrain and water rendering—but it can be used in similar meshes.

The main point to bear in mind is that we can use other techniques to calculate the tessellation factor, but we always have to be aware of the tessellation factor and mipmap levels with all the patches to avoid lines between them.

In addition, we have seen that it is better if we can use functions to represent a mesh like the ocean, because the resolution can be as high as the tessellation factor sets. If we use a heightmap, as we do for the terrain, it would be possible to not have enough information in the texture, and we would have to interpolate between texels.

Bibliography

[Microsoft 09] Microsoft. "Windows DirectX Grapics Documentation." August, 2009.

[Szécsi and Arman 08] László Szécsi and Khashayar Arman. *Procedural Ocean Effects.* Hingham, MA: Charles River Media, 2008.

Practical and Realistic Facial Wrinkles Animation

Jorge Jimenez, Jose I. Echevarria, Christopher Oat, and Diego Gutierrez

Virtual characters in games are becoming more and more realistic, with recent advances, for instance, in the techniques of skin rendering [d'Eon and Luebke 07, Hable et al. 09, Jimenez and Gutierrez 10] or behavior-based animation.[1] To avoid lifeless representations and to make the action more engaging, increasingly sophisticated algorithms are being devised that capture subtle aspects of the appearance and motion of these characters. Unfortunately, facial animation and the emotional aspect of the interaction have not been traditionally pursued with the same intensity. We believe this is an important aspect missing in games, especially given the current trend toward story-driven AAA games and their movie-like, real-time cut scenes.

Without even realizing it, we often depend on the subtleties of facial expression to give us important contextual cues about what someone is saying, thinking, or feeling. For example, a wrinkled brow can indicate surprise, while a furrowed brow may indicate confusion or inquisitiveness. In the mid-1800s, a French neurologist named Guillaume Duchenne performed experiments that involved applying electric stimulation to his subjects' facial muscles. Duchenne's experiments allowed him to map which facial muscles were used for different facial expressions. One interesting fact that he discovered was that smiles resulting from true happiness utilize not only the muscles of the mouth, but also those of the eyes. It is this subtle but important additional muscle movement that distinguishes a genuine, happy smile from an inauthentic or sarcastic smile. What we learn from this is that facial expressions are complex and sometimes subtle, but extraordinarily important in conveying meaning and intent. In order to allow artists to create realistic, compelling characters, we must allow them to harness the power of subtle facial expression.

[1] Euphoria NaturalMotion technology

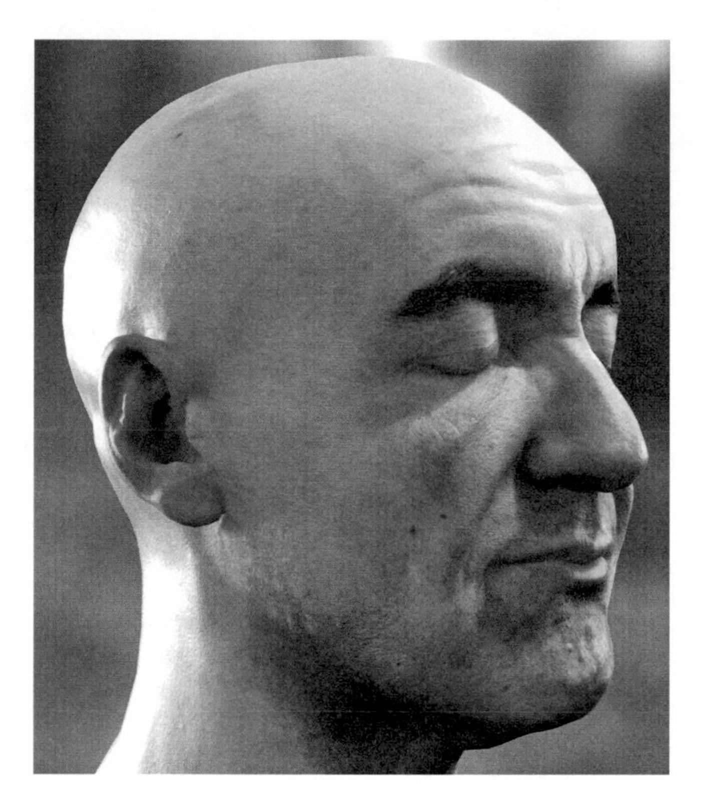

Figure 6.1. This figure shows our wrinkle system for a complex facial expression composed of multiple, simultaneous blend shapes.

We present a method to add expressive, animated wrinkles to characters' faces, helping to enrich stories through subtle visual cues. Our system allows the animator to independently blend multiple wrinkle maps across regions of a character's face. We demonstrate how combining our technique with state-of-the-art, real-time skin rendering can produce stunning results that enhance the personality and emotional state of a character (see Figures 6.1 and 6.2).

This enhanced realism has little performance impact. In fact, our implementation has a memory footprint of just 96 KB. Performance wise, the execution time of our shader is 0.31 ms, 0.1 ms, and 0.09 ms on a low-end GeForce 8600GT, mid-range GeForce 9800GTX+ and mid-high range GeForce 295GTX, respectively. Furthermore, it is simple enough to be added easily to existing rendering engines without requiring drastic changes, even allowing existing bump/normal textures to be reused, as our technique builds on top of them.

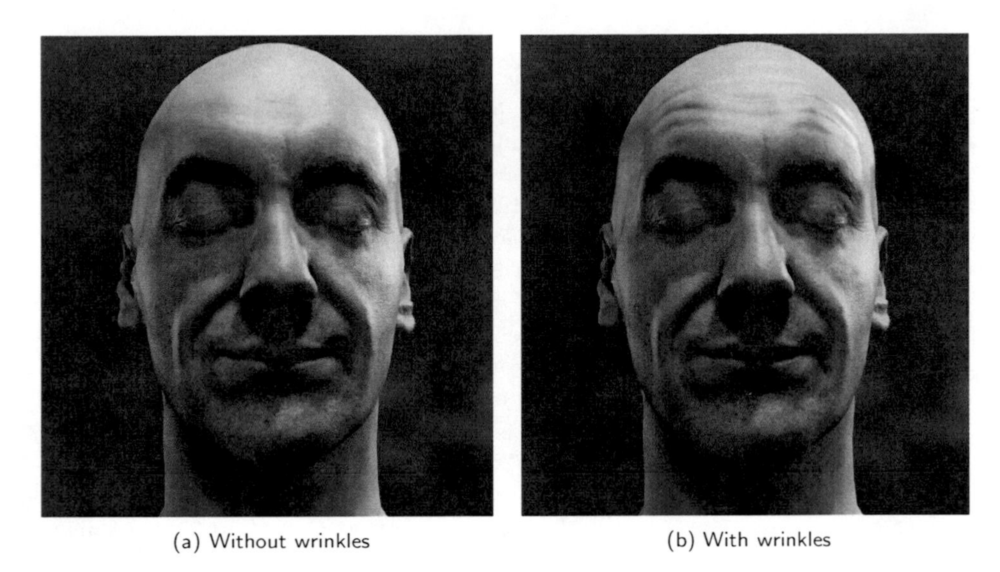

(a) Without wrinkles (b) With wrinkles

Figure 6.2. The same scene (a) without and (b) with animated facial wrinkles. Adding them helps to increase visual realism and conveys the mood of the character.

6.1 Background

Bump maps and normal maps are well-known techniques for adding the illusion of surface features to otherwise coarse, undetailed surfaces. The use of normal maps to capture the facial detail of human characters has been considered standard practice for the past several generations of real-time rendering applications. However, using static normal maps unfortunately does not accurately represent the dynamic surface of an animated human face. In order to simulate dynamic wrinkles, one option is to use length-preserving geometric constraints along with artist-placed wrinkle features to dynamically create wrinkles on animated meshes [Larboulette and Cani 04]. Since this method actually displaces geometry, the underlying mesh must be sufficiently tessellated to represent the finest level of wrinkle detail. A dynamic facial-wrinkle animation scheme presented recently [Oat 07] employs two wrinkle maps (one for stretch poses and one for compress poses), and allows them to be blended to independent regions of the face using artist-animated weights along with a mask texture. We build upon this technique, demonstrating how to dramatically optimize the memory requirements. Furthermore, our technique allows us to easily include more than two wrinkle maps when needed, because we no longer map negative and positive values to different textures.

6.2 Our Algorithm

The core idea of this technique is the addition of wrinkle normal maps on top of the base normal maps and blend shapes (see Figure 6.3 (left) and (center) for example maps). For each facial expression, wrinkles are selectively applied by using weighted masks (see Figure 6.3 (right) and Table 6.1 for the mask and weights used in our examples). This way, the animator is able to manipulate the wrinkles on a per-blend-shape basis, allowing art-directed blending between poses and expressions. We store a wrinkle mask per channel of a (RGBA) texture; hence, we can store up to four zones per texture. As our implementation uses eight zones, we require storing and accessing only two textures. Note that when the contribution of multiple blend shapes in a zone exceeds a certain limit, artifacts can appear in the wrinkles. In order to avoid this problem, we clamp the value of the summation to the $[0, 1]$ range.

While combining various displacement maps consists of a simple sum, combining normal maps involves complex operations that should be avoided in a time-constrained environment like a game. Thus, in order to combine the base

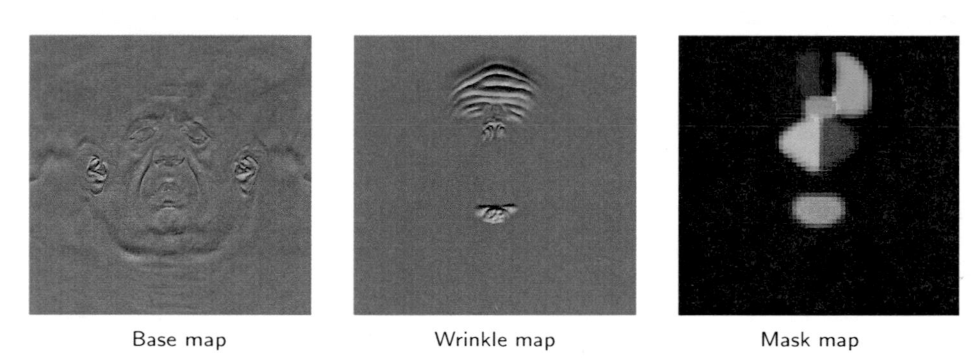

Figure 6.3. The wrinkle map is selectively applied on top of the base normal map by using a wrinkle mask. The use of partial-derivative normal maps reduces this operation to a simple addition. The yellowish look is due to the encoding and storage in the R and G channels that this technique employs. Wrinkle-zone colors in the mask do not represent the actual channels of the mask maps, they are put together just for visualization purposes.

	Red	Green	Blue	Brown	Cyan	Magenta	Orange	Gray
Joy	1.0	1.0	0.2	0.2	0.0	0.0	0.0	0.0
Surprise	0.8	0.8	0.8	0.8	0.0	0.0	0.0	0.0
Fear	0.2	0.2	0.75	0.75	0.3	0.3	0.0	0.6
Anger	-0.6	-0.6	-0.8	-0.8	0.8	0.8	1.0	0.0
Disgust	0.0	0.0	-0.1	-0.1	1.0	1.0	1.0	0.5
Sad	0.2	0.2	0.75	0.75	0.0	0.0	0.1	1.0

Table 6.1. Weights used for each expression and zone (see color meaning in the mask map of Figure 6.3).

and wrinkle maps, a special encoding is used: partial-derivative normal maps [Acton 08]. It has two advantages over the conventional normal map encoding:

1. Instead of reconstructing the z-value of a normal, we just have to perform a vector normalization, saving valuable GPU cycles;

2. More important for our purposes, the combination of various partial-derivative normal maps is reduced to a simple sum, similar to combining displacement maps.

```
float3 WrinkledNormal(Texture2D<float2> baseTex,
                      Texture2D<float2> wrinkleTex,
                      Texture2D maskTex[2],
                      float4 weights[2],
                      float2 texcoord) {
  float3 base;
  base.xy = baseTex.Sample(AnisotropicSampler16, texcoord).gr↩
     ;
  base.xy = -1.0 + 2.0 * base.xy;
  base.z = 1.0;

  #ifdef WRINKLES
  float2 wrinkles = wrinkleTex.Sample(LinearSampler,
                                      texcoord).gr;
  wrinkles = -1.0 + 2.0 * wrinkles;

  float4 mask1 = maskTex[0].Sample(LinearSampler, texcoord);
  float4 mask2 = maskTex[1].Sample(LinearSampler, texcoord);
  mask1 *= weights[0];
  mask2 *= weights[1];

  base.xy += mask1.r * wrinkles;
  base.xy += mask1.g * wrinkles;
  base.xy += mask1.b * wrinkles;
  base.xy += mask1.a * wrinkles;
  base.xy += mask2.r * wrinkles;
  base.xy += mask2.g * wrinkles;
  base.xy += mask2.b * wrinkles;
  base.xy += mask2.a * wrinkles;
  #endif

  return normalize(base);
}
```

Listing 6.1. HLSL code of our technique. We are using a linear instead of an anisotropic sampler for the wrinkle and mask maps because the low-frequency nature of their information does not require higher quality filtering. This code is a more readable version of the optimized code found in the web material.

This encoding must be run as a simple preprocess. Converting a conventional normal $n = (n_x, n_y, n_z)$ to a partial-derivative normal $n' = (n'_x, n'_y, n'_z)$ is done by using the following equations:

$$n'_x = \frac{n_x}{n_z} \qquad n'_y = \frac{n_y}{n_z}.$$

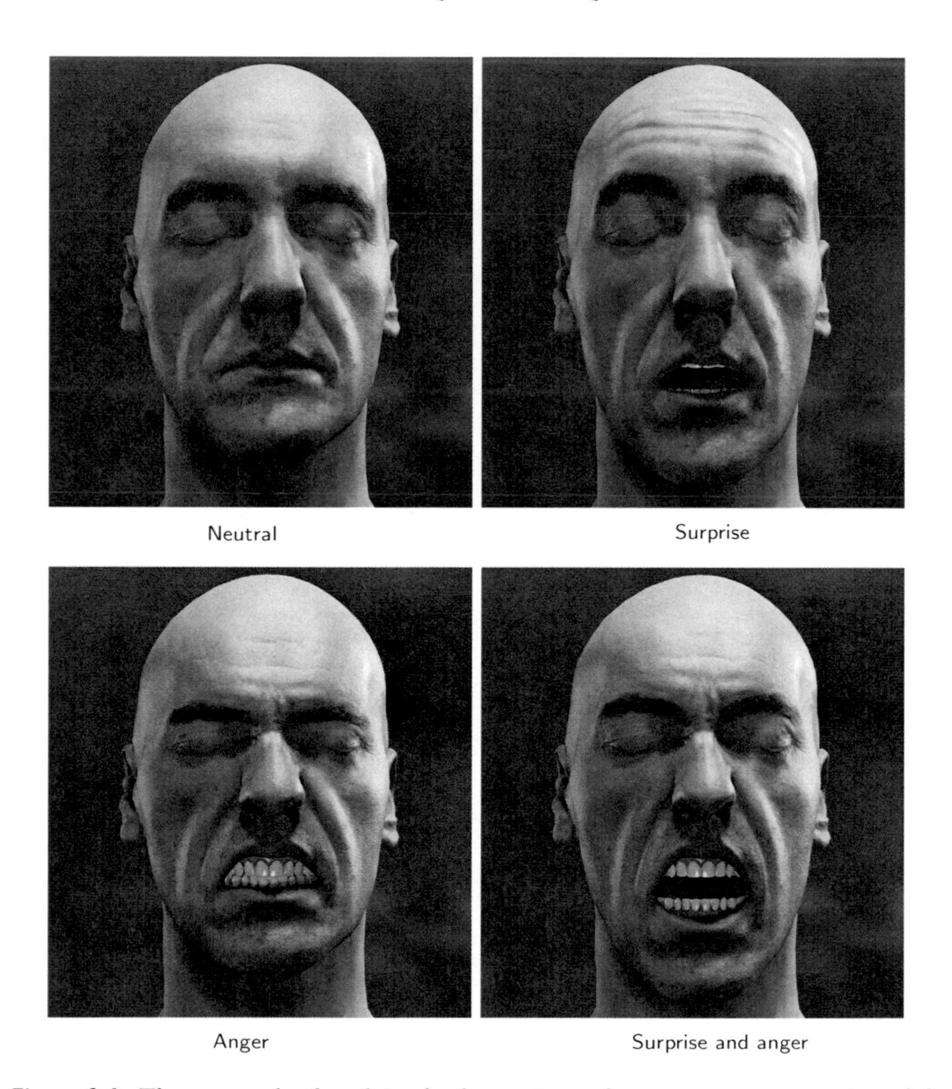

Figure 6.4. The net result of applying both surprise and anger expressions on top of the neutral pose is an unwrinkled forehead. In order to accomplish this, we use positive and negative weights in the forehead wrinkle zones, for the surprise and angry expressions, respectively.

In runtime, reconstructing a single partial-derivative normal n' to a conventional normal $\hat{n}$ is done as follows:

$$\begin{aligned} n &= (n'_x, n'_y, 1), \\ \hat{n} &= \frac{n}{\|n\|}. \end{aligned}$$

Note that in the original formulation of partial-derivative normal mapping there is a minus sign both in the conversion and reconstruction phases; removing it from both steps allows us to obtain the same result, with the additional advantage of saving another GPU cycle.

Then, combining different partial-derivative normal maps consists of a simple summation of their (x, y)-components before the normalization step. As Figure 6.3 reveals, expression wrinkles are usually low frequency. Thus, we can reduce map resolution to spare storage and lower bandwidth consumption, without visible loss of quality. Calculating the final normal map is therefore reduced to a summation of weighted partial-derivative normals (see Listing 6.1).

A problem with facial wrinkle animation is the modeling of compound expressions, through which wrinkles result from the interactions among the basic expressions they are built upon. For example, if we are surprised, the frontalis muscle contracts the skin, producing wrinkles in the forehead. If we then suddenly became angry, the corrugator muscles are triggered, expanding the skin in the forehead, thus causing the wrinkles to disappear. To be able to model these kinds of interactions, we let mask weights take negative values, allowing them to cancel each other. Figure 6.4 illustrates this particular situation.

6.2.1 Alternative: Using Normal Map Differences

An alternative to the use of partial-derivative normal maps for combining normal maps is to store differences between the base and each of the expression wrinkle maps (see Figure 6.5 (right)) in a manner similar to the way blend-shape interpolation is usually performed. As differences may contain negative values, we perform a scale-and-bias operation so that all values fall in the $[0, 1]$ range, enabling storage in regular textures:

$$d(x, y) = 0.5 + 0.5 \cdot (w(x, y) - b(x, y)),$$

where $w(x, y)$ is the normal at pixel (x, y) of the wrinkle map, and $b(x, y)$ is the corresponding value from the base normal map. When DXT compression is used for storing the differences map, it is recommended that the resulting normal be renormalized after adding the delta, in order to alleviate the artifacts caused by the compression scheme (see web material for the corresponding listing).

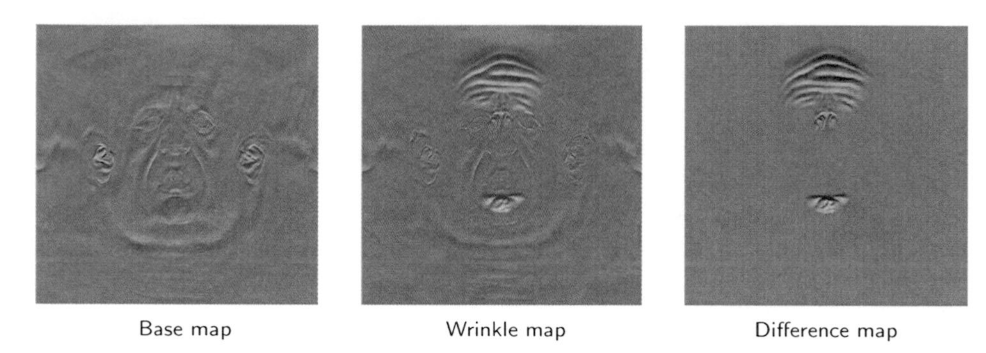

Figure 6.5. We calculate a wrinkle-difference map by subtracting the base normal map from the wrinkle map. In runtime, the wrinkle-difference map is selectively added on top of the base normal map by using a wrinkle mask (see Figure 6.3 (right) for the mask). The grey color of the image on the right is due to the bias and scale introduced when computing the difference map.

Partial-derivative normal mapping has the following advantages over the differences approach:

- It can be a little bit faster because it saves one GPU cycle when reconstructing the normal, and also allows us to add only two-component normal derivatives instead of a full (x, y, z) difference; these two-component additions can be done two at once, in only one cycle. This translates to a measured performance improvement of $1.12x$ in the GeForce 8600GT, whereas we have not observed any performance gain in either the GeForce 9800GTX+ nor in the GeForce 295GTX .

- It requires only two channels to be stored vs. the three channels required for the differences approach. This provides higher quality because 3Dc can be used to compress the wrinkle map for the same memory cost.

On the other hand, the differences approach has the following advantages over the partial-derivative normal mapping approach:

- It uses standard normal maps, which may be important if this cannot be changed in the production pipeline.

- Partial-derivative normal maps cannot represent anything that falls outside of a $45°$ cone around $(0, 0, 1)$. Nevertheless, in practice, this problem proved to have little impact on the quality of our renderings.

The suitability of each approach will depend on both the constraints of the pipeline and the characteristics of the art assets.

6.3 Results

For our implementation we used DirectX 10, but the wrinkle-animation shader itself could be easily ported to DirectX 9. However, to circumvent the limitation that only four blend shapes can be packed into per-vertex attributes at once, we used the DirectX 10 stream-out feature, which allows us to apply an unlimited number of blend shapes using multiple passes [Lorach 07]. The base normal map has a resolution of 2048×2048, whereas the difference wrinkle and mask maps have a resolution of 256×256 and 64×64, respectively, as they contain only low-frequency information. We use 3Dc compression for the base and wrinkle maps, and DXT for the color and mask maps. The high-quality scanned head model and textures were kindly provided by XYZRGB, Inc., with the wrinkle maps created manually, adding the missing touch to the photorealistic look of the images. We used a mesh resolution of 13063 triangles, mouth included, which is a little step ahead of current generation games; however, as current high-end systems become mainstream, it will be more common to see such high polygon counts, especially in cinematics.

To simulate the subsurface scattering of the skin, we use the recently developed screen-space approach [Jimenez and Gutierrez 10, Jimenez et al. 10b], which transfers computations from texture space to screen space by modulating a convolution kernel according to depth information. This way, the simulation is reduced to a simple post-process, independent of the number of objects in the scene and easy to integrate in any existing pipeline. Facial-color animation is achieved using a recently proposed technique [Jimenez et al. 10a], which is based on *in vivo* melanin and hemoglobin measurements of real subjects. Another crucial part of our rendering system is the Kelemen/Szirmay-Kalos model, which provides realistic specular reflections in real time [d'Eon and Luebke 07]. Additionally, we use the recently introduced filmic tone mapper [Hable 10], which yields really crisp blacks.

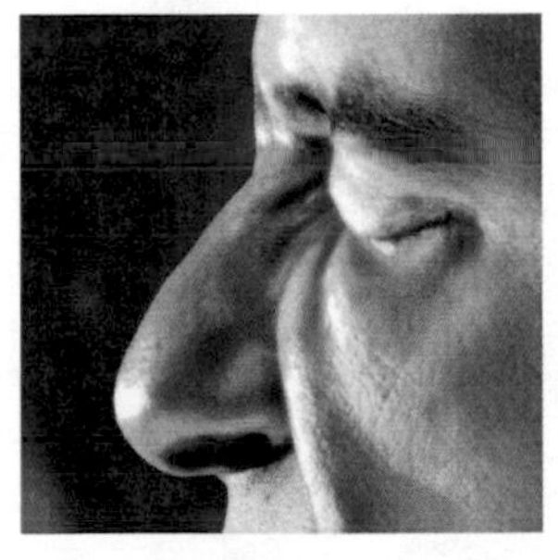

Nasalis

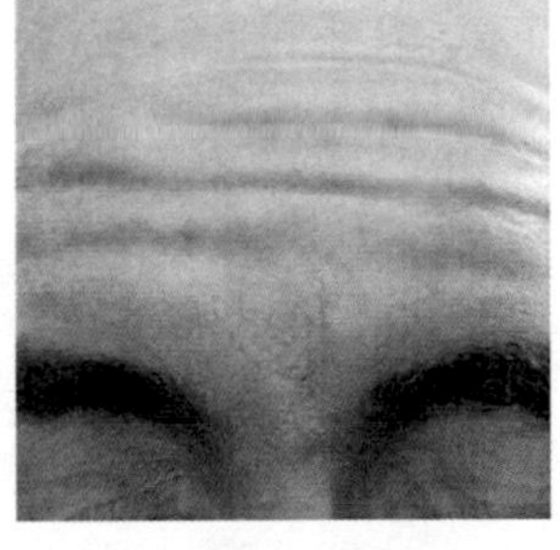

Frontalis

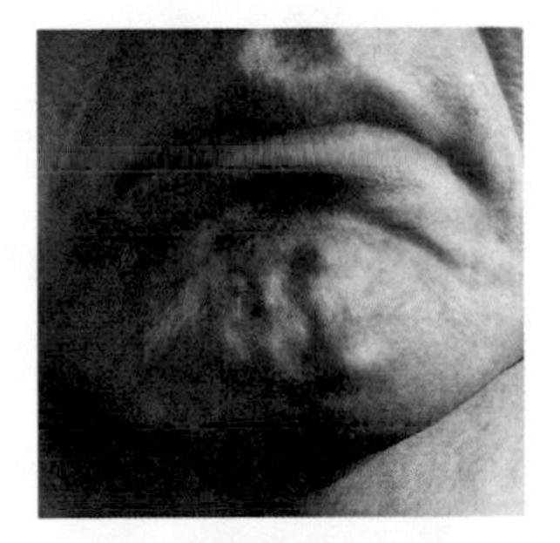

Mentalis

Figure 6.6. Closeups showing the wrinkles produced by nasalis (nose), frontalis (forehead), and mentalis (chin) muscles.

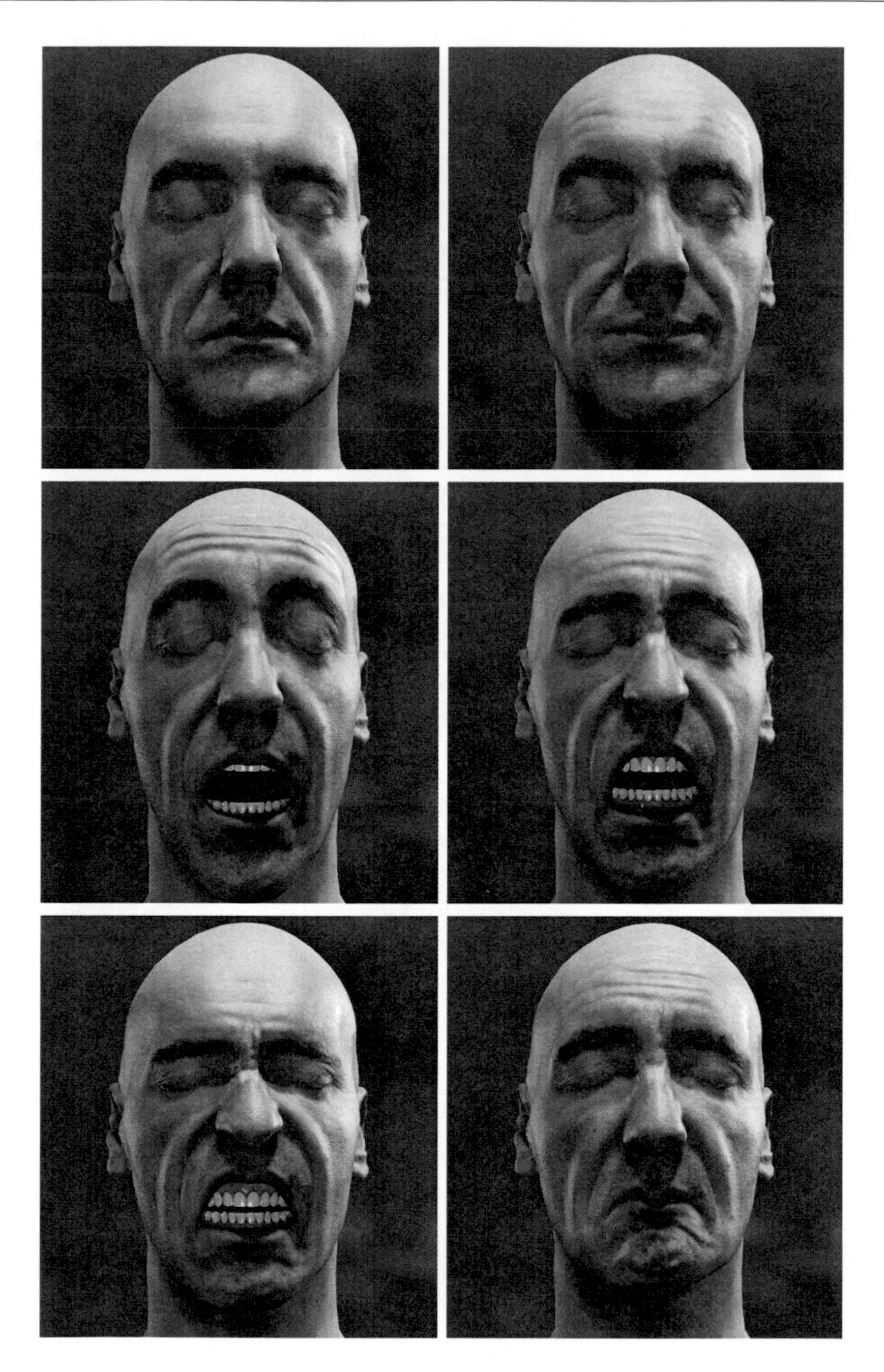

Figure 6.7. Transition between various expressions. Having multiple mask zones for the forehead wrinkles allows their shape to change according to the animation.

	Shader execution time
GeForce 8600GT	0.31 ms
GeForce 9800GTX+	0.1 ms
GeForce 295GTX	0.09 ms

Table 6.2. Performance measurements for different GPUs. The times shown correspond specifically to the execution of the code of the wrinkles shader.

For the head shown in the images, we have not created wrinkles for the zones corresponding to the cheeks because the model is tessellated enough in this zone, allowing us to produce geometric deformations directly on the blend shapes.

Figure 6.6 shows different close-ups that allow appreciating the wrinkles added in detail. Figure 6.7 depicts a sequential blending between compound expressions, illustrating that adding facial-wrinkle animation boosts realism and adds mood to the character (frames taken from the movie are included in the web material).

Table 6.2 shows the performance of our shader using different GPUs, from the low-end GeForce 8600GT to the high-end GeForce 295GTX. An in-depth examination of the compiled shader code reveals that the wrinkle shader adds a per-pixel arithmetic instruction/memory access count of 9/3. Note that animating wrinkles is useful mostly for near-to-medium distances; for far distances it can be progressively disabled to save GPU cycles. Besides, when similar characters share the same (u, v) arrangement, we can reuse the same wrinkles, further improving the use of memory resources.

6.4 Discussion

From direct observation of real wrinkles, it may be natural to assume that shading could be enhanced by using techniques like ambient occlusion or parallax occlusion mapping [Tatarchuk 07]. However, we have found that wrinkles exhibit very little to no ambient occlusion, unless the parameters used for its generation are pushed beyond its natural values. Similarly, self-occlusion and self-shadowing can be thought to be an important feature when dealing with wrinkles, but in practice we have found that the use of parallax occlusion mapping is most often unnoticeable in the specific case of facial wrinkles.

Furthermore, our technique allows the incorporation of additional wrinkle maps, like the lemon pose used in [Oat 07], which allows stretching wrinkles already found in the neutral pose. However, we have not included them because they have little effect on the expressions we selected for this particular character model.

6.5 Conclusion

Compelling facial animation is an extremely important and challenging aspect of computer graphics. Both games and animated feature films rely on convincing characters to help tell a story, and a critical part of character animation is the character's ability to use facial expression. We have presented an efficient technique for achieving animated facial wrinkles for real-time character rendering. When combined with traditional blend-target morphing for facial animation, our technique can produce very compelling results that enable virtual characters to accompany both their actions and dialog with increased facial expression. Our system requires very little texture memory and is extremely efficient, enabling true emotional and realistic character renderings using technology available in widely adopted PC graphics hardware and current generation game consoles.

6.6 Acknowledgments

Jorge would like to dedicate this work to his eternal and most loyal friend Kazán. We would like to thank Belen Masia for her very detailed review and support, Wolfgang Engel for his editorial efforts and ideas to improve the technique, and Xenxo Alvarez for helping to create the different poses. This research has been funded by a Marie Curie grant from the Seventh Framework Programme (grant agreement no.: 251415), the Spanish Ministry of Science and Technology (TIN2010-21543), and the Gobierno de Aragón (projects OTRI 2009/0411 and CTPP05/09). Jorge Jimenez was additionally funded by a grant from the Gobierno de Aragón. The authors would also like to thank XYZRGB Inc. for the high-quality head scan.

Bibliography

[Acton 08] Mike Acton. "Ratchet and Clank Future: Tools of Destruction Technical Debriefing." Technical report, Insomniac Games, 2008.

[d'Eon and Luebke 07] Eugene d'Eon and David Luebke. "Advanced Techniques for Realistic Real-Time Skin Rendering." In *GPU Gems 3*, edited by Hubert Nguyen, Chapter 14, pp. 293–347. Reading, MA: Addison Wesley, 2007.

[Hable et al. 09] John Hable, George Borshukov, and Jim Hejl. "Fast Skin Shading." In *ShaderX*^{7}, edited by Wolfgang Engel, Chapter II.4, pp. 161–173. Hingham, MA: Charles River Media, 2009.

[Hable 10] John Hable. "Uncharted 2: HDR Lighting." Game Developers Conference, 2010.

[Jimenez and Gutierrez 10] Jorge Jimenez and Diego Gutierrez. "Screen-Space Subsurface Scattering." In *GPU Pro*, edited by Wolfgang Engel, Chapter V.7. Natick, MA: A K Peters, 2010.

[Jimenez et al. 10a] Jorge Jimenez, Timothy Scully, Nuno Barbosa, Craig Donner, Xenxo Alvarez, Teresa Vieira, Paul Matts, Veronica Orvalho, Diego Gutierrez, and Tim Weyrich. "A Practical Appearance Model for Dynamic Facial Color." *ACM Transactions on Graphics* 29:6 (2010), Article 141.

[Jimenez et al. 10b] Jorge Jimenez, David Whelan, Veronica Sundstedt, and Diego Gutierrez. "Real-Time Realistic Skin Translucency." *IEEE Computer Graphics and Applications* 30:4 (2010), 32–41.

[Larboulette and Cani 04] C. Larboulette and M. Cani. "Real-Time Dynamic Wrinkles." In *Proc. of the Computer Graphics International*, pp. 522–525. Washington, DC: IEEE Computer Society, 2004.

[Lorach 07] T. Lorach. "DirectX 10 Blend Shapes: Breaking the Limits." In *GPU Gems 3*, edited by Hubert Nguyen, Chapter 3, pp. 53–67. Reading, MA: Addison Wesley, 2007.

[Oat 07] Christopher Oat. "Animated Wrinkle Maps." In *SIGGRAPH '07: ACM SIGGRAPH 2007 courses*, pp. 33–37. New York: ACM, 2007.

[Tatarchuk 07] Natalya Tatarchuk. "Practical Parallax Occlusion Mapping." In *ShaderX*^{5}, edited by Wolfgang Engel, Chapter II.3, pp. 75–105. Hingham, MA: Charles River Media, 2007.

Procedural Content Generation on the GPU

Aleksander Netzel and Pawel Rohleder

7.1 Abstract

This article emphasizes on-the-fly procedural creation of content related to the video games industry. We demonstrate the generating and rendering of infinite and deterministic heightmap-based terrain utilizing fractal Brownian noise calculated in real time on a GPU. We take advantage of a thermal erosion algorithm proposed by David Cappola, which greatly improves the level of realism in heightmap generation. In addition, we propose a random tree distribution algorithm that exploits previously generated terrain information. Combined with the natural-looking sky model based on Rayleigh and Mie scattering, we achieved very promising quality results at real-time frame rates. The entire process can be seen in our DirectX10-based demo application.

7.2 Introduction

Procedural content generation (PCG) refers to the wide process of generating media algorithmically. Many existing games use PCG techniques to generate a variety of content, from simple, random object placement over procedurally generated landscapes to fully automatic creation of weapons, buildings, or AI enemies. Game worlds tend to be increasingly rich, which requires a lot of effort that we can minimize by utilizing PCG techniques. One of the basic PCG techniques in real-time computer graphics applications is the heightmap-based terrain generation [Olsen 04].

7.3 Terrain Generation and Rendering

Many different real-time, terrain-generation techniques have been developed over the last few years. Most of them utilize procedurally generated noise for creating a heightmap. The underlying-noise generation method should be fast enough to get at least close to real time and generate plausible results. The most interesting technique simulates 1/f noise (called "pink noise"). Because this kind of noise occurs widely in nature, it can be easily implemented on modern GPUs and has good performance/speed ratio. In our work, we decided to use the approximation of fractal Brownian motion (fBm).

Our implementation uses spectral synthesis, which accumulates several layers of noise together (see Figure 7.1). The underlying-noise generation algorithm is simple Perlin noise, which is described in [Green 05]. Although its implementation relies completely on the GPU, it is not fast enough to be calculated with every frame because of other procedural generation algorithms. We therefore used a system in which everything we need is generated on demand. The terrain is divided into smaller nodes (the number of nodes has to be a divider of heightmap size, so that there won't be any glitches after generation), and the camera is placed in the center of all nodes. Every node that touches the center node has a bounding box. Whenever collision between the camera and any of the bounding boxes is detected, a new portion of the heightmap is generated (along with other procedural content). Based on the direction of the camera collision and the position of nodes in world space, we determine new UV coordinates for noise

Figure 7.1. Procedurally generated terrain with random tree distribution.

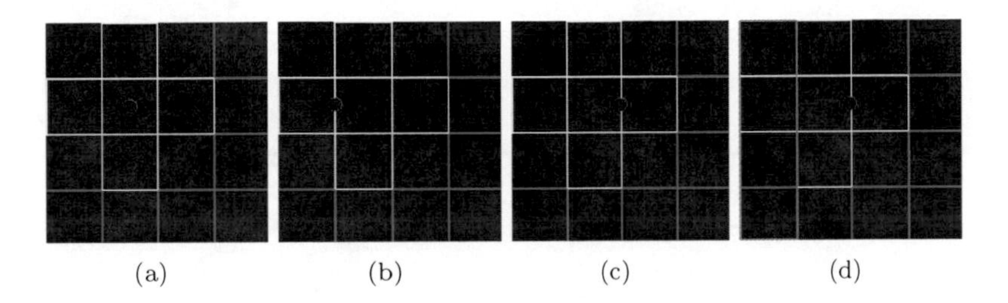

Figure 7.2. Red: camera, yellow: AABB, green: different patches in the grid. (a) Camera in the middle. (b) Collision detected with AABB. (c) New generation of procedural content, new AABB. (d) New row that is generated.

generation. This operation guarantees, with respect to the heightmap generation algorithm, that terrain will be continuous, endless, and realistic.

We could also optimize processing power by half, knowing that at most there are two rows of heightmaps that should be regenerated. We simply copy specified rows or columns and generate one additional row/column via the pixel shader (in Figure 7.2, all patches with a blue outline). When a camera collision occurs in the corner where two AABBs overlap, we have to generate a new row and column of data. To overcome the problem with camera collision and the newly generated AABB, we have to make the bounding boxes a little bigger or smaller, so there will be free space between them.

Our algorithm also utilizes a *level-of-detail* (LOD) system. Since we are assured of the relative positions of all nodes, we can precalculate all variations (there are only a small number) of index buffers. The center node has to have the lowest LOD (highest triangle count). Since we know that the camera will always be placed in the center node, we don't have to switch index buffers for different terrain LOD because the distance between the patch containing the camera and other patches is always the same (that's basically how our system works).

Another natural phenomenon we decided to simulate is erosion. Erosion is defined as the transporting and depositing of solid materials elsewhere due to wind, water, gravity, or living organisms. The simplest type of erosion is thermal erosion [Marak 97], wherein temperature changes cause small portions of the materials to crumble and pile up at the bottom of an incline. The algorithm is iterative and is as follows for each iteration: for every terrain point that has an altitude higher than the given threshold value (called *talus angle* (T)), some of its material will be moved to neighboring points. In the case that many of the neighbors' heights are above the talus angle, material has to be properly distributed.

The implementation is fairly simple. We compare every point of the heightmap with the heights of neighboring points and calculate how much material has to be moved. Since pixel shader limitation restricts us from scattering data (before UAVs in DirectX11), we use an approach proposed by David Cappola [Capolla 08]. To overcome pixel shader limitation, we need to make two passes. In the first pass, we evaluate the amount of material that has to be moved from a given point to its neighbors; in the second pass, we actually accumulate material, using information provided by the first pass.

To improve the visual quality of our terrain, we generate a normal map, which is pretty straightforward. We use a given heightmap and apply the Sobel filter to it. Using the sun position, we can calculate shading (Phong diffuse). Since we target DX10 and higher we use texture arrays to render terrain with many textures [Dudask 07].

Collision detection is managed by copying a GPU-generated heightmap onto an offline surface and reading the height data array. Then we can check for collision using a standard method on the CPU and determine an accurate position inside an individual terrain quad using linear interpolation. Or, without involving the CPU, we could render to a 1×1 render target instead; however, since our tree placement technique is not "CPU free" (more on that later), we need the possibility of being able to check for height/collision without stressing the GPU.

7.4 Environmental Effects

Since we want to simulate a procedurally-rich natural environment, we have to be able to render more than just terrain with grass texture—we want plants. In the natural environment, plant distribution is based on many conditions such as soil quality, exposure (wind and sun), or the presence of other plants. In our work, we used a heightmap-based algorithm to determine tree placement along generated terrain.

Our tree-placement technique consists of two separate steps. First, we generate a tree-density map, which characterizes tree placement. The next step involves rendering instanced tree meshes. We use a density map to build a stream with world matrices. The height and slope of the terrain are the most common factors on which plant growth is dependent; both are easy to calculate in our case since we already have a heightmap and a normal map.

Using a simple pixel shader, we calculate density for every pixel as a mix of slope, height, and some arbitrarily chosen parameters. As a result, we get texture with tree distribution where value 1.0 corresponds to the maximum number of trees and 0.0 corresponds to no tree. To give an example, we present one of our versions (we tried many) in Listing 7.1 and the result in Figure 7.3. There are no strict rules on how to implement the shader. Anything that suits your needs is acceptable.

When the tree density map is ready to go, we have to build a stream with world matrices for instancing. Because the tree-density map texel can enclose a huge range in world space, we cannot simply place one tree per texel because the

```
float p = 0.0f;
//Calculate slope.
float f_slope_range = 0.17;
float f_slope_min = 1.0 - f_slope_range;
float3 v_normal = g_Normalmap.Sample( samClamp, IN.UV ).xyz * 2.0
               - 1.0;
float f_height = g_Heightmap.Sample( samClamp, IN.UV ).x * 2.0
               - 1.0;
float f_slope = dot( v_normal, float3(0,1,0) );
f_slope = saturate(f_slope - f_slope_min);
float f_slope_val = smoothstep(0.0, f_slope_range, f_slope);
//Get relative height.
float f_rel_height_threshold = 0.002;
float4 v_heights = 0;

v_heights.x = g_Heightmap.Sample( samClamp, IN.UV
                + float2( -1.0 / f_HM_size, 0.0) ); //Left
v_heights.y = g_Heightmap.Sample( samClamp, IN.UV
                + float2(  1.0 / f_HM_size, 0.0) ); //Right
v_heights.z = g_Heightmap.Sample( samClamp, IN.UV
                + float2(  0.0, -1.0 / f_HM_size) ); //Top
v_heights.w = g_Heightmap.Sample( samClamp, IN.UV
                + float2(  0.0,  1.0 / f_HM_size) ); //Down

v_heights = v_heights * 2.0 - 1.0;
v_heights = abs(v_heights - f_height);
v_heights = step(v_heights, f_rel_height_t);
p = dot(f_slope_val, v_heights) * 0.25;

return p;
```

Listing 7.1. Tree-density map pixel shader.

generated forest will be too sparse. To solve this issue, we have to increase the trees-per-texel ratio; therefore, we need one more placement technique.

We assign each tree type a different radius that determines how much space this type of tree owns (in world space). It can be compared to the situation in a real environment when bigger trees take more resources and prevent smaller trees from growing in their neighborhood. Also, we want our trees to be evenly but randomly distributed across a patch corresponding to one density map texel.

Our solution is to divide the current patch into a grid wherein each cell size is determined by the biggest tree radius in the whole patch. The total number of cells is a mix of the density of the current texel and the space that the texel encloses in world space. In the center of every grid cell, we place one tree and move its position using pseudorandom offset within the grid to remove repetitive patterns.

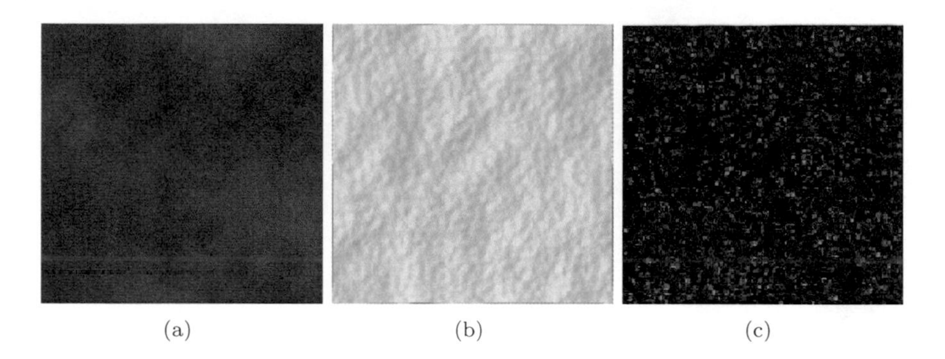

Figure 7.3. (a) Heightmap, (b) normal map, and (c) tree-density map.

Using camera frustum, we process only visible texels to form a density map. Based on the distance to the camera, the tree's LOD is determined so that only trees close to the camera will be rendered with a complete list of faces.

After all these steps, we have a second stream prepared and we are ready to render all the trees. We also input the maximum number of trees that we are about to render because there can easily become too many, especially when we have a large texel-to-world-space ratio. Care has to be taken with this approach since we cannot simply process a tree-density map row by row; we have to process texels close to the camera first. If we don't, we might use all "available" trees for the farthest patches, and there won't be any close to the camera. In this situation, we may use billboards for the farthest trees, or use a smooth transition into the fog color.

The last part of our procedural generation is the Rayleigh-Mie atmospheric scattering simulation [West 08, O'Neil 05]. Our implementation follows the technique described in *GPU Gems 2*. We first calculate optical depth to use it as a lookup table for further generating Mie and Rayleigh textures. Mie and Rayleigh textures are updated in every frame (using rendering to *multiple render targets* (MRT)) and are then sampled during sky-dome rendering. This method is efficient, fast, and gives visually pleasing results.

7.5 Putting It All Together

In Figure 7.4 we present what our rendering loop looks like. As described earlier, we perform only one full generation of procedural content whenever a collision of the camera with one of the bounding boxes is detected.

When collision is detected, we transform camera position into UV position for heightmap generation. After generating the heightmap, we calculate the erosion and the normal map. The last part of this generation step is to calculate new AABB.

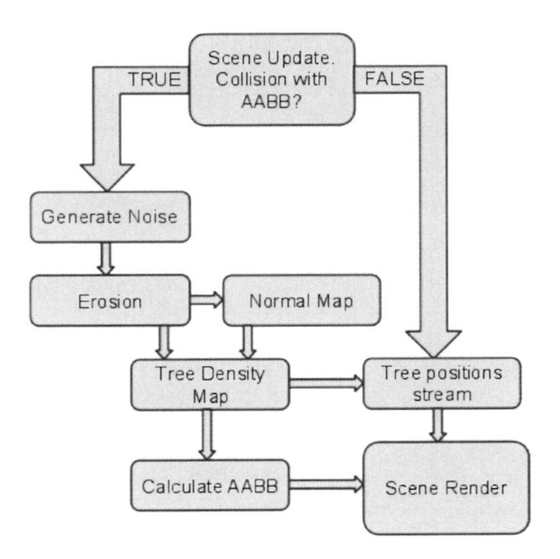

Figure 7.4. Rendering loop.

The tree-position stream is calculated with each frame since it depends on the camera orientation (see Figure 7.5).

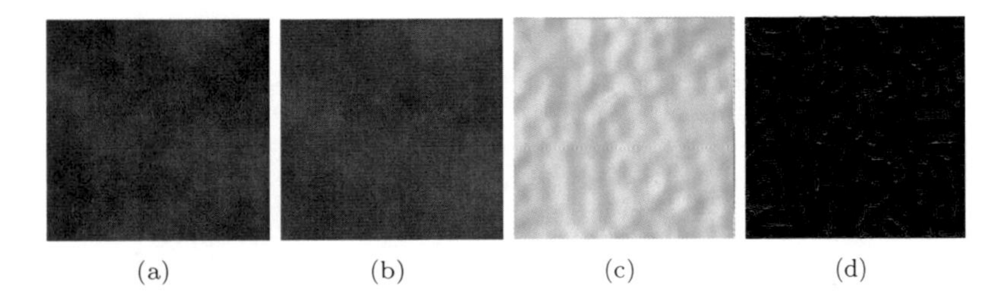

Figure 7.5. (a) Heightmap, (b) erosion map, (c) normal map, and (d) tree-density map.

7.6 Conclusions and Future Work

We implemented all of the techniques described in this article using Microsoft DirectX 10. All parameters controlling algorithm behavior can be changed during real time. Table 7.1 shows the minimum, maximum, and average number of *frames per second* (fps) in our framework, with 200 iterations of the erosion algorithm.

As we can see, the average number of frames is close to the number of maximum frames because usually only tree stream is generated. So a decrease of fps by a factor of 10 in the minimum is due to the generation of procedural content.

	NVIDIA GTX 260	ATI Radeon HD 5900
fps average	351	352
fps minimum	66	179
fps maximum	385	373

Table 7.1. Minimum, maximum, and average number of frames per second.

Therefore, we could procedurally generate content in every frame and keep up with real time, but our system doesn't require that.

The obvious optimization is to put the generation of the texture into another thread. For instance, erosion iterations can be divided into several frames.

Possibilities for developing further procedural generations are almost endless. We mention the techniques we will be developing in the near future. Tree-density mapping can be used not only for tree generation but also for other plant seeding systems like grass and bushes. Since we can assume that bushes can grow close to the trees, we can place some of them into the grid containing one tree. Of course, we must take into account that bushes are unlikely to grow directly in a tree's shadow. Also, the same density map can be used for grass placement. Since grass is likely to grow further away (because natural conditions for grass growth are less strict), we can blur the density map to get a grass density map. Therefore, without much more processing power, we can generate grass and bushes.

The next step could involve generating rivers and even whole cities, but cities put some conditions on generated terrain. Terrain under and around a city should be almost flat. The best solution is to combine artist-made terrain with procedural terrain. The only challenge is to achieve a seamless transition between the two.

Another interesting idea is to use other types of erosion (i.e., erosion described in *ShaderX*^{7}) or different noise generators. For heightmap generation, since every function that returns height for a given x, y can be used, options are practically limitless.

For example, one might take advantage of the possibilities provided by the latest version of DirectX 11 Compute Shaders. It provides many new features that make procedural generation easier.

In conclusion, PCG offers the possibility to generate virtual worlds in a fast and efficient way. Many different techniques may be used to create rich environments, such as terrain with dense vegetation with a very small amount of artist/level designer work. Our application could be easily extended or integrated into an existing game editor. Since our techniques offer interactive frame rates it could be used for current games, like flight simulators or any other open-space game with a rich environment.

Bibliography

[Capolla 08] D. Capolla. "GPU Terrain." Available at http://www.m3xbox.com/GPU_blog, 2008.

[Dudask 07] B. Dudask. "Texture Arrays for Terrain Rendering." Available at http://developer.download.nvidia.com/SDK/10.5/direct3d/Source/TextureArrayTerrain/doc/TextureArrayTerrain.pdf, 2007.

[Green 05] S. Green. "Implementing Improved Perlin Noise." In *GPU Gems 2*, edited by Hubert Nguyen, pp. 73–85. Reading, MA: Addison-Wesley, 2005.

[Marak 97] I. Marak. "Thermal Erosion." Availalbe at http://www.cg.tuwien.ac.at/hostings/cescg/CESCG97/marak/node11.html, 1997.

[Olsen 04] J. Olsen. "Real-Time Procedural Terrain Generation." Available at http://oddlabs.com/download/terrain_generation.pdf, 2004.

[O'Neil 05] S. O'Neil. *Accurate Atmospheric Scattering*. Reading, MA: Addison-Wesley, 2005.

[West 08] M. West. "Random Scattering: Creating Realistic Landscapes." Available at http://www.gamasutra.com/view/feature/1648/random_scattering_creating_.php, 2008.

Vertex Shader Tessellation

Holger Gruen

8.1 Overview

This chapter presents a method to implement tessellation using vertex shaders only. It only works on DX10 class hardware and above. The technique itself doesn't need any data in addition to the already available vertex data in contrast to instanced tessellation techniques like those given in [Gruen 05,Tatarinov 08]. It solely relies on the delivery of `SV_VertexID` and the usage of the original vertex and index buffers as input shader resources to the vertex shader that performs the tessellation. Furthering the work in [Gruen 05,Tatarinov 08], we present here a way to support edge-based fractional tessellation.

This method not only enables a new tessellation technique on DX10 class hardware, but it also can be used in combination with the DX11 tessellation pipeline to reach overall tessellation factors that are higher than 64. This is important as often triangles in typical game scenes are too large to apply techniques like tessellated displacement mapping to them with a maximum tessellation factor of "only" 64. Although the approach described here is used in the context of tessellation, the same basic ideas can be used to do any sort of geometry amplification. In theory, using these insights, the geometry shader stage introduced with DirectX 10 can be made largely redundant.

8.2 Introduction

Having a low polygon-count (low poly) object representation that can later on be refined and smoothed on-the-fly has always been a very desirable goal.

The downside of a low poly mesh is obvious; it simply does not look as good as a high polygon-count (high poly) object. Some aspects of this low poly look can be improved by using normal maps to display fine details using per-pixel lighting. Still, the silhouettes of objects rendered using normal mapping look crude.

Tessellation is a refinement technique that can be used to generate a high number of output triangles per input triangle; it can thus smooth silhouettes to generate an organic look. Relatively low poly objects can then be used to save precious video memory, something that is especially relevant on game consoles.

One approach to generate nice silhouettes for low poly objects is to replace triangles of the object with curved surface patches. Alternatively, one can represent the object completely by the control points necessary to define these patches.

Ideally, one wants to evaluate curved surfaces over objects completely on-the-fly on the GPU, and this is where tessellation comes into play. With the introduction of DX11 class hardware, tessellation has become a mainstream hardware feature. This article tries to simplify the use of tessellation on DX10 class hardware and to make it a "first-class citizen" for all game-development efforts.

8.3 The Basic Vertex Shader Tessellation Algorithm

In order to understand the basic algorithm, please consider the following steps:

1. Set up an empty vertex input layout. As a consequence, no data will be delivered to the vertex shader from any vertex buffer.

2. Unbind all index and vertex buffers from the device.

3. Define `SV_VertexID` (or the appropriate OpenGL identifier) as the only input to the vertex shader. Only a vertex-ID will be delivered to the vertex shader.

4. Issue a draw call that delivers enough vertices (however, only `SV_VertexID` will be delivered) to the vertex shader to create a tessellation.

5. Bind all vertex and index buffers necessary to draw the current mesh as shader resources for the vertex stage.

A tessellation with a tessellation factor of N generates $N^2 - 1$ subtriangles as shown in Figure 8.1. In order to generate such a tessellation, it is now necessary to issue a draw call that delivers enough vertex-IDs to the vertex shader to output $N^2 - 1$ triangles. As each triangle in the tessellation needs three vertices, the overall number of vertex IDs to be issued is $3 \times (N^2 - 1)$.

So how does one then use the delivered vertex-IDs in the vertex shader to output a tessellation as shown in Figure 8.1? The answer lies in assigning an ID to each subtriangle of the tessellation. Figure 8.1 shows how to assign these IDs.

The vertex shader now needs to perform the following steps:

1. Compute the ID of the current subtriangle from `SV_VertexID`.

2. From the subtriangle-ID, compute an initial barycentric coordinate for the output vertex of the current subtriangle.

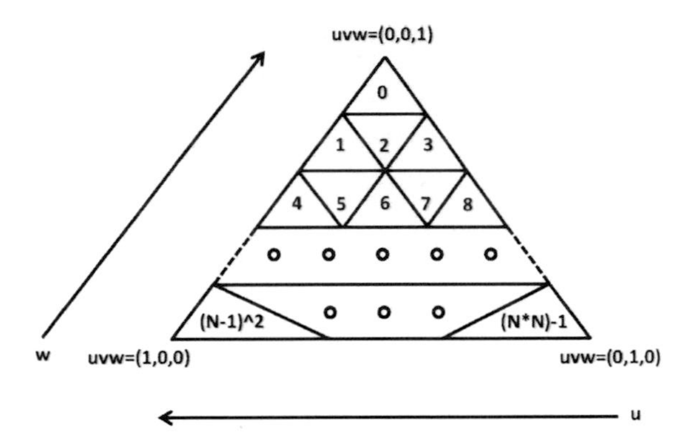

Figure 8.1. Assigning IDs to every subtriangle of a tessellation with tessellation factor N.

3. Compute the index of the current mesh triangle from `SV_VertexID`.
4. Fetch the three indices for the current mesh triangle from the index buffer using Load().
5. Fetch the vertex data for all three vertices of the current mesh triangle from the vertex buffers using the indices from Step 4.
6. Modify the barycentric coordinates from Step 2 to move to the right vertex of the subtriangle based on `SV_VertexID`.
7. Use the barycentric coordinates from Step 7 to compute the output vertex from three sets of vertex attributes from mesh vertices fetched in Step 5.

Listing 8.1 shows a real-world implementation of Steps 1–7.

```
// original vertex input structure
struct VS_RenderSceneInput
{
     float3 f3Position    : POSITION;
     float3 f3Normal      : NORMAL;
     float2 f2TexCoord    : TEXCOORD;
};

// ib and vb bound as shader resource by the vertex shader
Buffer<uint>         g_bufIndices              : register( t0  );
Buffer<float4>       g_bufVertices             : register( t1  );

Vs_to_ps VS_Tessellate( uint in_vertexID  : SV_VertexID )
{
   Vs_to_ps O;

   // g_uSubTriangles holds the number of subtriangles for the
   // current tessellation factor e.g., (N*N)-1
```

```
// compute what subtriangle we're in at the moment
uint subTriID = ( in_vertexID / 3 ) % g_uSubTriangles;

// subTriID to float
float fSubTriID = float(subTriID );

// compute barycentric coords u, v, and w
float fRow  = floor( sqrt( fSubTriID ) );  // how far we are along u
float incuv = 1.0f / g_fTessFactor;
float u     = ( 1.0f + fRow ) / g_fTessFactor;

// compute how far along the row we are
float fCol   = fSubTriID - ( fRow * fRow );
uint  uCol   = uint( fCol );
float v      = incuv * floor( fCol * 0.5f );

// correct u
u -= v;

// setup w
float w = 1.0f - u - v;

// compute the vertex index in the current mesh
uint vertexID = ( ( in_vertexID / 3 ) / g_uSubTriangles ) * 3
              + ( in_vertexID % 3 );

// compute offsets for fetching indices from index buffer
uint  uOffset1 = ( vertexID / uint(3) ) * 3;
uint3 uOffset3 = uint3( uOffset1, uOffset1 + 1, uOffset1 + 2 )
               + g_StartIndexLocation.xxx; // start offset for idxs

// fetch indices
uint3 indices;

indices.x = g_bufIndices.Load( uOffset3.x ).x;
indices.y = g_bufIndices.Load( uOffset3.y ).x;
indices.z = g_bufIndices.Load( uOffset3.z ).x;

// add base vertex location from constant buffer
indices += g_BaseVertexLocation.xxx;

// compute offset for vertices into the vertex buffer
uint3 voffset3 = indices * g_VertexStride.xxx + g_VertexStart.xxx;

// load vertex data for u=1.0 - vertex 0 of the mesh triangle
float4 dataU0  = g_bufVertices.Load( voffset3.x );
float4 dataU1  = g_bufVertices.Load( voffset3.x + 1 );
float3 f3PosU  = dataU0.xyz;
float3 f3NormU = float3( dataU0.w, dataU1.xy );
float2 f2TexU  = dataU1.zw;

// load vertex data for v=1.0 - vertex 1 of the mesh triangle
float4 dataV0  = g_bufVertices.Load( voffset3.y );
float4 dataV1  = g_bufVertices.Load( voffset3.y + 1 );
float3 f3PosV  = dataV0.xyz;
float3 f3NormV = float3( dataV0.w, dataV1.xy );
float2 f2TexV  = dataV1.zw;

// load vertex data for w=1.0 - vertex 2 of the mesh triangle
float4 dataW0  = g_bufVertices.Load( voffset3.z );
float4 dataW1  = g_bufVertices.Load( voffset3.z + 1 );
float3 f3PosW  = dataW0.xyz;
float3 f3NormW = float3( dataW0.w, dataW1.xy );
float2 f2TexW  = dataW1.zw;
```

```
    // compute output vertex based on vertexID
    // shift uvw based on this
    uint uVI = vertexID % uint( 3 );

    [flatten] if( uVI == uint( 0 ) ) // vertex at u==1
    {
       [flatten] if( ( uCol & uint(1) ) != 0 )
          v += incuv, u -= incuv; // constant w line
    }
    else [flatten] if( uVI == uint( 1 ) ) // vertex at v == 1
    {
       [flatten] if( ( uCol & uint(1) ) == 0 )
          v += incuv, u -= incuv; // constant w line
       else
       {
          v += incuv, u -= incuv; // constant w line
          w += incuv, u -= incuv; // constant v line
       }
    }
    else // vertex at w ==1
    {
       [flatten] if( ( uCol & uint(1) ) == 0 )
          u -= incuv, w += incuv; // constant v line
       else
          w += incuv, u -= incuv; // constant v line
    }

    // Pass through world space position
    O.f3Position = ( mul( float4( u * f3PosU + v * f3PosV
                     + w * f3PosW, 1.0f ), g_f4x4World ) ).xyz;

    O.f3Normal = u * f3NormU + v * f3NormV + w * f3NormW;
    O.f3TexC   = u * f3TexU + v * f3TexV + w * f3TexW;

    return O;
}
```

Listing 8.1. A tessellation vertex shader.

We have described how to tessellate in the triangular domain. In a similar way, we can implement tessellation for quadrangular domains. In the following section, we present a technique to implement fractional tessellation factors for a quadrangular tessellation.

8.4 Per-Edge Fractional Tessellation Factors

The only way to implement per-edge fractional tessellation factors is to send enough vertices to the vertex-shader stage to support the maximum-per-edge tessellation factor. Actually, one even needs to send two additional vertices to allow for the phasing in of the new fractional vertices on the left- and right-hand sides of the edge center to make sure that a fully symmetrical tessellation is created. Such a symmetrical tessellation is necessary to make sure that there are no gaps between meeting edges from adjacent triangles.

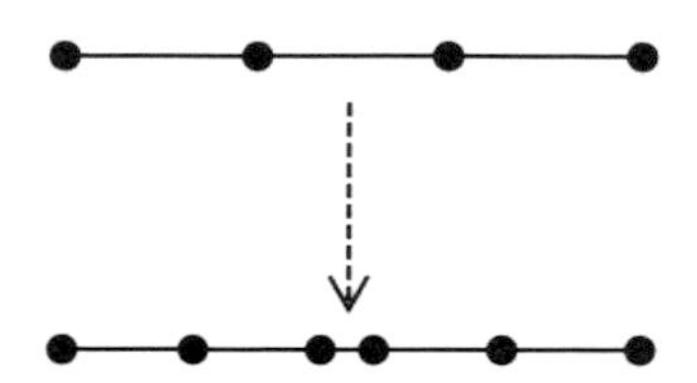

Figure 8.2. Fractional transitioning between a tessellation factor 3.0 and a tessellation factor 5.0 of an edge.

Figure 8.2 shows the tessellation that a fractional tessellation factor of roughly 3.5 would create.

For edges of the tessellated triangles and for interior vertices that need a lower than maximum fractional tessellation factor, the vertex shader must ensure that a number of zero-area triangles get created.

This is not the most efficient way of creating a tessellation, especially when there are big differences between the tessellation factors on the three edges of a triangle or the four edges of a quad. Nevertheless, it is the only way to deal with this problem on DX10-class hardware.

In general, however, this does not pose a severe problem, since one can, for example, use one fractional tessellation factor for an object or a mesh. This will work nicely as long as the objects or meshes that get tessellated are not too large.

If objects or meshes get large, for example, when the mesh represents a large terrain, then one tessellation factor is not enough. In such a case, it makes sense to break up the mesh into a number of patches that have similar maximum-per-edge fractional tessellation factors. Each set of patches can then be sent to the vertex shader for tessellation in one draw call.

Listing 8.2 shows an implementation of a vertex shader that supports per-edge fractional tessellation factors for quadrangular terrain patches. For the quad domain, one needs to issue six vertices per subquad (generated by the tessellation) since the vertex shader needs to output two triangles per quad.

Please note that the vertex shader in Listing 8.2 relies on the use of instanced draw calls to kick off the tessellating vertex shader. The instanced draw calls issue six vertices per subquad and one instance per quad of the untessellated terrain.

```
static const float2 uv_shift[4] = {
                                    float2( 0.0f, 0.0f ),
                                    float2( 1.0f, 0.0f ),
                                    float2( 1.0f, 1.0f ),
                                    float2( 0.0f, 1.0f ),
                                  };

VS_OUTPUT VSTessellateEdges( uint in_vertId : SV_VertexID,
                             uint in_instID : SV_InstanceID )
```

```
{
    VS_OUTPUT O;

    int i;

    // get the four tess factors for the four edges of the quad
    float4 f4EdgeFactor = g_f4EdgeFactor;

    // get the maximum tess factor (needs to be 2 + max edge factor)
    float  tf             = float( g_uVSTessFactor.x );

    // use vertex id to compute the id of the subquad
    // and to compute the row/col in the quad tessellation
    uint  uSubQuadID     = in_vertId / 6;
    float fSubQuadID     = float( uSubQuadID );
    float inv_tf         = 1.0f / tf;
    float fRow           = float( uSubQuadID / g_uVSTessFactor.x );
    float fCol           = floor( fSubQuadID - fRow * tf );

    // now compute u and v
    float v      = fRow * inv_tf;
    float u      = fCol * inv_tf;

    // compute offset into vertex buffer that holds
    // the vertices of the terrain
    uint offset = ( in_instID ) * 4;

    // fetch data for the 4 corners of the quad
    float4 data0 = g_bufVertices.Load( offset );
    float4 data1 = g_bufVertices.Load( offset + 1 );
    float4 data2 = g_bufVertices.Load( offset + 2 );
    float4 data3 = g_bufVertices.Load( offset + 3 );

    float3 f3Pos[4];

    // use data to init the corner positions in f3Pos[4]

    . . .

    // update u, v based on what vertex of the 6 vertices
    // that are used to output the 2 triangles that make
    // up a subquad (==quad created by the tessellation)
    uint vID = in_vertId % 6;
    vID = vID < 3 ? ( vID == 2 ? 3 : vID ) :
                    ( vID >  4 ? 3 : ( vID - 2 ) );

    float2 uv_off = inv_tf * uv_shift[ vID ];

    u += uv_off.x;
    v += uv_off.y;

    // compute screen space adjusted tessellation factors for each
    // edge of the original input quad
    // please note that on the CPU the same tessefactors
    // get computed in order to bin terrain quads into bins
    // with comparable maximum tessellation factors
    f4EdgeFactor = computeEdgeTessFactors( f3Pos[0], f3Pos[1],
                                           f3Pos[2], f3Pos[3] );

    // combine tessellation factors to a single one that depends
    // on the current uv position in the highest res tessellation
    float4 f4EdgeFlag = ( float4( 1.0f - v, 1.0f - u , v, u ) +
                        ( 1.0f/128.0f ).xxxx ) >= (1.0f).xxxx ?
                        (1.0f).xxxx : (0.0f).xxxx;
```

```
float   fDotEf      = dot( f4EdgeFlag, (1.0f).xxxx );
float   fInterpTF   = max( max( f4EdgeFactor.x, f4EdgeFactor.y ),
                          max( f4EdgeFactor.z, f4EdgeFactor.w ) );

[flatten]if( fDotEf != 0.0f )
{
    fInterpTF = f4EdgeFlag.x != 0.0f ? f4EdgeFactor.x : fInterpTF;
    fInterpTF = f4EdgeFlag.y != 0.0f ? f4EdgeFactor.y : fInterpTF;
    fInterpTF = f4EdgeFlag.z != 0.0f ? f4EdgeFactor.z : fInterpTF;
    fInterpTF = f4EdgeFlag.w != 0.0f ? f4EdgeFactor.w : fInterpTF;
}

// now we need to compute the closest uv position in the next
// closest (lower but even) tessfactor
float fLowTF  = float( ( uint( floor( fInterpTF +
                         ( 1.0f / 128.0f ) ) ) >> 1 ) ) * 2.0f;
float fHighTF = fLowTF + 2.0f;
float fHighU  = 0.5f + ( u > 0.5f ?
                 floor( ( u - 0.5f ) * fHighTF +( 1.0f/ 128.0f ) ) :
               - floor( ( 0.5f - u ) * fHighTF +( 1.0f / 128.0f ) ) )
                / fHighTF ;
float fHighV  = 0.5f + ( v > 0.5f ?
                 floor( ( v - 0.5f ) * fHighTF +( 1.0f / 128.0f ) ) :
               - floor( ( 0.5f - v ) * fHighTF +( 1.0f / 128.0f ) ) )
                / fHighTF ;
float fLowU   = 0.5f + ( fHighU > 0.5f ?
                 floor((fHighU - 0.5f) * fLowTF + (1.0f / 128.0f) ) :
               - floor((0.5f - fHighU) * fLowTF + (1.0f / 128.0f) ) )
                / fLowTF ;
float fLowV   = 0.5f + ( fHighV > 0.5f ?
                 floor((fHighV - 0.5f) * fLowTF  + (1.0f / 128.0f)) :
               - floor((0.5f - fHighV) * fLowTF  + (1.0f / 128.0f)))
                / fLowTF ;

// now we need to update the real uv morphing between low tess and
// high tess UV
float fLerp = 1.0f - ( ( fHighTF - fInterpTF ) / 2.0f );

// lerp uv - this will generate zero area triangles for
// triangles that are not needed
u = ( 1.0f - fLerp ) * fLowU + fLerp * fHighU;
v = ( 1.0f - fLerp ) * fLowV + fLerp * fHighV;

// compute bilinear lerp of corners
O.vPosition.xyz = ( 1.0f - u ) * ( ( 1.0f - v ) * f3Pos[0] +
                                           v    * f3Pos[1] ) +
                         u    * ( ( 1.0f - v ) * f3Pos[3] +
                                           v    * f3Pos[2] );
O.vPosition.w    = 0.0f;

// do whatever you need to do to make the terrain look
// interesting
. . .
. . .

// transform position
O.vPosition = mul( float4( O.vPosition.xyz, 1.0f ),
                   g_mWorldViewProjection );

return O;
```

Listing 8.2. A vertex shader that implements quad domain.

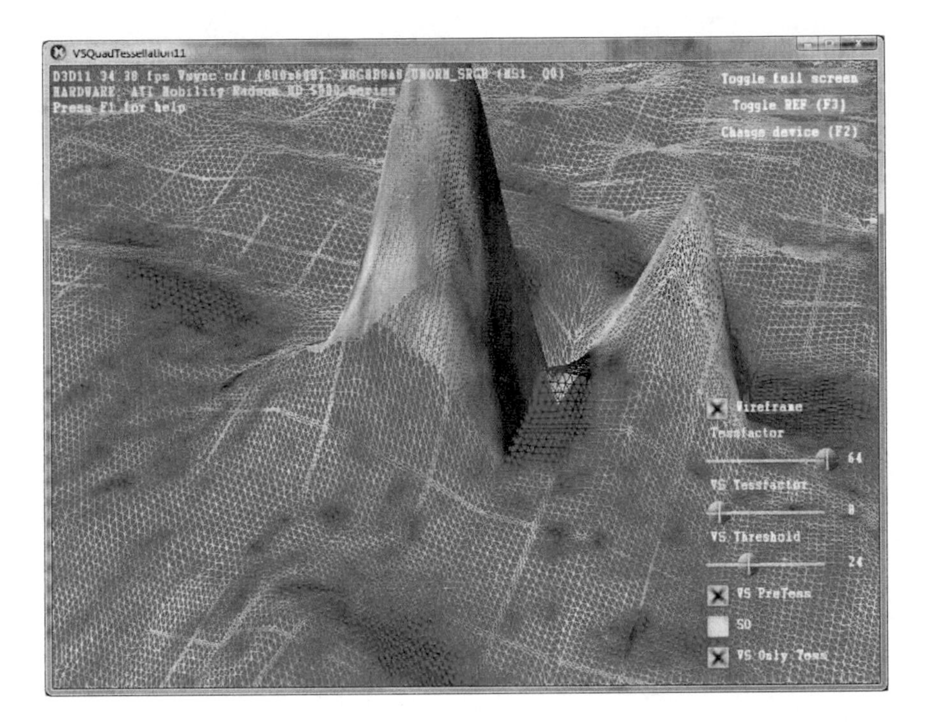

Figure 8.3. Wireframe view showing fractional tessellation factors implemented in a vertex shader.

Figure 8.3 shows the vertex shader in action. Although the sample application that was used to generate the screenshot is a genuine DX11 application, the shot shown uses the vertex shader only to generate the tessellation shown. The fractional tessellation factors can clearly be observed as seams between patches.

8.5 Conclusion

This chapter has presented a simplified way to implement tessellation in a vertex shader. The method is simplified in that it does not need any data other than what is already present in the vertex and index buffers. In addition, we present a way to implement fractional tessellation factors.

The concepts shown in this chapter can easily be used to get around using a geometry shader under many circumstances. As geometry shader performance relies on the sizing of internal buffers, using the concepts presented here may lead to faster and parallel performance.

Bibliography

[Gruen 05] H. Gruen. "Efficient Tessellation on the GPU through Instancing." *Journal of Game Development* 1:3 (2005), pp. 5–21.

[Tatarinov 08] A. Tatarinov. "Instanced Tessellation in DirectX 10." Available at http://developer.download.nvidia.com/presentations/2008/GDC/Inst_Tess_Compatible.pdf, 2008.

Optimized Stadium Crowd Rendering

Alan Chambers

The video games industry has benefited immensely from the wealth of information on instanced rendering and crowd simulation to have emerged in recent years [Dudash 07]. It has allowed developers to create a much more immersive gaming experience with hoards of visible characters that render at interactive frame rates. However, while these techniques form the basis of efficient crowd rendering within a stadium environment, they often fail to realize the unique set of problems and optimization opportunities that exist within this field. This chapter aims to extend on previous work in the field of crowd rendering, focusing predominantly on the math, techniques, and optimizations that can be made for stadium crowds.

9.1 Introduction

Many sports titles feature stadia that require some degree of crowd rendering technology. The geometric detail required for both the seats and the crowd characters is a problem on current commercial consoles, especially when trying to reproduce an 80,000-seat stadium. This is made more difficult by the fact that stadium crowds are a peripheral feature that we want to spend as little time on as possible. Indeed, many video games end up having to go with a simple solution, prerendered flip book animations. These generally suffer from inconsistent lighting and a flat overall appearance. Instanced crowd simulation provides the basis for a system that is not only visually much better but is also able to react to in-game events, which can greatly enhance the game play experience. However, even some of the latest games that use this technique suffer from perspective issues and completeness problems. This can lead to whole sections of the crowd being incorrectly oriented at times or undesirable gaps appearing when viewing them from certain angles, both of which can detract from the realism of the experience.

The system presented in this chapter tried to address many of these problems in *Rugby Challenge* on PlayStation® 3, Xbox 360, and PC. It begins with a tour of the data pipeline that explains how we can accurately place seats around the stadium and populate them with characters at runtime. A discussion of the real-time rendering process and its problems are then laid out together with performance evaluations that highlight the bottlenecks and potential areas for improvement. In addition to this, we also reveal the tricks for achieving colored "writing" in the stands, ambient occlusion that darkens the upper echelons, and crowd density that can be controlled live in-game. Following this, we are able to focus some discussion on the optimizations that can reduce the cost of the system even further. The symmetrical nature of stadium architectures and their fixed seating structure allow us to make specific optimizations that we could not do for a generic crowd system. Essentially, a complete pipeline for a fast and flexible stadium crowd simulation is presented, together with a discussion of the problems you can expect to face when implementing this type of system.

9.2 Overview

This chapter explores the use of instancing technology, deferred rendering [Policarpo 05], and imposters [Schaufler 95] in reducing the cost of rendering huge crowds on the GPU without compromising the color consistency in the scene.

The system is split into two content pipelines and three rendering phases. Each content pipeline produces data offline that is then loaded at runtime and fed into a specific rendering phase. The results of these two renders are then used in the final phase to shade and light the crowd area of the scene (see Figure 9.1).

The Model Content Pipeline is responsible for creating the character and seat geometry that will be featured in the stadium. It uses our existing model conditioner tools and exports data in an optimal format for the target platform.

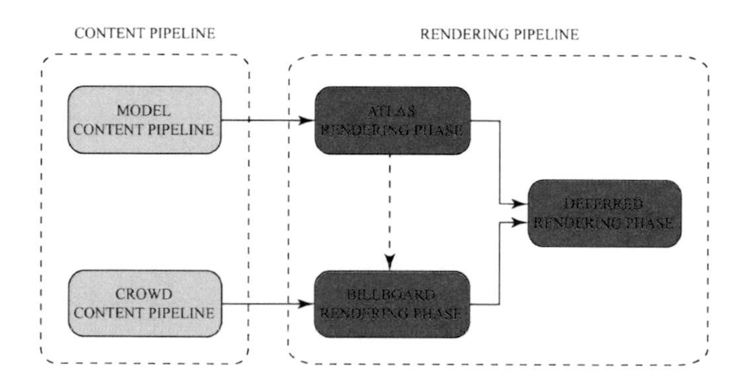

Figure 9.1. Overview of the crowd rendering pipeline.

The Crowd Content Pipeline produces a mesh that ultimately determines where our seats and characters will be instanced in the scene. Each vertex in the crowd mesh represents a real-world seat position from which we will instantiate our imposters.

The Atlas Rendering Phase takes the model set produced in the Model Content Pipeline and generates a texture atlas at runtime using our normal model rendering pipeline. Each model is rendered into the atlas from a specific viewpoint to create an imposter image that we can use multiple times.

The Billboard Rendering Phase instances each position in the crowd mesh to produce a view plane-aligned quadrilateral [Möller et al. 08] that contains one of the imposter images we generated in the Atlas Rendering Phase. It is an intermediate step in producing the final shaded crowd scene.

The Deferred Rendering Phase is the final stage in the pipeline. It uses the buffers produced in the Atlas and Billboard Rendering Phases to shade the crowd in a manner that is consistent with the rest of the scene. These colors are then written into the framebuffer to display our final crowd image.

9.3 Content Pipeline

Accurate seat and character placement is vital for creating a convincing stadium crowd representation. In this section we discuss the model data that we use to generate our imposters and the platform for accurately placing each one around the scene.

9.3.1 Model Content Pipeline

The first content pipeline produces the character and seat data for the Atlas Rendering Phase. Each model we produce needs to be performance friendly, i.e., have a fairly low polygon count with per-pixel lighting disabled. Note that we still require surface normal information to perform our per-pixel lighting calculations later in the pipeline.

The pipeline produces a list of character models, some female and some male, that we can dynamically load at runtime, usually before the start of a match. The geometry is skinned so we can control their gestures at runtime and make them respond to in-game events in a fluid and dynamic manner. If we use an atlas to map a variety of standard clothes for each character, we can swap it out at load time for a dedicated team color strip to introduce home and away supporters. This contributes to the general color in the stands and immediately relates the crowd to the teams on the playing surface.

The seat can be a static piece of geometry that is tailored to each particular stadium. However, this item is a special case since it is not colored directly. Instead, we use a grayscale texture that will act as a mask and a light map. The mask information is contained in the alpha channel so we can identify seat pixels

Figure 9.2. The source models used to create imposters.

and swap out the gray color for a stadium-specific one. Normally this procedure would leave us with a flat color, but if we apply the original grayscale value as a light map, then we will be able to create authentic seats and colorize them on an individual basis (see Figure 9.2).

9.3.2 Crowd Content Pipeline

The second content pipeline provides the data for accurately positioning our seat and character imposters around the stadium. Since the seats have a fixed position in the stadium and the crowd characters occupy the same space, we can use each vertex in a standard mesh to specify the position of both elements. These vertices can then be instanced into a quadrilateral at runtime to map a desired imposter image. Furthermore, we can encode elements of data into an associated color stream to influence the work done in the shader. This makes the rendering pipeline very different than that of a normal mesh. For this reason, we chose to decouple the crowd mesh from the stadium mesh and have it exist as a separate entity, despite the obvious position and orientation dependencies. This decoupling gave us the freedom to make optimizations that would have been awkward to integrate into a generic model rendering pipeline (see Figure 9.3).

The crowd mesh contains four data streams: vertex, billboard UVs, ambient occlusion UVs, and color. Note that since all the vertices are unique and we are instancing each one in our draw calls, we can actually omit the index buffer in our setup to save memory. Each data stream has a special role to play in the system. The vertex stream contains the seat positions that we extrapolate into billboards in the shader to create our imposters. In order to do this, the billboard

Figure 9.3. The crowd mesh is separate from the stadium. Each blue dot is a vertex in the mesh that specifies a real-world seat position.

UV stream needs to be populated with the four unit corners of a quadrilateral so we can map each instanced vertex to a specific corner from within the shader. The ambient occlusion UV stream is used to map each vertex into a texture that describes the ambient light information for the stadium. Finally, we have a four-component color stream for smuggling vertex attributes into the shader. We use this to connect some important values to the vertex data:

- X - density,
- Y - seat ID,
- Z - dude ID,
- W - shirt ID [optional].

The *density* value refers to the probability of someone actually sitting in that seat. It is used by the vertex shader to determine whether the seat is occupied. If the value at that vertex is higher than the required crowd density, then we will draw an empty seat. The *seat ID* is simply an index into a 256 $\times$ 1 color palette that will define the final color for the seat. We use the *dude ID* to index into a specific slot of the atlas so we can specify exactly which person is seated in a particular seat. Finally, we can choose to set up a *shirt ID* to control the appearance of our characters in the same way as the seat ID [Maïm 08]. All of these values are generated offline with the exception of the dude ID, which we create randomly at runtime to avoid patterns in the crowd from appearing. This is not strictly necessary, since the inclusion of various stadia with different crowd

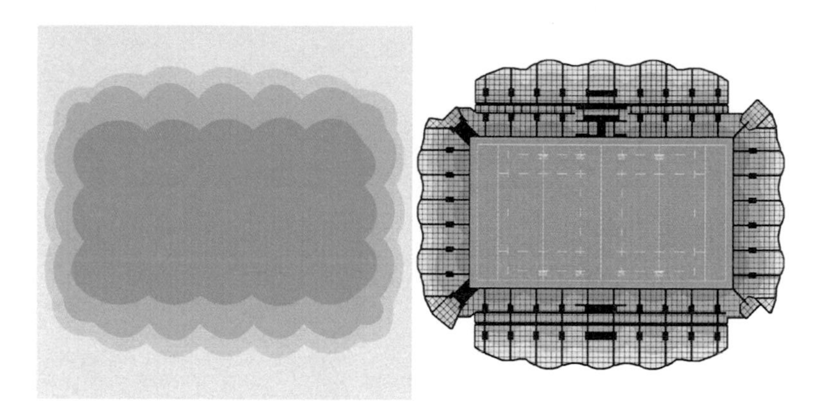

Figure 9.4. Example map that generates our seat color palette and crowd mesh color indices.

meshes, densities, and team colors can be enough to make them seem different every time.

Manually setting density and color indices for individual seats can be a very mundane process, so to help the artists we created a script that allowed them to take hand-painted maps and apply them directly to a crowd mesh, as shown in Figure 9.4. The color range can then be extracted from the texture and stored in a separate palette to determine individual seat colors at runtime.

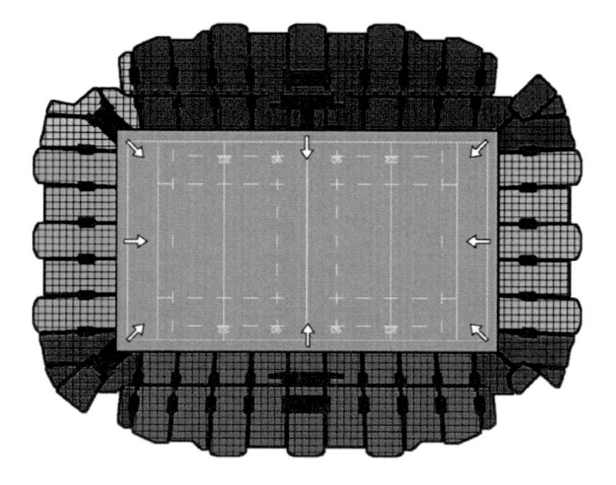

Figure 9.5. Dividing our crowd mesh into eight sections is enough to cover every aspect of the stadium.

The crowd mesh needs to be divided up into sections so we can control the view direction of the characters. To do this, we have each section contain a world-space transform that influences the orientation of our models at render time. Each section will trigger a render of the entire model set to ensure our characters oriented correctly, so we must keep these to a minimum if we are to remain fast at runtime. Most stadia naturally divide up into eight distinct sections, whose spectators coarsely share the same view direction as shown in Figure 9.5. Of course, this simple approach does mean there will be slight discrepancies in the viewing angle for some subsections of the crowd. However, these are almost always unnoticeable. Overall, we found this level of granularity worked well in creating a seamless camera panning experience around the whole arena, even for circular stadia.

Note that each section is set to look at the center of the stadium along the ground plane so each character faces the on-field action.

9.4 Rendering

The problems associated with drawing a large number of characters in a stadium environment can be solved efficiently with three separate draw phases. The first phase renders each of the models into a texture atlas to effectively cache the results and avoid the expense of repeated triangulation setups, transforms, and rasters that we would have to endure with fully fledged models. In this form we can reduce the complexity of the data to one quadrilateral per crowd member and not lose any fidelity. The primary drawback with the process is that rendering large amounts of billboards in such close proximity can, from certain camera angles, create huge amounts of overdraw. This means we are going to compute a lot of pixels that end up being discarded. To minimize the impact of overdraw we can defer the color and lighting to a later stage and use a simple fragment shader to cut down the cost of the actual pixel work. Each phase plays an important part in creating our crowd at an interactive frame rate, and this section describes them in detail.

9.4.1 Atlas Rendering Phase

We need to generate a texture atlas at runtime that contains all the viewpoint renders of our characters and seats so our billboards can reference various imposter images. Essentially, the data produced in the Model Content Pipeline is rendered out to two buffers containing color and normal information. With this data we can apply lighting and bump mapping techniques at a later stage to give our imposters a dynamic appearance. Each of the models needs to be rendered from a specific viewpoint that compensates for the flat 2D billboard projection of the imposters. To do this we have to set up a separate scene that contains the target model at the origin and an offscreen camera that orbits the model at a set distance, as shown in Figure 9.6.

Figure 9.6. The offscreen camera orbit around our models.

The position and orientation of the offscreen camera is fundamental to creating the imposter effect. We need to make the camera orbit the character in a way that reflects the orientation of both the world camera and the crowd section if our imposter is to be displayed correctly. To do this we have three control parameters that can influence the final camera position: pitch, relative angle, and distance.

The *pitch* of the main scene camera can be mapped directly to the offscreen camera to ensure that it shares the same view incline. The *relative angle* refers to the angle between the main scene camera and the crowd section forward vectors. It is a key element to rendering from the correct perspective, essentially providing the orbiting influence around the y-axis. These two parameters allow us to construct a unit vector that defines the ideal line on which to position our camera. A scalar *distance* value can then be applied to determine the final position of our offscreen camera and ensure we orbit an arbitrary focus point at a set distance (see Figure 9.7).

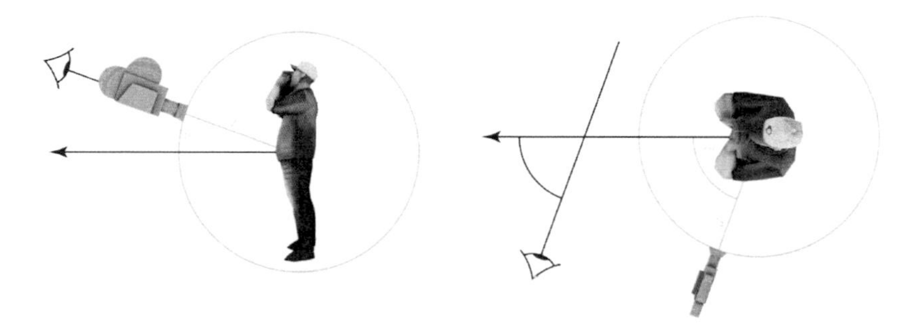

Figure 9.7. The control parameters that influence the orbit.

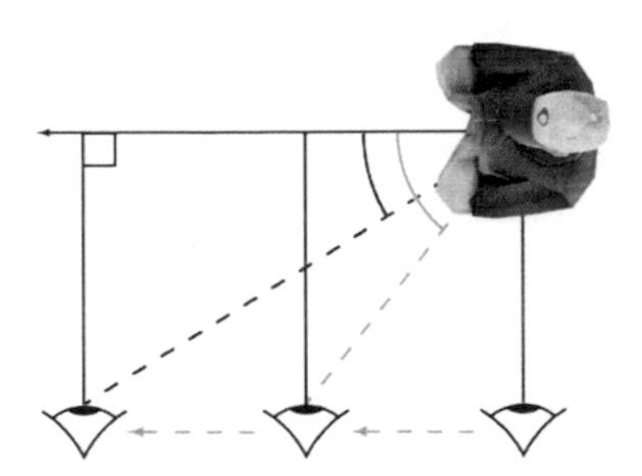

Figure 9.8. The translational influence of the world camera.

However, there is also a relationship between the world-camera position and the crowd section that we must account for if our models are to be correctly oriented within the billboards at all times. Consider the case in Figure 9.8, where the world-camera view direction is perpendicular to that of the crowd section. Here we can clearly see that the offscreen camera view direction needs to change as we translate away from the crowd section.

To compensate for this, we deploy a function to scale the relative angle according to the world-camera position (p_{wc}), the crowd section position (p_{cs}), and the crowd section forward vector (v_{cs}):

$$f(x) = \frac{0.5}{x+1} + 0.5,$$

where $x = (p_{wc} - p_{cs}) \cdot \hat{v}_{cs}$.

If we retrieve the vector between the world camera and the crowd section, we can calculate the scalar product along the crowd vector and feed it into our function to retrieve the amount of compensation required. Note, we made p_{cs} a point on the front edge of the bound of the crowd section so our relative-angle scalar was tuned to the first row of the section. This is important since the game is usually rendered from a position on the pitch, so we need to ensure crowd members closest to the on-field action are rendered with minimal discrepancies.

Finally, we calculate a look-at matrix for the camera with our newly computed orbit position, as shown in Listing 9.1. Generally, this should be focused on a fixed point in the scene so that all the models are framed correctly and consistently.

Since texture size has a direct impact on performance, we need to carefully consider the dimensions of our atlas to ensure it is never a bottleneck. To do this reliably we divide the atlas up into slots that have equal dimensions. Each slot contains a different model render and is mapped through a section ID and a model ID. These values identify the row and column indices that we use to access the character data from a specific slot within the atlas. This setup makes it easy to control the final texture size through the slot resolution while also enabling us to automatically adjust the atlas size when our model set changes.

```
void Section::computeCameraPosition( const Camera& world_camera )
const
{
    Vector3 camera_zaxis( -world_camera.getForwardAxis() );
    Vector3 crowd_xaxis = crowd_matrix.getRightAxis();
    Vector3 crowd_zaxis = crowd_matrix.getForwardAxis();

    camera_zaxis.setY( 0.0f );
    camera_zaxis = normalize( camera_zaxis );

    float x_dot = dot( camera_zaxis, crowd_xaxis );
    float z_dot = dot( camera_zaxis, crowd_zaxis );

    float angle = fast_acos( z_dot );

    angle = x_dot >= 0 ? angle : angle * -1;

    Vector3 crowd_edge_pos = Vector3( 0.0f, 0.0f,
        crowd_bounds.getZ() ) * crowd_matrix;
    Vector3 camera_wpos = world_camera.getPosition();
    camera_wpos.setY( 0.0f );
    Vector3 scale_vector = camera_wpos - crowd_edge_pos;
    float scalar = dot( scale_vector, crowd_zaxis );
    angle *= ( 0.5f / ( abs( scalar ) + 1.0f ) ) + 0.5f;
    float pitch = world_camera.getPitch();

    Vector3 camera_fwd = Matrix3::rotationX( -pitch ).
        getForwardAxis();
    Vector3 camera_dir = Matrix3::rotationY( angle ) * camera_fwd;
    Vector3 camera_pos = camera_dir * MIN_CAMERA_DIST;

    return camera_pos;
}
```

Listing 9.1. Computes an offscreen camera position.

The best resolution to use depends on the number of models and sections in your scene. We found a slot size of 64 × 128 in an RGBA format gave us a good detail to performance ratio on PlayStation® 3 for eight sections and ten models. Increasing this in powers of two saw significant performance hits as shown in Table 9.1.

Slot Width × Height	PlayStation® 3
64 × 128	0.16 ms
128 × 256	0.38 ms
256 × 512	1.31 ms

Table 9.1. The costs associated with rendering one section containing ten models into the atlas. Each section occupies a separate row of the atlas and each model occupies a separate column.

Figure 9.9. Base and associated normal map atlas.

Since we will be deferring all the color and lighting calculations until later in the pipeline, we need to set the pass up to take advantage of multiple render targets and have the fragment shader output two components: a flat color and a surface normal.

We can now render the model set into the atlas for each visible section. The first model we draw is the seat and it is always rendered into the first slot of the appropriate row in the atlas. This marks out the seat area with our alpha mask value as well as storing its light map information. Since the seat is the same for all the characters, we can copy this image across to the other slots in the row. Each character is then set in a specific pose and rendered over the top from the same viewpoint, as shown in Figure 9.9. Now we only have to perform one texture lookup in the shader to get both the seat and character information. Note that we must take care to ensure the character image is never clipped because of wild cheering animations that exceed the viewport scissor rectangle.

When rendering our texture atlas, care must be given to the use of bilinear filtering as it can cause problems for us in the later parts of the rendering pipeline. The alpha mask values along the edges of our characters will become polluted when they are sampled since the edge texels will be averaged with untouched texels. This can cause visible fringes to appear since our seat texels cannot be detected and swapped out for the correct color. To solve this we can implement a region map that stores the alpha information in a separate A8 texture. We can

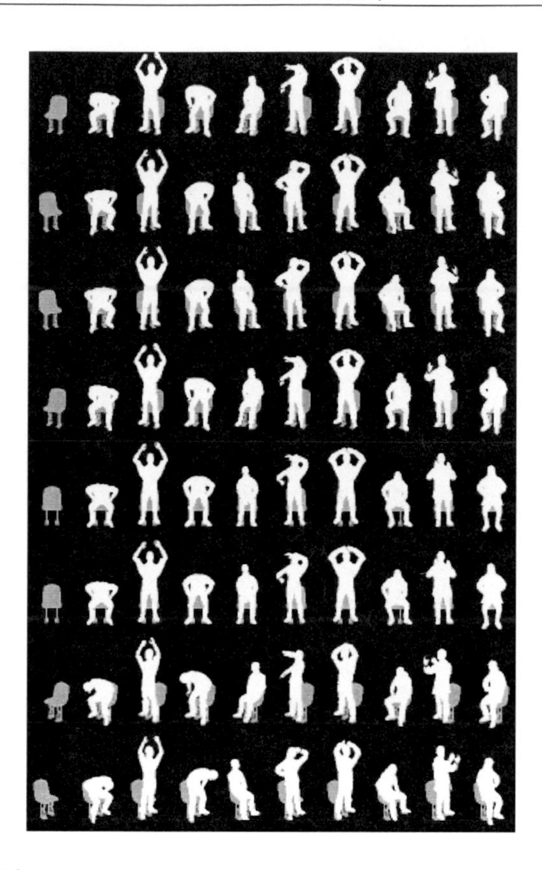

Figure 9.10. Region map identifies seat pixels from character pixels.

then use nearest filtering to sample the region map and safely identify our actual crowd pixels while using bilinear filtering to retrieve our color values from the atlas (see Figure 9.10).

One of the most important techniques we need to deploy as a postprocess over the texture atlas is real-time mip-mapping. Both the Billboard Rendering Phase and the Deferred Rendering Phase require fast texture lookups to reach maximum efficiency. The use of mip-maps improves texture cache performance and essentially speeds up texture fetching, which is vital to make the system run fast. The actual solution we adopted was straightforward. We simply loop through and downsample each level into the next by rendering a quad with the current level as the source texture. Note that the use of bilinear filtering in this step can again be problematic. The alpha value pollution leads to a reduction in the alpha test coverage, i.e., the proportion of pixels that pass the alpha test decreases with each level [Castao 10]. This causes each level to become progressively more transparent and so the image appears to thin out with distance.

Slot Width × Height	PlayStation® 3
64 × 128	0.17 ms
128 × 256	0.67 ms
256 × 512	3.01 ms

Table 9.2. The costs associated with mip-mapping one row of ten slots in the atlas.

The use of mip-mapping at runtime has an associated cost. Usually the decrease in texture-fetching latency when reading from the atlas makes the process worthwhile. However, the size of the atlas once again becomes a vital consideration. As the resolution of the atlas increases, the cost of performing the mip-mapping in real time increases dramatically as shown in Table 9.2. This further illustrates the importance of selecting the right atlas resolution for your application.

9.4.2 Billboard Rendering Phase

The objective of this phase is to instance each of our crowd quadrilaterals and stencil out the crowd region in the scene. The primary bottleneck is overdraw. To minimize its effect, we render out each visible section of the crowd mesh into several intermediate buffers, as shown in Figure 9.11.

We defer as much computation as possible to the later stages in the pipeline to avoid heavy pixel-processing work that will never end up in the final image. This is done over several steps. First, we set up three fullscreen 2 × 16-bit floating-point render targets together with the main scene depth-stencil buffer. These

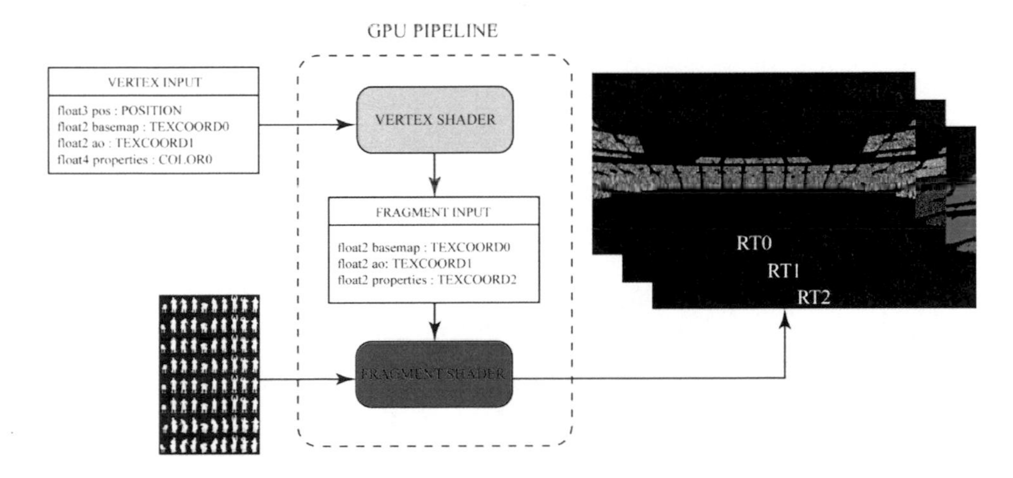

Figure 9.11. An overview of the billboard shader pipeline.

render targets will store information relating to the following:

- RT0 - interpolated base map UVs,
- RT1 - ambient occlusion UVs,
- RT2 - seat ID and section ID.

The stencil buffer provides an easy and efficient means of marking out our crowd region. If we set it up to always pass on write and replace its contents with a custom reference value, we only have to draw into our offscreen render targets to start marking out the area that will be covered by our crowd. The binding of the main scene depth-stencil buffer is important. Despite rendering into an offscreen buffer, we are still able to populate the main scene with depth information and stencil out the region of crowd for further pixel processing (see Figure 9.12).

The crowd mesh is submitted as quad data for rendering. We bind each of the four data streams in the crowd mesh, modulating the billboard UV stream and dividing the others by a factor of four to instance each corner of our quadrilaterals. Each repeated position can then be mapped onto an appropriate corner using the billboard UV information that comes into the vertex shader. This leaves us needing to determine which image in the atlas we will map on to the quadrilateral. The outcome depends on whether we want to render a character or an empty seat at that position. If we have a target density value for a particular match, we can do a step in the shader to see if this is greater than the density value in the color stream and select an appropriate UV set. Dynamically adjusting population levels live in-game then becomes only a matter of changing the target density value that gets sent to the shader. In this way, we can progressively empty the stadium near the end of a match if the home team is losing or populate it with more people over the course of a season if the team is doing well (see Figure 9.13).

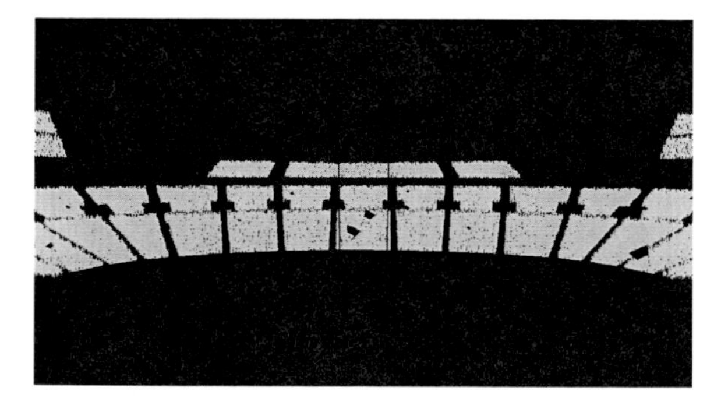

Figure 9.12. The stenciled region of the crowd.

Figure 9.13. Crowd density can change in-game.

The fragment shader for this phase needs to be as cheap as possible. The vast amount of layered quad data can create huge amounts of overdraw, which eats up GPU time when we are trying to render large crowds. To reduce this, we must have the shader perform a minimal number of texture lookups and defer as much of the pixel processing as possible. Texture lookups have an unpredictable cost and since the atlas is rendered in real time, it cannot benefit from swizzling optimizations to speed up neighboring memory accesses. This means we want to avoid them as much as possible to prevent overdraw becoming prohibitively expensive. Note that the application of per-pixel lighting is costly at this stage and can cause explosions in render time when viewing the crowd from certain camera angles. Indeed, any situation that causes the imposters to stack up on top of each other will generate huge layers of work over the same pixel, most of which will end up being discarded (see Figure 9.14).

The performance problems associated with overdraw are highlighted in Figure 9.15. In typical scenes, the crowd occupies between 0 and 25% of our total screen-space area. In these cases, each crowd character usually covers only a

Figure 9.14. The overdraw associated with crowd imposters.

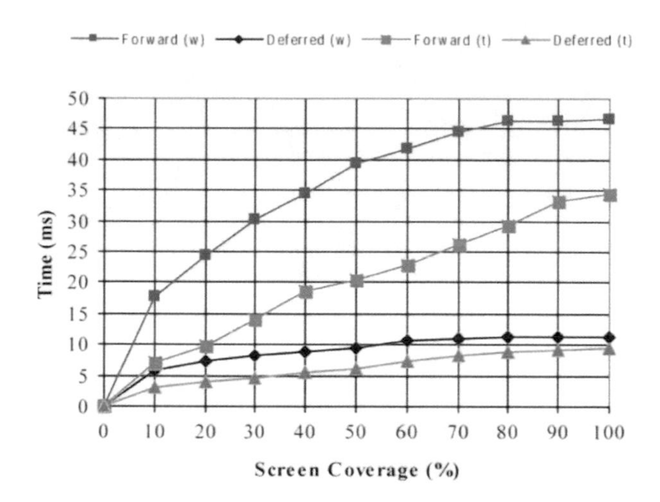

Figure 9.15. Metrics for a section of stadium crowd with 18,600 characters. The (w) timings are worst case. These were taken with a camera that looked down the whole length of the section to maximize overdraw. The (t) timings are typical case. These were taken from a typical in-game perspective that suffers less overdraw.

few pixels, which makes overdraw much less expensive. However, there are times when the crowd can occupy as much 40% or more of the framebuffer. If the camera is zoomed in on the crowd, the pixel space each character occupies will increase and this will drive up the cost of overdraw.

The fact that overdraw is camera dependent means we can control it to a certain extent. However, performance can swing wildly if we are not careful. Figure 9.15 shows us that by deferring the majority of our pixel processing in this phase, we become much less susceptible to swings in performance. The constant nature of the results means we can afford to give our in-game cameras greater freedom without fear of an unforeseen impact on frame rate.

To simplify our fragment shader, we perform just one lookup from the region map. This will determine if we have a crowd texel or a dead space texel. If the alpha value is zero we have a dead space texel and execute a pixel kill in the shader to prevent us from writing into the stencil buffer. However, if the alpha value is nonzero we have a crowd texel and immediately write out the three values that were passed into the shader. This forces a stencil write to the main depth-stencil buffer, which marks out the crowd region while deferring as much computation as possible. Note that because of the simplicity of the shader, we do not submit this phase to our Z-Pre Pass system since it will actually duplicate work and increase the load on the GPU.

9.4.3 Deferred Rendering Phase

The final phase of the rendering pipeline generates a final output color using the outputs from the previous stages (see Figure 9.16). We submit a fullscreen quadrilateral so our code is executed over all the framebuffer pixels and we can draw into our previously stenciled crowd region. Note that the trivial vertex processing in this phase means we are always going to be fragment-shader bound, so changing the GPR allocation accordingly can help speed up the pass on hardware with unified memory architectures. The work is split into three components: color, lighting, and shadows. However, other stages such as a fog and tone mapping [Hable 10] can be added to keep the look and feel completely congruous with the rest of the scene.

The base color contribution is determined from the maps we produced in the previous two phases. The UVs that enter the shader are used to perform a lookup into each of the maps rendered out from the Billboard Rendering Phase. This gives us the RT0, RT1, and RT2 data for a particular fragment. The RT0 value contains our base map UVs and is used to fetch a color value from the texture atlas and an alpha value from the region map. This is going to yield one of three things: a seat texel, a character texel, or a dead space texel. We don't care about dead space texels since they will end up being eliminated by the stencil test in the ROP. However, we do need to determine if we have a seat texel so we can apply the color in the palette associated with the current stadium. We can do this by testing our region map texel for the mask value we used on the seat model. If the seat value is present, we use the index in the RT2 data to fetch a color texel from the palette before applying the light map value in the RGB atlas texel. This gives us a final base-color contribution for the seat and allows us to create convincing effects such as writing in the stands or wave-like patterns as shown in Figure 9.17. If the mask value is not present then we can assume we have a character pixel and simply use the texel value directly as our base-color contribution.

The lighting solution heavily dictates the performance of this phase. We need it to be fast while also reflecting the light setup in the main scene to prevent

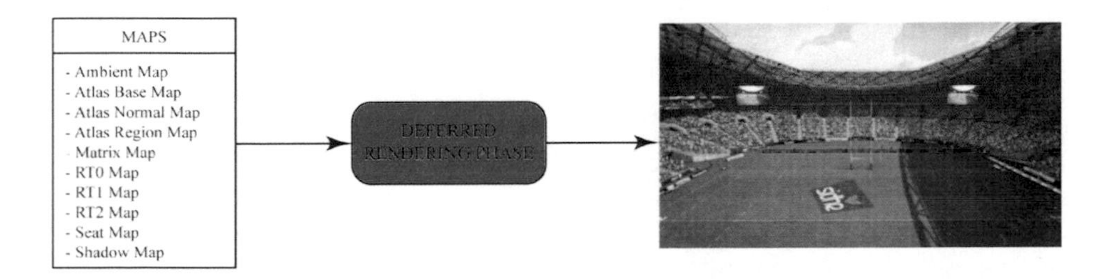

Figure 9.16. The shader pipeline for the final phase.

Figure 9.17. Individual seats can be assigned specific colors to create effects such as writing in the stands.

the crowd looking out of place. Although it would be nice to implement specular lighting using BRDF models [Kelemen et al. 01] and skin shading with subsurface scattering [Green 03], the computation across the whole crowd would be enormous. The main aim here is to produce a system that is fast and efficient while also giving the appearance of high fidelity and consistency with the main scene. To do this we need to cut down the amount of work in the fragment shader by using a small set of lights and cheaper equations that approximate the main scene conditions. We found that supporting one ambient light and one directional light was sufficient for illuminating the entire crowd in a manner that was consistent with the rest of the scene:

$$L_O(v) = (c_{\text{amb}}(c_{\text{ao}})) + (c_{\text{shdw}}(c_{\text{dir}}(n \cdot l_{\text{dir}})),$$

where c_{amb} is the ambient color, c_{ao} is the ambient occlusion contribution, c_{shdw} is the shadow contribution, c_{dir} is the directional light color, n is the surface normal, and l is the directional light vector.

The ambient light is traditionally thought of as a base-color constant that represents light coming from all directions. However, spherical harmonics have enabled us to take into account the color emission from surrounding surfaces as well [Sloan et al. 02]. This can be an expensive runtime process that often requires the computation of multiple coefficients in several directions. However, we can approximate this at load time by computing the light from the stadium in two or three directions and taking an average before it is supplied to the shader. This allows us to coarsely match the ambient light of the scene without impeding performance in the fragment shader.

Furthermore, we know that some areas of the stadium will also be occluded from the main light sources in the scene because of their architectural design.

Figure 9.18. Ambient occlusion that darkens the upper echelons of the stands.

In these cases, the amount of ambient light received will reduce as the field of incoming light decreases. This leads to the far, upper areas of stadia becoming progressively darker as the roof and surrounding walls shield more of the light source. To account for this, we can apply a cheap ambient occlusion solution [Christensen 03] using a map that describes the ambient light levels in the stadium. This allows us to retrieve the proportion of ambient light we need to apply at any point in the scene. The UVs stored in the RT1 buffer can then be used to sample the ambient map for a value that modulates the ambient light color component for the crowd pixel (see Figure 9.18).

The directional light information is used to compute the diffuse component of our lighting equation. However, the values in the normal map are in model local space so we need some way of getting them into the same space as our crowd section before we can evaluate the equation. Ideally we would fetch the transform directly from the crowd section using the section ID but since we cannot access elements of a constant buffer as an array from inside the shader, we have to use a different solution. We created a small RGBA texture and packed the upper 3×3 matrix of all our section transforms into a separate row during an initialize step. Each pixel then contains one vector from the 3×3 matrix in a compressed form, which means all eight sections can be stored in a 3×8 texture. The shader then only has to do three texel lookups to construct a matrix that can transform the values in our normal map into the correct space for a particular section. Note that the size of the texture makes this a cache-friendly operation that can be performed in parallel with ALU operations. Each normal we extract from the map is then transformed and fed into the standard n dot l lighting equation [Shirley et al. 05] to modulate our directional light color and produce an initial diffuse value.

The lighting model also contains a shadow component that needs to be applied to the diffuse. Since the computation of real-time shadows [Diamand 10] is often

Figure 9.19. A well approximated dynamic lighting solution will keep the crowd looking consistent with the rest of the scene.

expensive, we have to aggressively reduce the complexity for the crowd to ensure it runs fast. We decided upon two things:

1. The crowd would not cast shadows.

2. The crowd would receive shadows through an existing shadow map.

Calculating the shadow component now becomes only a case of looking up the shadow map value for the pixel. This is cheap to perform and effectively controls our diffuse input, ensuring no crowd pixels are lit from an obscured directional light source. The result is a convincing shadowing effect that is consistent with the main scene (see Figure 9.19).

A final color is then computed by adding the ambient and diffuse components to produce the lighting value, $L_0(v)$, that modulates our base color contribution. Note that the cost of this phase was fixed to 3.03 ms for a 720 p render target on PlayStation® 3, but this can be easily reduced using early rejection features such as stencil cull, as described in Section 9.5.2.

9.5 Further Optimizations

The technique presented in this chapter outlines an efficient model for real-time generation of stadium crowds. However, it can be optimized even further. This section focuses on areas that can be improved and offers solutions to help boost performance.

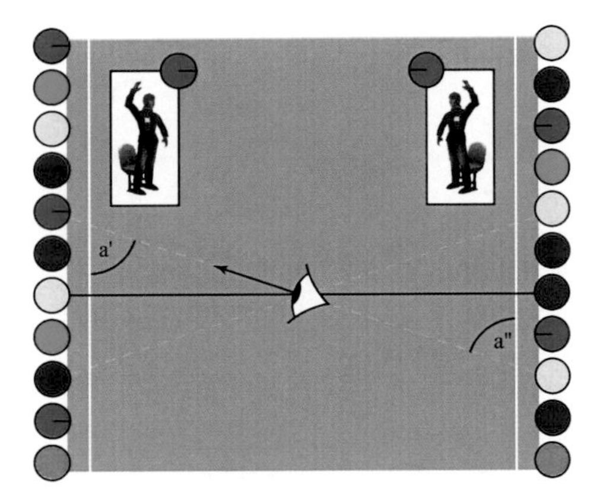

Figure 9.20. Each dot represents a vertex in the crowd mesh; the color corresponds to a particular crowd member to use in the atlas. The symmetry of the scene allows us to eliminate viewpoint renders but it does not mean we have to use the same configuration of crowd members in the opposing section.

9.5.1 Symmetry

The nature of stadium architectures presents us with a unique way in which we can reduce render time. The symmetry that exists between opposing stands allows us to effectively share the images produced in the Atlas Rendering Phase. Since each crowd member has a fixed world position and view direction, we are guaranteed to have congruent camera angles while we remain in the confines of the pitch (see Figure 9.20).

This relationship allows us to render with respect to one section and merely mirror the result for the opposing section in the Billboard Rendering Phase. We can see in Figure 9.9 that there is a mirror image of the models in each row elsewhere in the atlas. By sharing the appropriate row we can effectively halve the number of crowd section renders required for the simulation in the Atlas Rendering Phase. As a result, we only need to render each model from four viewpoints to populate an entire stadium full of people. This reduced the cost of the phase on PlayStation® 3 to a consistent and affordable time of around 1 ms as seen in Table 9.3.

	PlayStation® 3
8 sections	2.87 ms
4 sections	1.20 ms

Table 9.3. The performance benefits of symmetry optimizations. Timings were taken with each section rendering ten models.

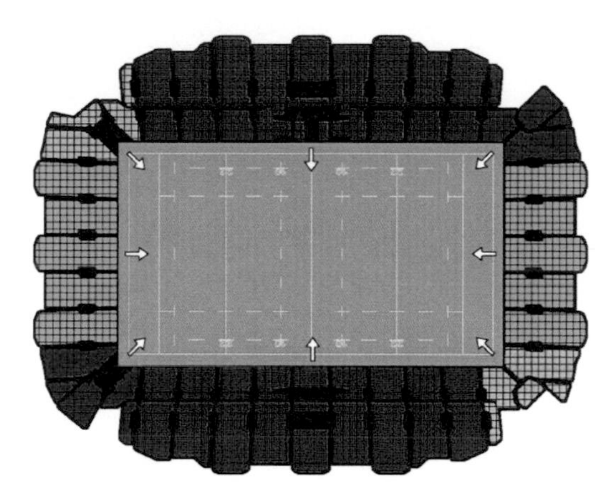

Figure 9.21. The group allocation of opposing sections.

The reduction in size of the atlas is significant for performance. Since we are only rendering half the number of sections we can save both memory and render time. This is significant for two reasons:

1. Sampling is less likely to suffer cache misses.

2. There is less data to mip-map in real time.

To implement the system we need to assign each section in the crowd mesh to one of four groups as shown in Figure 9.21. Each group contains only the sections that are diametrically opposed so we can track the symmetrical relationship and easily share their imposter images. Of course, not every stadium will have opposing stands, or even corner sections. In these cases, we can simply populate the symmetrical group with only one section or remove the group completely to avoid drawing them altogether.

To take advantage of the symmetry between opposing stands we need to modify the code in Listing 9.1. If we use the absolute on the scalar product before calculating the relative angle, we will always get a positive input value that mirrors the angle of the opposing crowd section. We then only have to flip the atlas UVs in the vertex shader to be able to use the image with mirrored crowd sections. Note that opposing crowd sections are still free to have a completely different character and seat configuration since only the atlas itself is mirrored and not the crowd section geometry.

The integration of mirrored rendering has an important consequence for the lighting calculations we do in the Deferred Rendering Phase. Although the image is horizontally flipped for mirrored sections, the actual normal values will remain

Figure 9.22. The need to flip the normals for mirrored sections.

in the same relative space as the original section. If we do not account for this, the lighting equation will not be solved correctly for the mirrored sections. This is easily resolved by adapting some of the matrices we pack into the 3×8 texture. Since these matrices are only used for lighting calculations, we can simply flip the x-axis for the mirrored sections when we pack in the data. The shader will then automatically flip the normal when it applies the lighting transform with no extra overhead incurred (see Figure 9.22).

To be able to detect mirrored sections we need to identify in which half of the stadium the crowd section lies. Fortunately, this can be easily computed, as shown in Listing 9.2.

```
void Section::isMirror() const
{
    Vector3 xAxis( 1.0f, 0.0f, 0.0f );
    Vector3 zAxis( 0.0f, 0.0f, 1.0f );
    Vector3 dir = m_worldTransform.getForwardAxis();

    dir.setY( 0.0f );
    dir = normalize( dir );
    float dp = dot( dir, zAxis ) + dot( dir, xAxis );
    m_isMirror = ( dp >= 0.0f );
};
```

Listing 9.2. Computes the symmetry state for a section.

If we take the dot product between the crowd-section forward axis and each unit x- and z-axis, we can add the two results together to determine the side of the stadium the crowd section lies in (see Figure 9.23). Any section that lies in the second half of the arena is then identified as a mirror. Furthermore, we can integrate this functionality into our content build to automatically detect mirrored sections and set up the correct lighting transform to use at runtime.

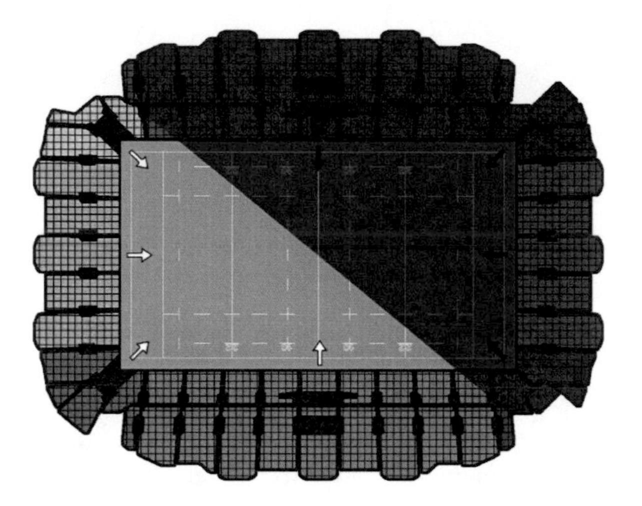

Figure 9.23. Detecting mirrored sections with the dot product.

9.5.2 Stencil Cull

The use of the stencil buffer presented so far only ensures that our crowd system draws to the parts of the framebuffer that we need it to; it does very little for performance. This is because the hardware only rejects pixels based on the stencil test during the ROP stage of the GPU pipeline, which occurs after our main bottleneck, fragment processing, has been executed. Essentially, work will still be carried out for pixels that are never actually displayed unless we take advantage of the stencil cull features in modern hardware. Stencil culling hardware can reject whole pixel quads before they reach the fragment shader provided all the pixels in the quad are set to CULL. However, it can only do this if the results of the stencil buffer are flushed into the stencil cull system. This means we must populate the stencil buffer before we can utilize the early rejection hardware to cull pixel data. Since the Billboard Rendering Phase is already populating the stencil buffer, we can easily instruct the hardware to set the corresponding stencil cull bit to PASS when we stencil out our crowd area. With a properly populated stencil cull region we can set up the Deferred Rendering Phase to recycle the results and early reject all the pixel quads that are still set to CULL. This can avoid the cost of pixel processing for large areas of the framebuffer since our crowd rarely occupies a significant portion of the final image. We found this to provide a very necessary win on both PlayStation® 3 and Xbox 360, as shown in Table 9.4.

	PlayStation® 3
STENCIL CULL OFF	3.03 ms
STENCIL CULL ON	0.57 ms

Table 9.4. The performance benefits of stencil cull on the scene shown in Figure 9.12 at 720 p.

9.5.3 Variation

The technique can be made even cheaper if we minimize the number of model renders in each crowd section. Currently each crowd section triggers a redraw of the entire model set in its associated row of the atlas. While this maximizes the dispersion of models throughout the crowd, it also maximizes the cost of the Atlas Rendering Phase since we have to accommodate all of the available models. The large atlas size increases the cost of the real-time mip-mapping and the chance of enduring texture cache misses. To make it run fast, we need to keep the model set at a minimum without destroying the illusion that there are lots of different people in the crowd.

The optimization is simple; each section renders a subset of the available models. Although this cuts down on the variety within each section itself, it still allows for every model to be used across the whole stadium. Provided there is an overlap of models across neighboring sections, the dispersion can appear seamless to the point where the reduced model set goes unnoticed. This can be taken even further if we assume the corner sections are going to contain less people than the main stands. In this case, we can render half the amount of models for those sections and pack the results into one row of the atlas. Combining this with the mirrored rendering approach means we can effectively drop the number of rows in the atlas to just three with little noticeable reduction in character variety. The performance gains from this are evident in Table 9.5.

Sections × Models	PlayStation® 3
8 × 10	2.87 ms
8 × 6	1.6 ms
3 × 6	0.62 ms

Table 9.5. Minimizing the number of models we render in each crowd section keeps the cost low without adversely affecting perceived variety.

9.5.4 Nighttime Lighting

At night the lighting characteristics are very different from daytime. During the day there is a strong directional light source from the sun but at night, large flood

Figure 9.24. A nighttime scene with pure ambient lighting on the crowd closely resembles real-world conditions.

lights tend to produce a more ambient distribution, with light coming from all directions. If we use a specialized shader for night games, we can take advantage of the conditions and make our crowd run even faster. We can do this by completely removing the computation for the directional light contribution. By performing fewer texture lookups and ALU calculations in the Deferred Rendering Phase, we can save up to 1 ms on PlayStation® 3 and have the final result still closely resemble the conditions of a live match (see Figure 9.24).

9.5.5 Multicore CPUs

The CPU essentially controls the amount of work done by the GPU. The more work the CPU sends to the GPU, the more of its time we will consume. If we are smart, we will avoid sending the GPU work wherever we can, especially if it will never end up being displayed. Modern multicore CPUs present some interesting opportunities for offloading work from the GPU. In some cases, postprocessing techniques such as MLAA [Reshetov 09] have been completely lifted off the GPU and put onto a multicore configuration [De Pereyra 11]. These types of postprocessing techniques usually mean that the GPU has to stall at the end of a frame until the work has finished and it can flip the buffers. However, instead of letting the GPU stay idle for this time, we can give it some work to get on with for the next frame. By hiding work behind some other processing, we effectively get it done for free.

The Atlas Rendering Phase is an ideal candidate. It has to be refreshed every frame and re–mip-mapped. If we can do this while the GPU is normally idle then we can save the cost of the whole phase. Of course, this will lead to the atlas being rendered one frame behind as the results are based on a previous camera configuration. Ordinarily this is not a problem because the discrepancy in viewing angle is so small, it is unnoticeable. However, it can prove to be a problem when the camera suddenly switches direction or position. In this case, the atlas will need a total refresh to avoid a one frame "pop" caused from using stale data. As long as we can detect this camera change, we can re-render the geometry on demand for the affected frame and ensure we never use stale data. A one frame spike every time the camera cuts to a different area is usually more desirable than a permanent slice of GPU-time being lost every frame.

Another area in which we can use the CPU is vertex culling. If we farm a job out to another thread, we can have the CPU point-test every vertex in the crowd mesh against our camera frustum and build up a command buffer of visible vertices to send to the GPU. This can help reduce the cost of triangulation setup and vertex processing since we are essentially rejecting redundant data that will only end up being clipped.

9.6 Limitations

Although the technique is fast and flexible, it is not without limitation. It has been optimized for use within the confines of the stadium and, as such, if the camera is put in obscure places, the illusion can be broken. Often this can be easily rectified by adding new methods for calculating the offscreen camera position that better suit the individual needs of the title. Indeed, the system can be made much more flexible in this respect if we have a selection of specialized routines that each camera can hook into. With the ability to tailor each viewpoint calculation, we can resolve almost any camera complication.

9.7 Conclusion

Realistic stadium crowds are becoming increasingly important for selling a convincing sports arena in video games. However, creating a fast and efficient system to solve this is a complex problem that touches on a wide range of graphical areas.

This article has presented a complete pipeline for implementing realistic crowds on current commercial consoles. It has drawn attention to the main bottleneck of the process—overdraw—and offered a solution that minimizes the GPU workload. This was strengthened with several suggestions to optimize the system even further with symmetry, early-rejection hardware, and multicore CPUs.

Despite the ground made in the article, there is still plenty of room for improvement. In particular, the technique does not consider close-quarters rendering of the crowd. If the camera frustum includes a section of crowd at close range, then implementing a LOD system that can generate higher-resolution renders would be advantageous.

In summary, the technique has several benefits. It provides a way to create crowds that look and move in a fluid manner while also maintaining the color consistency in the scene. The system is fast and having been broken down into several phases, there is scope to hide the processing of whole sections behind other work on the CPU or GPU. This leads us to conclude that rendering large crowds in a stadium environment is even more viable on today's commercial consoles than those in an open air environment where there are less optimization opportunities.

9.8 Acknowledgments

I would like to extend an extra special thanks to Kester Maddock for being such a brilliant source of graphical knowledge on the project and always making himself available to discuss ideas. I also want to thank Wolfgang Engel, Robert Higgs, and Stuart Sharpe for giving me some great feedback on the article. I'm also grateful to Duncan Withers for producing some of the art that you see in the paper. Finally, I'd like to thank Mario Wynands and Tyrone McAuley at Sidhe for supporting my work and allowing me to publish the material.

Bibliography

[Castao 10] Ignacio Castao. "Computing Alpha Mipmaps." Available at http://the-witness.net/news/2010/09/computing-alpha-mipmaps, 2010.

[Christensen 03] Per H. Christensen. "Global Illumination and All That." In *ACM SIGGRAPH Course 9 (RenderMan, Theory and Practice)*. New York: ACM, 2003.

[De Pereyra 11] Alexandre De Pereyra. "MLAA: Efficiently Moving Antialiasing from the GPU to the CPU." Intel Labs, 2011.

[Diamand 10] Ben Diamand. "Shadows in God of War III" Paper presented at the Game Developer's Conference, San Francisco, March 9–13, 2010.

[Dudash 07] Bryan Dudash. "Animated Crowd Rendering." In *GPU Gems 3*, edited by Hubert Nguyen, pp. 39–52. Reading, MA: Addison Wesley, 2007.

[Green 03] Simon Green. "Real-Time Approximations to Subsurface Scattering." In *GPU Gems*, edited by Randima Fernando, pp. 263–278. Reading, MA: Addison Wesley, 2003.

[Hable 10] John Hable. "Uncharted 2—HDR Lighting." Paper presented at the Game Developer's Conference, San Francisco, March 9–13, 2010.

[Kelemen et al. 01] Csaba Kelemen and Lázló Szirmay-Kalos "A Microfacet Based Coupled Specular-Matte BRDF Model with Importance Sampling." In *Proceedings Eurographics*, pp. 25–34. Aire-la-Ville, Switzerland: Eurographics Association, 2001.

[Maïm 08] Jonathan Maïm and Daniel Thalman. "Improved Appearance Variety for Geometry Instancing." In *ShaderX*^{6}, edited by Wolfgang Engel, pp. 17–28. Hingham, MA: Charles River Media, 2008.

[Möller et al. 08] Thomas Akenine-Möller, Eric Haines, and Naty Hoffman. *Real-Time Rendering*. Wellesley, MA: A K Peters, 2008.

[Policarpo 05] Fabio Policarpo and Francisco Fonseca. "Deferred Shading Tutorial." http://bat710.univ_Tyon1.fr/~jciehl/Public/educ/GAMA/2007/Deferred_Shading_Tutorial_SBGAMES2005.pdf, 2005.

[Reshetov 09] Alexander Reshetov. "Morphological Antialiasing." Intel Labs, 2009.

[Schaufler 95] Gernot Schaufler. "Dynamically Generated Imposters." In *Modeling Virtual Worlds—Distributed Graphics, MVD Workshop*, pp. 129–135. Infix Verlag, GOP, Austria, 1995.

[Shirley et al. 05] Peter Shirley, Michael Ashikhmin, Michael Gleicher, Stephen R. Marschner, Erik Reinhard, Kelvin Sung, William B. Thompson, and Peter Willemsen. *Fundamentals of Computer Graphics*, Second Edition. Wellesley, MA: A K Peters, 2005.

[Sloan et al. 02] Peter-Pike Sloan, Jan Kautz, and John Snyder. "Precomputed Radiance Transfer for Real-Time Rendering in Dynamic, Low-Frequency Lighting Environments." *Proc. SIGGRAPH '02, Transactions on Graphics* 21:3 (2002), 527–536.

Geometric Antialiasing Methods

Emil Persson

10.1 Introduction and Previous Work

Recently a number of techniques have been introduced for performing antialiasing as a postprocessing step, such as morphological antialiasing (MLAA) [Reshetov09, Jimenez et al. 11], fast approximate antialiasing (FXAA) [Lottes 11], and subpixel reconstruction antialiasing (SRAA) [McGuire and Luebke 11]. These techniques operate on the color buffer and/or depth buffer and in the case of SRAA on super-resolution buffers. Another approach is to use the actual geometry information to accomplish the task [Malan 10]. This method relies on shifting vertices to cover gaps caused by rasterization rules and approximates the edge distances using gradients.

In this chapter, we discuss geometric postprocess antialiasing (GPAA) [Persson 11a], which is an alternative approach that operates on an aliased image and applies the antialiasing post-step using geometry information provided directly by the rendering engine. This results in a very accurate smoothing that has none of the temporal aliasing problems seen in MLAA or the super-resolution buffers needed for SRAA or the traditional MSAA. Additionally, we will discuss geometry buffer antialiasing (GBAA) [Persson 11b], which is based on a similar idea, but is implemented by storing geometry information to an additional buffer. This technique is expected to scale better with dense geometry and provides additional benefits, such as the ability to antialias alpha-tested edges.

10.2 Algorithm

Two geometric antialiasing methods will be discussed here. The first method operates entirely in a postprocess step and is called geometric postprocessing antialiasing (GPAA). This method draws lines over the edges in the scene and applies the proper smoothing. The second method is more similar to traditional MSAA in that it lays down the required information during main scene rendering and does the smoothing in a final resolve pass. This has scalability advantages over GPAA with dense geometry and provides additional opportunities for smoothing alpha-tested edges, geometric intersection edges, etc. On the down side, it requires another screen-sized buffer to hold the geometric information. Hence, it is called geometry buffer antialiasing (GBAA).

10.2.1 Geometric Postprocessing Antialiasing

Overview. Provided there is an aliased image and geometric information available in the game engine, it is possible to antialias the geometric edges in the scene. The algorithm can be summarized as follows:

1. Render the scene.

2. Copy the backbuffer to a texture.

3. Overdraw aliased geometric edges in a second geometry pass and blend with a neighbor pixel to produce a smoothed edge.

Steps 1 and 2 are straightforward and require no further explanation. Step 3 is where the GPAA technique is applied. For each edge in the source geometry a line is drawn to the scene overdrawing that edge in the framebuffer. Depth testing is enabled to make sure only visible edges are considered for antialiasing. A small depth-bias will be needed to make sure lines are drawn on top of the regular triangle-rasterized geometry. For best results, it is better to bias the scene geometry backwards instead of pushing the GPAA lines forward since you can apply slope depth-bias for triangles. However, if the edges are constructed from a geometry shader, it is possible to compute the slope for the primitive there, as well as for the adjacent primitives across the edge.

The antialiasing process is illustrated in Figure 10.1. Here, a geometric edge is shown between two primitives. It may be adjacent triangles sharing an edge or a foreground triangle somewhere over the middle of a background surface.

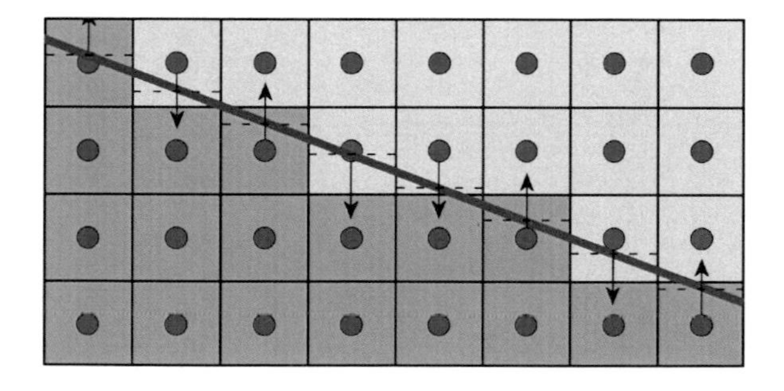

Figure 10.1. GPAA sample direction and coverage computation.

The first step in smoothing an edge is determining the major direction of its line in screen space, i.e., if it is more horizontal than vertical or vice versa. In Figure 10.1 the line is considered to be mostly horizontal, in which case we will choose neighbors vertically. Whether to go up or down depends on which side of the pixel center the edge is. If the edge is below the pixel center, we sample down, otherwise up. The same logic applies for the vertical case, except we choose between left and right.

Vertex shader. In processing an edge, the vertex shader first transforms the vertices to clipspace and then converts to screen-space positions in pixel units. A direction vector from first to second vertex is computed. If the absolute value of the x-component is larger than the absolute value of the y-component, then the major direction is horizontal, otherwise vertical. The line equation for the edge as well as a flag indicating the major direction is passed down to the pixel shader. Listing 10.1 shows the vertex shader.

```
struct VsIn {
    float3 Position0 : Position0;
    float3 Position1 : Position1;
};

struct PsIn {
    float4 Position : SV_Position;

    // The parameters are constant across the line, so
    // use the nointerpolation attribute. This way the
    // vertex shader can be slightly shorter.
    nointerpolation float4 KMF : KMF;
};

float4x4 ViewProj;
float4 ScaleBias;

PsIn main(VsIn In) {
    PsIn Out;

    float4 pos0 = mul(ViewProj, float4(In.Position0, 1.0));
    float4 pos1 = mul(ViewProj, float4(In.Position1, 1.0));

    Out.Position = pos0;

    // Compute screen-space position and direction of line
    float2 pos = (pos0.xy / pos0.w) * ScaleBias.xy
                 + ScaleBias.zw;
    float2 dir = (pos1.xy / pos1.w) * ScaleBias.xy
                 + ScaleBias.zw - pos;

    // Select between mostly horizontal or vertical
    bool x_gt_y = (abs(dir.x) > abs(dir.y));

    // Pass down the screen-space line equation
    if (x_gt_y) {
        float k = dir.y / dir.x;
        Out.KMF.xy = float2(k, -1);
```

```
    } else {
        float k = dir.x / dir.y;
        Out.KMF.xy = float2(-1, k);
    }
    Out.KMF.z = -dot(pos.xy, Out.KMF.xy);
    Out.KMF.w = asfloat(x_gt_y);

    return Out;
}
```

Listing 10.1. The GPAA vertex shader.

We are passing down a line equation of the form $y = k \cdot x + m$, or $x = k \cdot y + m$ in the vertical case. Depending on the major direction, we pass the slope factor k in the x- or y-component of the KMF output. In the pixel shader, we compute the signed distance from the pixel center to the line, so we want to subtract the pixel's y in the horizontal case and x in the vertical case. Thus we pass down -1 in the other component so that a simple dot product can be used regardless of the major direction of the line. The m parameter is computed to the z-component KMF output and the w gets the direction flag.

Pixel shader. Provided the line equation and a copy of the backbuffer, the antialiasing can be done with a fairly simple pixel shader (see Listing 10.2).

```
Texture2D BackBuffer;
SamplerState Filter;
float2 PixelSize;

float4 main(PsIn In) : SV_Target {
    // Compute the distance from pixel to the line
    float diff = dot(In.KMF.xy, In.Position.xy) + In.KMF.z;

    // Compute the coverage of the neighboring surface
    float coverage = 0.5f - abs(diff);
    float2 offset = 0;

    if (coverage > 0) {
        // Select direction to sample a neighbor pixel
        float off = (diff >= 0)? 1 : -1;
        if (asuint(In.KMF.w))
            offset.y = off;
        else
            offset.x = off;
    }

    // Blend with neighbor pixel using texture filtering.
    return BackBuffer.Sample(Filter,
        (In.Position.xy + coverage * offset.xy) * PixelSize);
}
```

Listing 10.2. The GPAA pixel shader.

First, we evaluate the distance to the line and use that to compute the coverage. Note that the coverage computed here is that of the other surface from which we sample a neighboring pixel, not the pixel's own coverage. If the edge cuts through the pixel center, the distance will be zero and consequently the coverage 0.5, i.e., the pixel is split evenly between the current pixel and the neighbor. If the edge touches the edge of the pixel, the distance is −0.5 or 0.5 depending on if it is the top or bottom edge, and the coverage consequently goes to zero, meaning the neighbor pixel does not contribute to the final antialiased value of this pixel. The observant reader notes that the neighbor coverage never goes above 0.5. One might initially think that it should range from 0 to 1, but a quick look at Figure 10.1 illustrates why this is not the case. Had the coverage been over 0.5 the neighboring surface would have covered the pixel center, and consequently no longer have been a "neighbor surface" but the pixel's own surface. In other words, push the line in Figure 10.1 up and across a pixel center and that pixel goes from belonging to the green surface to the orange surface.

Results. Given that we operate on the backbuffer with the actual geometry at hand, the antialiasing results are very accurate in terms of coverage calculation. Ideally the pixel should blend with the background surface sampled at the same pixel center. That information is not available in the backbuffer; however, a neighbor pixel is in most practical circumstances sufficiently close. It should be noted, however, that recent development using DirectX 11 hardware has made order-independent translucency somewhat practical [Thibieroz 11]. Using such a method, it is possible to have the underlying surface available and sample down

Figure 10.2. Aliased scene before applying GPAA.

Figure 10.3. Scene antialiased with GPAA.

in the depth order instead of a neighbor pixel for slightly more accurate results (see Figures 10.2 and 10.3).

One notable property of GPAA is that it performs really well in all edge angles. Interestingly, it is the most accurate in the near horizontal and near vertical case, unlike MSAA and most other antialiasing methods where this is the worst case. GPAA has its worst case at diagonal edges, where other methods tend to work the best. But it is worth noting that even in this case, the quality is good.

Figures 10.4(a) and 10.4(b) show a standard pixel-zoomed antialiasing comparison. Before applying GPAA we have the regular stair-stepped edge. After applying GPAA the edge is very smooth.

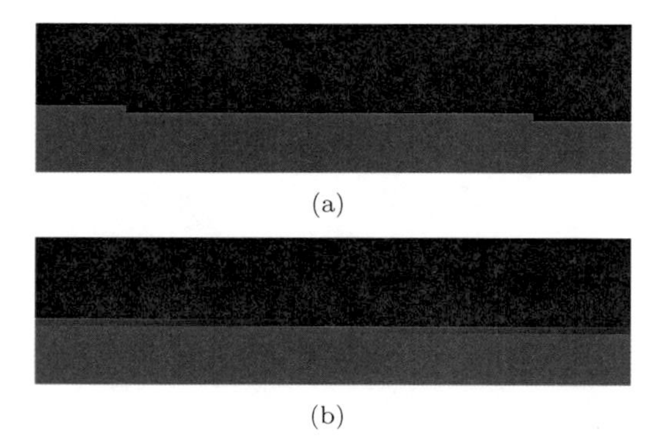

(a)

(b)

Figure 10.4. (a) Zoomed pixels before GPAA. (b) Zoomed pixels after GPAA.

While the GPAA approach works and produces high-quality output, it has some notable drawbacks. There is a scalability problem. The demo application for this method shows great performance for GPAA, but on the other hand it has a quite low polygon count. With increasingly dense geometry the cost will increase, especially since we rely on line rasterization, something for which regular consumer-level GPUs are not particularly optimized. Once a triangle is sufficiently small it is faster just to fill it than render the wireframe. Also, the overhead of having a second geometry pass for antialiasing may prove costly with a high vertex count.

10.2.2 Geometry Buffer Antialiasing

Overview. This method is similar in spirit to GPAA, but with the key difference being that the geometry information is stored to an additional render target during main scene rendering. Hence, no second geometry pass is required. A geometry shader is used to extract the geometric information on the fly and pass to the pixel shader. In the end, a fullscreen resolve pass is applied to the buffer to apply antialiasing.

Geometry shader. Given an existing vertex shader and pixel shader combination in the main scene, a geometry shader is inserted that adds another interpolator to pass down to the pixel shader. This interpolator contains three floats indicating the signed distance to each edge of the triangle. The fourth component contains flags indicating the major direction in screen space for the three edges. Using an interpolator instead of passing line equations is done so that only one interpolator is needed instead of three, and so that we can move the majority of the work to the geometry shader instead of the pixel shader. The geometry shader is presented in Listing 10.3.

```
struct PsIn {
    // Original pixel shader inputs go here
    // ...

    // Interpolator added for GBAA
    noperspective float4 Diff : Diff;
};

float ComputeDiff(const float2 pos0, const float2 pos1,
                  const float2 pos2, out uint major_dir)
{
    float2 dir = normalize(pos1 - pos0);
    float2 normal = float2(-dir.y, dir.x);
    float dist = dot(pos0, normal) - dot(pos2, normal);

    // Check major direction
    bool x_gt_y = (abs(normal.x) > abs(normal.y));

    major_dir = x_gt_y;
    return dist / (x_gt_y? normal.x : normal.y);
```

```
}

[maxvertexcount(3)]
void main(in triangle GsIn In[3], inout TriangleStream<PsIn> TS)
{
    float2 pos0 = (In[0].Position.xy / In[0].Position.w)
        * ScaleBias.xy + ScaleBias.zw;
    float2 pos1 = (In[1].Position.xy / In[1].Position.w)
        * ScaleBias.xy + ScaleBias.zw;
    float2 pos2 = (In[2].Position.xy / In[2].Position.w)
        * ScaleBias.xy + ScaleBias.zw;

    uint3 major_dirs;
    float diff0 = ComputeDiff(pos0, pos1, pos2, major_dirs.x);
    float diff1 = ComputeDiff(pos1, pos2, pos0, major_dirs.y);
    float diff2 = ComputeDiff(pos2, pos0, pos1, major_dirs.z);

    // Pass flags in last component. Add 1.0f (0x3F800000)
    // and put something in LSB bits to give the interpolator
    // some slack for precision.
    float major_dir = asfloat((major_dirs.x << 4) |
        (major_dirs.y << 5) | (major_dirs.z << 6) | 0x3F800008);

    TS.Append( Vertex(In[0], float4(0, diff1, 0, major_dir)) );
    TS.Append( Vertex(In[1], float4(0, 0, diff2, major_dir)) );
    TS.Append( Vertex(In[2], float4(diff0, 0, 0, major_dir)) );
}
```

Listing 10.3. The GBAA geometry shader.

First, we compute the screen-space position of the input vertices. To set up an interpolator for the line between pos0 and pos1 we first compute the signed distance of pos2 to this line. However, what we are really interested in is the distance along the major direction axis. This is just different by a scale factor, which is the reciprocal of the line normal's magnitude in that direction. Figure 10.5 illustrates how an interpolator is set up for a particular edge.

In Figure 10.5 the distance from the top vertex to the bottom edge of the triangle is computed as d. Given that the edge's major direction is horizontal we

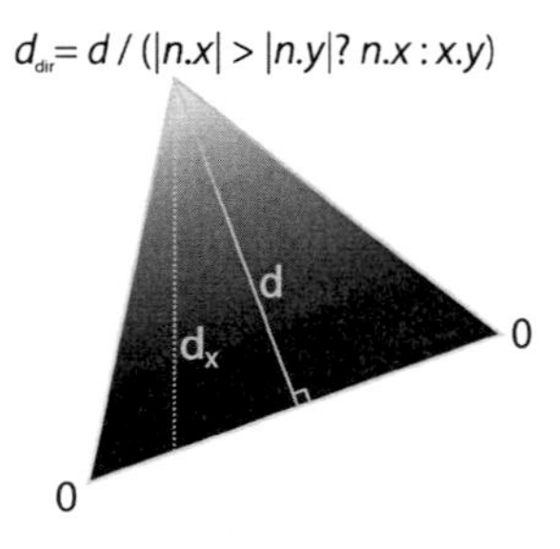

Figure 10.5. Setting up an interpolator to return the distance to the edge along the major direction.

are computing d_x, i.e., the vertical distance from the vertex to the edge, which is done using the equation for d_{dir} in Figure 10.5. This distance, which is returned from `ComputeDiff()` in the code, is set to that vertex, whereas the two vertices on the line naturally get a distance of exactly zero. So for the first line, the first two vertices get their x-component set to zero and the third vertex gets the edge distance. The same logic is applied to the other two edges and passed down in the y- and z-components, respectively. Note that the interpolator has to be declared with the `noperspective` attribute to make sure interpolation is done in screen space.

Given that the interpolator is a `float4` the w-component will be a floating point, whereas we really want to submit a bit-field. We accomplish this by encoding the bits into the mantissa of the float. To be friendly with the interpolator we have to give it a valid float; just throwing a smallish integer into the bottom bits will give us a denormalized float and all bets are off. In our tests we just got all zeros. Using 1.0f as the base is a reasonable choice, although any other valid float with enough zero bits in it will get the task done. The bitwise representation of 1.0f is 0x3F800000. We could throw our bits in the bottom of that number, and this appears to work in practice on the hardware we tried it on; however, it could be wise to not rely on the interpolator being precise to the last bits, even if all inputs are identical. So, we rather arbitrarily use bits 4 to 6, and enable bit 3 to make sure the lowest four bits can be imprecise in both directions without affecting the bits we use. So the bits we finally add in addition to our flags are 0x3F800008.

Pixel shader. In Listing 10.4 we take a look at the work we add to the pixel shader. In addition to whatever the shader was doing before, we add a snippet to take the input edge distances as provided by the interpolator and select the edge that is closest to the pixel. Using the major direction of the closest edge, we output a two-component distance value indicating horizontal or vertical distance to the edge.

```
PsOut main(PsIn In)
{
    // Other outputs of the shader go here
    // ...

    // Select the smallest distance
    float diff = In.Diff.x;
    int major_dir = asuint(In.Diff.w) & (1 << 4);

    if (abs(In.Diff.y) < abs(diff))
    {
        diff = In.Diff.y;
        major_dir = asuint(In.Diff.w) & (1 << 5);
    }
    if (abs(In.Diff.z) < abs(diff))
    {
```

```
        diff = In.Diff.z;
        major_dir = asuint(In.Diff.w) & (1 << 6);
    }

    float2 offset = 0;

    // Select direction to sample a neighbor pixel
    if (major_dir)
        offset.x = diff;
    else
        offset.y = diff;

    Out.Diff = offset;
}
```

Listing 10.4. The GBAA pixel shader.

Selecting the closest edge is merely a matter of selecting the component with the smallest absolute value. The corresponding direction flag is also extracted and the difference is supplied in either x or y. Note that we only ever shift along one axis, so we could easily pack the information into one component and just put a flag in the low-order bit indicating the direction, but keeping it as two values makes the shader output and the final resolve shader a bit simpler. Depending on the balance between ALU, bandwidth, memory, and buffer layout, however, it may sometimes be beneficial to squeeze it into one component.

Resolve. Once the scene has been rendered and we have a geometry buffer filled with edge distance information, we are ready to apply GBAA to antialias the image. The resolve pass is rendered as a fullscreen pass where we sample from a copy of the backbuffer and from the geometry buffer. Initially, we sample the

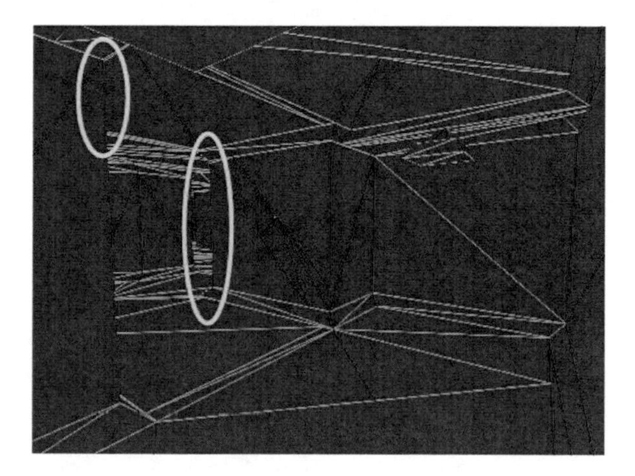

Figure 10.6. Gaps occurring in the geometry buffer along silhouette edges.

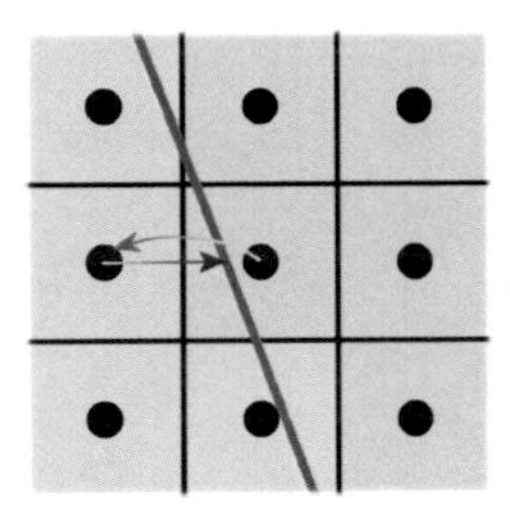

Figure 10.7. Patching the geometry buffer by considering neighbors' geometry information.

geometry buffer at the pixel location. Using the edge information, if any, we compute a texture-coordinate offset to sample the backbuffer texture using the texture filter to blend with a neighbor pixel. This will work for all edges where geometry information is present on both sides of the edge, i.e., for shared edges within a mesh. It will not work for silhouette edges of geometry in front of a background surface. In this case, only one side of the edge holds the relevant edge information. The background surface knows nothing about the geometry covering it. The problem is illustrated in Figure 10.6.

When visualizing the distance information there are noticeable gaps. The circled areas in Figure 10.6 show a couple of places where silhouette edges lack relevant geometry information on one side of the edge because those pixels were shaded by the background surface. To cover up the holes, we need to consider geometry information in neighboring pixels. If the pixel's own geometry information does not indicate that it is crossed by a geometric edge, we search the immediate neighborhood for pixels that might indicate otherwise. Figure 10.7 illustrates this process.

Here the green center pixel belongs to a background surface. It does not know that the cyan foreground surface is crossing over its territory. By checking its left neighbor, it can find that indeed there is an edge and it is indeed crossing over this pixel. The neighbor's information is then converted to the range of the pixel and used. Note that in Figure 10.7 there is also valid geometry information available in the pixel below it. Given that this edge is mostly vertical, it will point horizontally to the right, however, and will not be useful for the center pixel and

Neighbor	Range	Converted offset
Left	offset.x in $[0.5 \ldots 1.0]$	offset.x $-$ 1.0
Right	offset.x in $[-1.0 \ldots . -0.5]$	offset.x $+1.0$
Up	offset.y in $[0.5 \ldots 1.0]$	offset.y -1.0
Down	offset.y in $[-1.0 \ldots -0.5]$	offset.y $+1.0$

Table 10.1. Detecting and converting neighbors' geometry information.

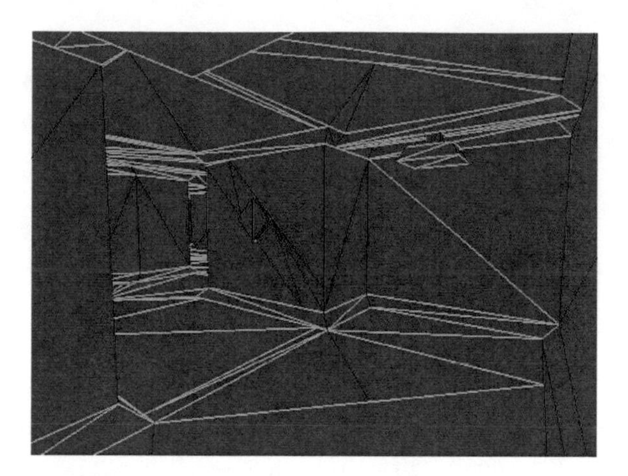

Figure 10.8. Patched geometry buffer with all gaps filled.

thus not used. The bottom-right pixel might look at it, but in this case it will be rejected because the distance is out of range, i.e., the edge does not cross that pixel. For patching up the geometry buffer we only need to check neighbors left, right, up, and down. Table 10.1 shows the offset ranges to look for in neighboring pixels and how to convert them. Figure 10.8 shows the final visualization of the geometry buffer after filling up the gaps.

Once we have patched the missing information from the geometry buffer, we can now blend with a neighbor pixel. As with GPAA, we use the texture filter to accomplish the blending. The final edge-distance information is thus converted to a sampling offset, and a single sample is taken from the backbuffer to get the final antialiased pixel. The complete GBAA resolve shader is presented in Listing 10.5.

```
Texture2D BackBuffer;
Texture2D <float2> GeometryBuffer;
SamplerState Linear, Point;
float2 PixelSize;

float4 main(PsIn In) : SV_Target
{
    float2 offset = GeometryBuffer.Sample(Point, In.TexCoord);

     // Check geometry buffer for an edge cutting through
     // the pixel, otherwise search neighbors
    [flatten]
    if (min(abs(offset.x), abs(offset.y)) >= 0.5f) {
        offset = 0.5f;

        float2 offset0 = GeometryBuffer.Sample(Point,
            In.TexCoord, int2(-1, 0));
        float2 offset1 = GeometryBuffer.Sample(Point,
            In.TexCoord, int2(+1, 0));
```

```
        float2 offset2 = GeometryBuffer.Sample(Point,
            In.TexCoord, int2(0, -1));
        float2 offset3 = GeometryBuffer.Sample(Point,
            In.TexCoord, int2(0, +1));

        if (abs(offset0.x - 0.75f) < 0.25f)
            offset = offset0 + float2(-1, 0.5f);

        if (abs(offset1.x + 0.75f) < 0.25f)
            offset = offset1 + float2(1, 0.5f);

        if (abs(offset2.y - 0.75f) < 0.25f)
            offset = offset2 + float2(0.5f, -1);

        if (abs(offset3.y + 0.75f) < 0.25f)
            offset = offset3 + float2(0.5f, 1);
    }

    // Convert distance to texture coordinate shift
    float2 off = (offset >= float2(0, 0))?
                 float2(0.5f, 0.5f) : float2(-0.5f, -0.5f);
    offset = off - offset;

    // Blend with neighbor pixel using texture filter
    return BackBuffer.Sample(Linear,
        In.TexCoord + offset * PixelSize);
}
```

Listing 10.5. The GBAA resolve shader.

Antialiasing alpha-tested edges. An important advantage of GBAA over GPAA is that the former is not strictly tied to geometric edges. Any form of edges that can be encoded in the geometry buffer will be properly antialiased. This means we can also antialias alpha-tested edges. All we need to do in the alpha-tested pass is to compute the distance to the edge, i.e., the cross-over point where the alpha value falls below the threshold value, in the same fashion as for geometric edges. A pretty straightforward way to accomplish this is presented in Listing 10.6. Essentially, we get the slope of alpha in x and y in screen space using gradient instructions and simply compute how far we need to go in either direction to get down to the alpha-reference value, and then we select the shortest direction.

```
float dx = ddx(alpha);
float dy = ddy(alpha);

float2 alpha_dist = 0.0f;

if (abs(dx) > abs(dy))
  alpha_dist.x = (alpha_ref - alpha) / dx;
else
  alpha_dist.y = (alpha_ref - alpha) / dy;
```

Listing 10.6. Alpha-tested edge distance.

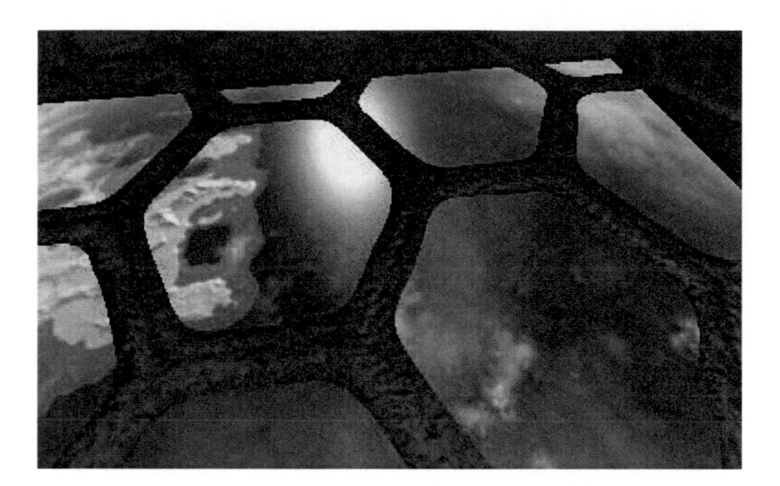

Figure 10.9. Alpha-tested edges before applying GBAA.

Other types of edges for which it might be interesting to use GBAA are bump edges internal to the surface when using parallax occlusion mapping [Brawley and Tatarchuk 04].

Results. Other than the ability to antialias alpha-tested edges, GBAA is for all practical purposes identical to GPAA in terms of image quality. Figures 10.2–10.4 are thus representative of the GBAA quality as well. For alpha-tested edges Figures 10.9 and 10.10 illustrate the quality of alpha-tested edge with and without antialiasing.

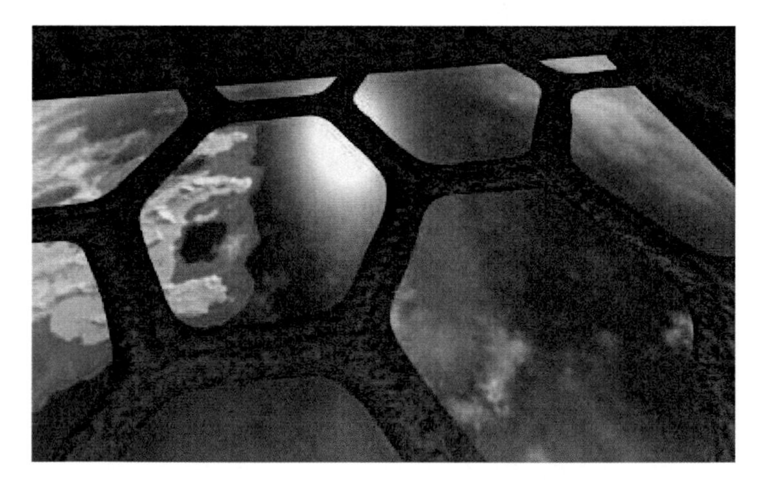

Figure 10.10. Alpha-tested edges antialiased with GBAA.

10.2.3 Performance and Optimizations

The main difference between GPAA and GBAA lies in performance characteristics and engine integration. With a small amount of geometry, such as in the provided demo applications, GPAA wins the performance race, although GBAA is not particularly slow in this case either. As geometry gets denser, GBAA gets more favorable, depending on hardware, resolution, and other factors.

For GBAA, where we have a fullscreen resolve pass, we can simply render the scene to a temporary buffer and then resolve to the backbuffer. GPAA, on the other hand, requires a copy of the backbuffer to be sampled as a texture. While this is a relatively small cost for a modern GPU, it is probably still worth optimizing when possible. Depending on what the engine is doing, it may already have a copy of the backbuffer at hand, for example, for refraction effects or post-effect purposes. In that case all is fine and the available copy can be used at no extra cost. In other cases, it should be noted that strictly speaking there is nothing in the technique that requires the use of a copy of the backbuffer rather than modifying it in place directly. However, reading from and writing to the same surface at the same time causes a read-after-write hazard and is generally considered a bad idea. Given the long pipelines of GPUs, as well as caching on the render backend, values tend to be written to memory long after a shader invocation produced its final outputs, and even if it is written to memory already, new shader invocations are not guaranteed to receive the updated value when sampling the texture since old values may still hang around in the texture cache.

For this reason, APIs generally declare this case as undefined. Some APIs, such as DirectX 10 and 11, actively prevent you from entering such a state. While undefined certainly gives you no guarantees of anything, in practice it generally means that you receive either the old or the new value. For these two techniques, these values are either the same, or both acceptable. So if you feel a bit adventurous, and your API does not prevent you, you may want to experiment with in-place modification. For fixed hardware platforms, such as consoles, where you don't have to worry about things breaking on a particular driver or hardware revision, it basically boils down to trying it out, and if it works it works. Unfortunately, in the case of the Xbox 360 the renderable memory (EDRAM) is separate from texture-sampleable memory (regular video memory), so on this platform it is impossible to do in-place modification for hardware reasons. In OpenGL there is the `GL_NV_texture_barrier` extension [Bolz 09], which relaxes the rules for read-after-write cases and thus allows for in-place modification. At the time of this writing, it is supported by both Nvidia and AMD. Strictly speaking, we do not fully comply with the conditions of that extension since we are sampling a neighbor pixel, but in practice it will almost certainly work. It is the closest we get to a defined behavior today. The important thing is that this extension also tells us that the driver will not actively prevent us from entering this "dangerous" state since it is now valid on the API level to have the same texture bound on both input and output.

An optimization for GPAA is to prefilter the mesh geometry and only draw lines for edges that are deemed necessary to antialias. For instance, shared edges between coplanar polygons can be excluded. It is also possible to store the polygon normals for each edge and only draw silhouette edges, i.e., where one normal points towards the camera and one points away. Using a geometry shader and adjacency, it is also possible to accomplish the same. In this case no additional memory is required for storing the vertices or indices.

10.3 Conclusion and Future Work

Two techniques have been presented that produce very high-quality antialiasing. However, there are several open opportunities for improvements and additional research. This includes efficient ways to remove internal edges from antialiasing when using a geometry shader, such as for GBAA. Currently these edges are needlessly antialiased. Typically this does not cause any problems, but ideally it should not be done.

Currently only one edge is considered per pixel. This could cause small artifacts in corners. Sometimes single pixels along an edge are off, because the method went by a different edge from the background. It may be possible to include several edges and blend with multiple neighbors, or alternatively, to prioritize edges based on depth or other heuristics.

Both techniques have been implemented on a PC using DirectX 10. It would be interesting to see how well these techniques would fit on current generation consoles. The consoles do not support a geometry shader, for instance, but there are ways around this limitation, although possibly at the cost of additional memory.

The current implementations of these techniques do not catch edges between intersecting meshes, such as for example, buildings placed on terrain where some parts of the buildings are bound to go some distance into the ground. For static geometry there is nothing preventing such edges from being detected in a preprocess step and stored, albeit this is not necessarily a small undertaking. For dynamic objects, it may not be very practical to attempt to detect such edges in real time. Theoretically, it ought to be possible to detect such edges by looking for creases in the depth buffer. Promising experiments have been done in this area.

Finally, it would be interesting to see what effect on quality blending with a background layer (OIT style) would have compared to blending with a neighbor pixel.

Bibliography

[Bolz 09] Jeff Bolz. "NV_texture_barrier." Available at http://www.opengl.org/registry/specs/NV/texture_barrier.txt, 2009.

[Brawley and Tatarchuk 04] Zoe Brawley and Natalya Tatarchuk. "Parallax Occlusion Mapping: Self-Shadowing, Perspective-Correct Bump Mapping Using Reverse Height Map Tracing." In *ShaderX*^{3}, edited by Wolfgang Engel, pp. 135–154. Hingham, MA: Charles River Media, 2004.

[Jimenez et al. 11] Jorge Jimenez, Belen Masia, Jose I. Echevarria, Fernando Navarro, and Diego Gutierrez. "Practical Morphological Anti-Aliasing." In *GPU Pro*^{2}, edited by Wolfgang Engel, pp. 95–114. Natick, MA: A K Peters, 2011.

[Lottes 11] Timothy Lottes. *NVIDIA FXAA*. Available at http://timothylottes.blogspot.com/2011/03/nvidia-fxaa.html, 2011.

[McGuire and Luebke 11] Morgan McGuire and David Luebke. "Subpixel Reconstruction Antialiasing." Available at http://research.nvidia.com/publication/subpixel-reconstruction-antialiasing, 2011.

[Malan 10] Hugh Malan. "Edge Anti-Aliasing by Post-Processing." In *GPU Pro*, edited by Wolfgang Engel, pp. 265–289. Natick, MA: A K Peters, 2010.

[Persson 11a] Emil Persson. "Geometric Post-process Anti-Aliasing." Available at http://www.humus.name/index.php?page=3D&ID=86, 2011.

[Persson 11b] Emil Persson. "Geometry Buffer Anti-Aliasing." Available at http://www.humus.name/index.php?page=3D&ID=87, 2011.

[Reshetov09] Alexander Reshetov. "Morphological Antialiasing." Available at http://visual-computing.intel-research.net/publications/papers/2009/mlaa/mlaa.pdf, 2009.

[Thibieroz 11] Nicolas Thibieroz. "Order-Independent Transparency using Per-Pixel Linked Lists." In *GPU Pro*^{2}, edited by Wolfgang Engel, pp. 409–432. Natick, MA: A K Peters, 2011.

GPU Terrain Subdivision and Tessellation

Benjamin Mistal

This paper presents a GPU-based algorithm to perform real-time terrain subdivision and rendering of vast detailed landscapes. This algorithm achieves smooth level of detail (LOD) transitions from any viewpoint and does not require any preprocessing of data structures on the CPU.

11.1 Introduction

There are a lot of existing terrain rendering and subdivision algorithms that achieve fantastic results, and fast frame rates. Some of these algorithms, however, are limited. They can require the preprocessing of large data sets, constant transferring of large data sets from the CPU to the GPU, limited viewing areas, or complex algorithms to merge together meshes in efforts to avoid cracks and seams as a result of various LOD subdivision techniques.

The GPU-based algorithm we developed addresses all of the above mentioned limitations, and presents a simple alternative to render highly detailed landscapes, without significant impact on the CPU. Figure 11.1 shows an example of this algorithm and also shows the generated underlying wire-frame mesh partially superimposed.

We describe a GPU-based algorithm to create a subdivided mesh with distance-based LOD that can be used for terrain rendering. Data amplification and multiple stream-out hardware capability are utilized to repeatedly subdivide an area to achieve a desired LOD. In addition, culling is also performed at each iteration of the algorithm, therefore avoiding a lot of unnecessary processing or subdivision of areas outside of the viewing frustum. Because the resulting data is retained and refined on the GPU, the CPU is mostly left available to perform other tasks. To show a practical use of this technique, we also utilize a smooth LOD transitioning scheme, and use a procedural terrain generation function to provide real-time rendering of a highly detailed and vast landscape.

Figure 11.1. Rendered terrain, with the underlying wireframe mesh partially superimposed.

Our algorithm was heavily inspired by two existing algorithms. The first inspiration came from the great desire to walk through the procedural mountains created by F. Kenton Musgrave [Ebert et al. 98], in real time. The second inspiration came from the visual beauty of the real-time water created by Claes Johanson, in his introduction of the projected grid concept [Johanson 04]. The concept helped form one of the ideas for the basis of our subdivision algorithm, by showcasing effective and efficient vertex placement to display a vast area of seascape.

11.2 The Algorithm

A few terms are used throughout this paper and are integral to understanding the general algorithm and the related descriptions. Section 11.2.1 will describe these terms and provide related calculations. An algorithm overview is provided in Section 11.2.2, and Sections 11.2.3 to 11.2.5 describe the main separate components of our algorithm.

11.2.1 Terms and Definitions

Viewable region. A *viewable region*, denoted by R, is defined as an axis aligned quadrilateral representing a region that is to be subdivided and/or rendered. This region is defined from a *center position* and an *applied offset* in both the positive and negative directions along each aligning axis (see Figure 11.2). The center position is denoted by R_C. The applied offset is denoted by R_λ. The bounding points of the region are denoted by P_1, P_2, P_3, and P_4.

Viewable region span. To quantify a *viewable region span*, denoted by θ, we defined the following calculation at point P and applied offset λ:

$$\theta(P, \lambda) = |P_{\text{ScreenR}} - P_{\text{ScreenL}}|,$$

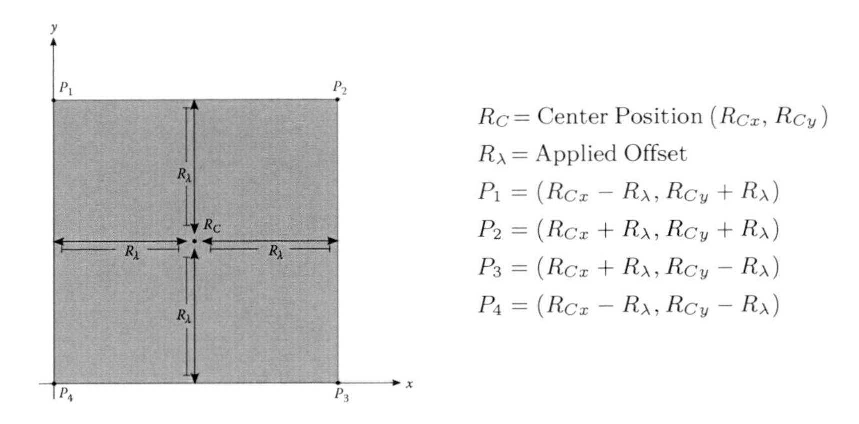

Figure 11.2. Example of a viewable region and its associated properties.

$$
\begin{aligned}
P_{\text{ScreenL}} &= (P_{\text{VProjL}xy})/P_{\text{ProjLw}},\\
P_{\text{ScreenR}} &= (P_{\text{ProjR}xy})/P_{\text{ProjRw}},\\
P_{\text{ProjL}} &= P_{WL} \times \text{matProjection},\\
P_{\text{ProjR}} &= P_{WR} \times \text{matProjection},\\
P_{WL} &= (P_{Wx} - \lambda, P_{Wy}),\\
P_{WR} &= (P_{Wx} + \lambda, P_{Wy}),\\
P_{W} &= P \times \text{matWorldView}.
\end{aligned}
$$

The above calculation for the viewable region span can be used at any position P and applied offset λ, and is used extensively within the LOD Transition Algorithm (explained in Section 11.2.4). We use this calculation instead of calculating the actual viewing surface area to avoid inconsistencies when viewable regions are viewed from different angles.

Maximum viewable region span. The *maximum viewable region span*, denoted by θ_{max}, is the maximum allowable viewable region span. This value, which is set by the user, is one of the main determining factors of the attainable LOD, and it plays a key part in both the Subdivision Algorithm (explained in Section 11.2.3) and the LOD Transition Algorithm (explained in Section 11.2.4).

Relative quadrant code. This code identifies the relative position of a split viewable region in relation to its parent viewable region. This code is utilized by the LOD Transition Algorithm (explained in Section 11.2.4), and calculated in the Subdivision Algorithm (explained in Section 11.2.3). Usually encoded as a 2-bit mask, this code becomes part of the definition of a viewable region.

11.2.2 Algorithm Overview

The algorithm operates in three stages:

Stage 1. Create the initial input stream of one or more viewable regions containing the area(s) to be viewed.

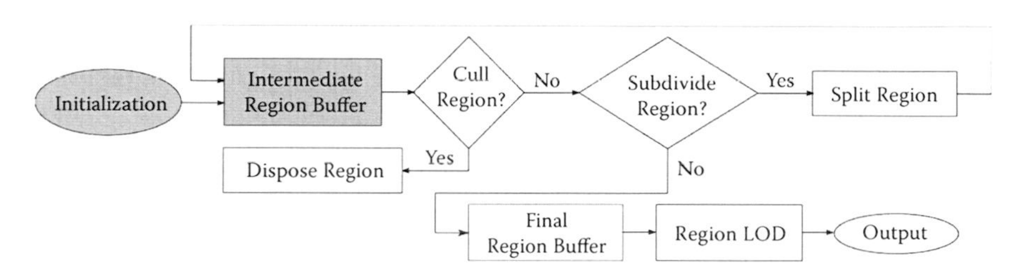

Stage 2. Take the initial input stream of viewable regions from Stage 1 and iteratively process each viewable region utilizing the Subdivision Algorithm (explained in Section 11.2.3).

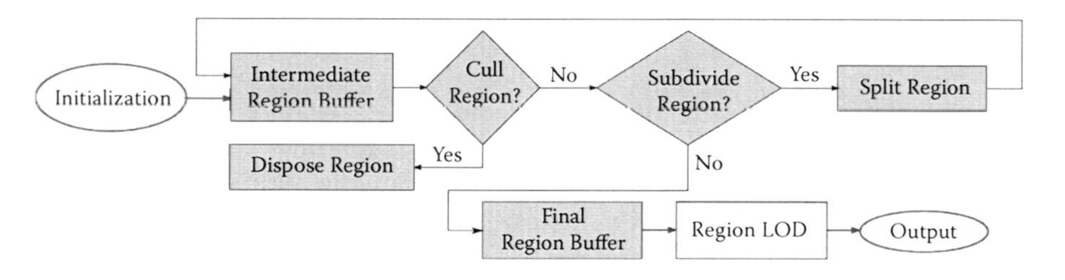

Stage 3. Render the resulting final output stream of viewable regions from Stage 2. At this stage, we utilize an LOD Transition Algorithm (explained in Section 11.2.4), as well as a Procedural Height Generation Algorithm (explained in Section 11.2.5) to render the resulting terrain.

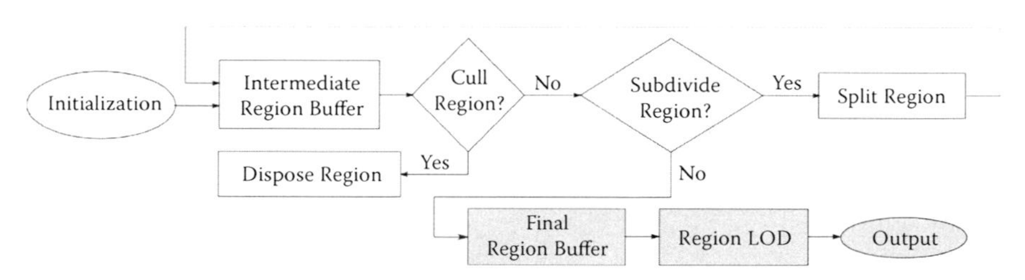

11.2.3 Subdivision Algorithm

In Stage 2 of the algorithm, for each iteration, we feed an input stream of viewable regions into the geometry shader stage of the rendering pipeline. The input stream for the first iteration is the initial input stream that was created in Stage 1. All subsequent iterations use the intermediate output stream from the preceding iteration as the input. Two output streams are utilized in the geometry shader. An intermediate output stream is used to hold viewable regions that are intended for further processing. A final output stream is used to hold viewable regions that require no further processing in the next iteration, and are to be rendered as part of Stage 3 of the algorithm.

Within the geometry shader, each viewable region is processed and one of the following three actions is performed:

Option 1. If the viewable region is determined to not intersect with the view frustum, it is culled. The culled viewable region is not added to either the intermediate or the final output stream.

When determining whether or not the viewable region intersects the view frustum, one must be mindful of any displacements performed during Stage 3 of the algorithm. When using this algorithm for terrain rendering, each viewable region is extruded into a volume, based on the maximum displacement that may be added in Stage 3. It is this extruded volume that is then tested for intersection with the view frustum.

Option 2. If the viewable region span θ, calculated at the viewable region R's center position R_C, and with the applied offset R_λ, written as $\theta(R_C, R_\lambda)$, is greater than the maximum viewable region span $\theta_{\max}$, then the viewable region R is split into four quadrants $(R1, R2, R3, R4)$. The quadrants, each of which become viewable regions themselves, are then added to the intermediate output stream to be reprocessed at the next iteration. A special code, unique to each quadrant, is also added to the output stream to identity the relative location of the split viewable regions to their parent viewable region. This extra piece of information, referred to as a *relative quadrant code*, and usually encoded into a 2-bit mask, is later utilized by the LOD Transition Algorithm (explained in Section 11.2.4) when rendering each viewable region.

To split the viewable region R (see Figure 11.3(a)) into the quadrants (see Figure 11.3(b)), we create four new viewable regions $(R1, R2, R3, R4)$ with their respective center positions $(R1_C, R2_C, R3_C, R4_C)$ and applied offsets $(R1_\lambda,\ R2_\lambda,\ R3_\lambda,\ R4_\lambda)$ defined as follows:

$$R1_C = \left(R_{Cx} - \frac{1}{2}R_\lambda, R_{Cy} + \frac{1}{2}R\lambda\right),$$

$$R2_C = \left(R_{Cx} + \frac{1}{2}R_\lambda, R_{Cy} + \frac{1}{2}R\lambda\right),$$
$$R3_C = \left(R_{Cx} + \frac{1}{2}R_\lambda, R_{Cy} - \frac{1}{2}R\lambda\right),$$
$$R4_C = \left(R_{Cx} - \frac{1}{2}R_\lambda, R_{Cy} - \frac{1}{2}R\lambda\right),$$

$$R1_\lambda = \frac{1}{2}R_\lambda,$$
$$R2_\lambda = \frac{1}{2}R_\lambda,$$
$$R3_\lambda = \frac{1}{2}R_\lambda,$$
$$R4_\lambda = \frac{1}{2}R_\lambda.$$

Option 3. If the viewable region span θ, calculated at the viewable region R's center position R_C, and with the applied offset R_λ, $\theta(R_C, R_\lambda)$, is less than or equal to the maximum viewable region span θ_{max}, then the viewable region R is added to final output stream of viewable regions.

Note that for the final iteration in Stage 2 of the algorithm, any viewable region not culled must be added to the final output stream, regardless of the viewable region span. This is to ensure that all remaining viewable regions have a chance

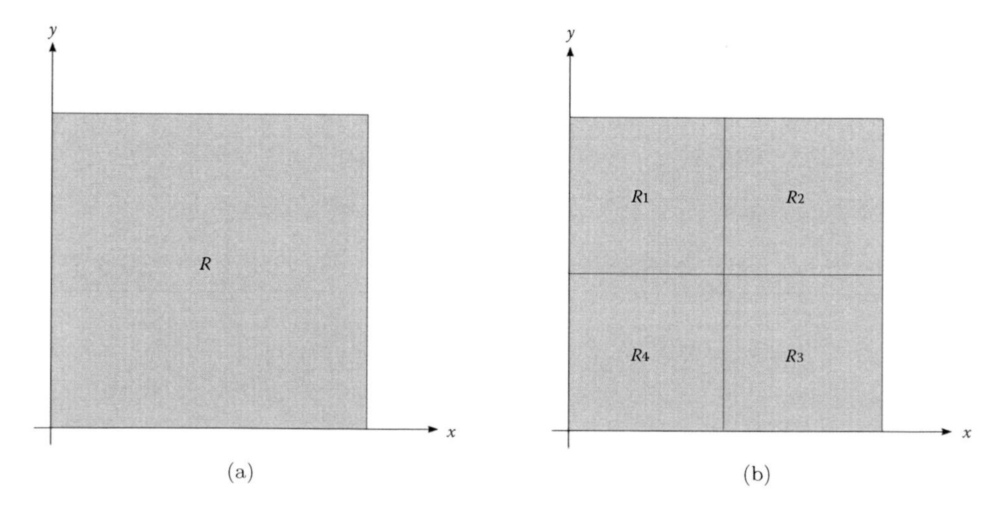

Figure 11.3. (a) The viewable region (R). (b) The split viewable regions ($R1, R2, R3, R4$).

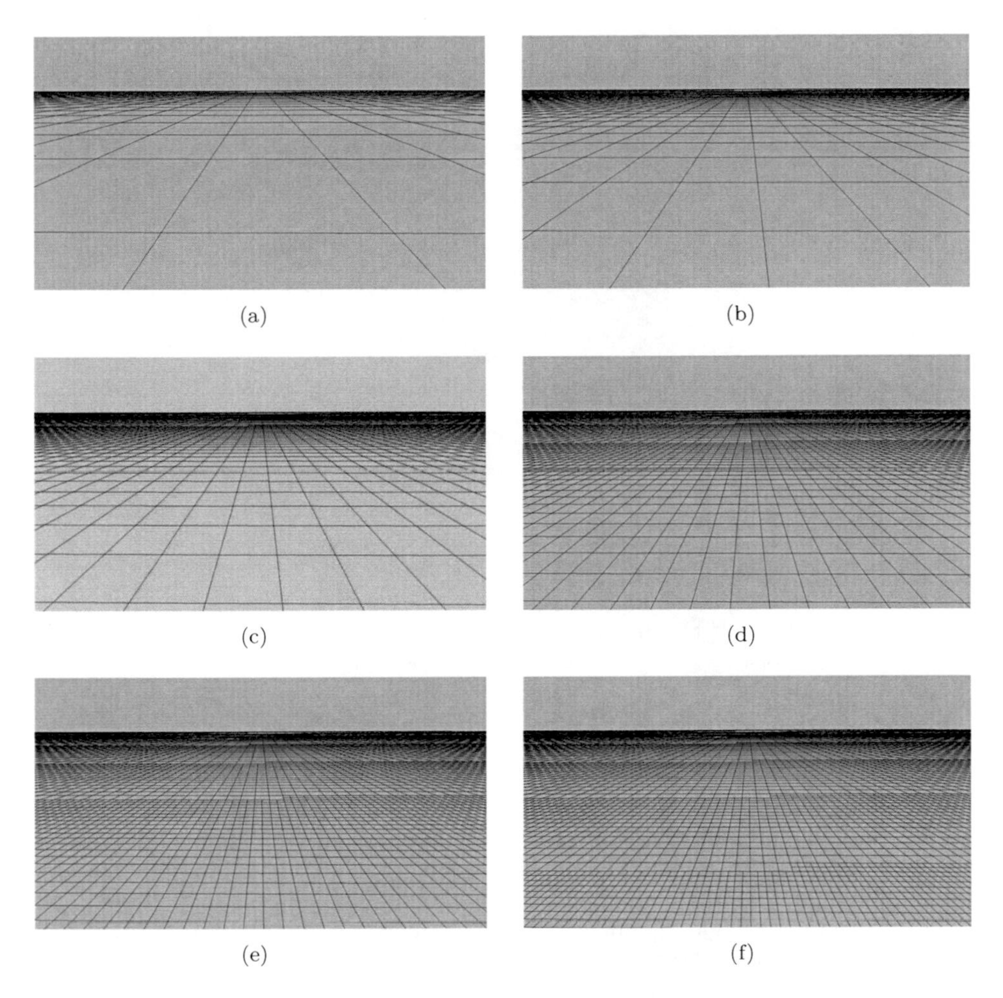

Figure 11.4. Viewable region after (a) N iterations, (b) $N+1$ iterations, (c) $N+2$ iterations, (d) $N+3$ iterations, (e) $N+4$ iterations, and (f) $N+5$ iterations.

to be rendered, and are not placed into an intermediate output stream that will not result in any further processing or rendering.

Figures 11.4(a)–11.4(f) show examples of a viewable region after a number of iterations through our subdivision algorithm.

11.2.4 LOD Transition Algorithm

In Stage 3 of the algorithm, a stream of viewable regions are rendered. One viewable region R with an applied offset of R_λ may be adjacent to another viewable

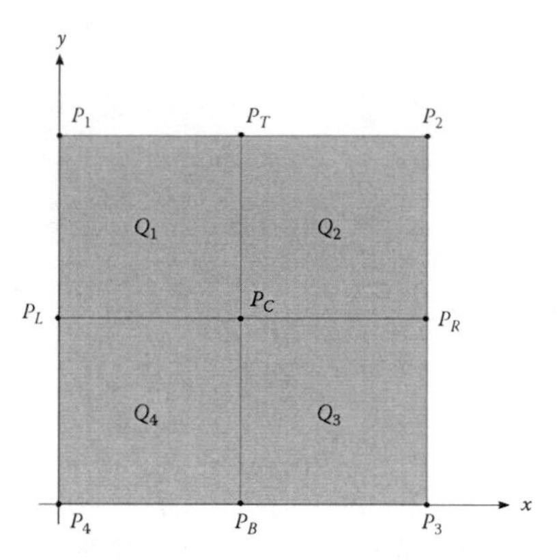

Figure 11.5. Example of the viewable region quadrilateral boundary points.

region with an applied offset that is either half or double the size of R_λ. Without performing a smooth transition between the two differently sized viewable regions, there would be visible discontinuities or other visual anomalies when rendering. We describe a method that offers a smooth transition between differently sized viewable regions. By rendering a viewable region as a set of quadrilaterals, we are able to morph the quadrilaterals in such as way to make the boundary between the larger and smaller viewable regions indistinguishable. This method eliminates seams, T-junctions, and visible boundaries between neighboring viewable regions of different sizes. A similar method was described by Filip Strugar [Strugar 10], although we have extended it to handle various boundary cases to ensure no cracks form anywhere within the mesh.

Each viewable region is rendered by splitting the viewable region into four quadrilaterals, denoted by Q_1, Q_2, Q_3, and Q_4 (see Figure 11.5). The boundary points for each quadrilateral, comprised from the collection of static boundary points (P_1, P_2, P_3, P_4) and the morphing boundary points (P_L, P_T, P_R, P_B, P_C), are defined as follows:

$$\begin{aligned}
Q_1 &= \{P_1, P_T, P_C, P_L\},\\
Q_2 &= \{P_T, P_2, P_R, P_C\},\\
Q_3 &= \{P_C, P_R, P_3, P_B\},\\
Q_4 &= \{P_L, P_C, P_B, P_4\}.
\end{aligned}$$

Collectively, these nonoverlapping quadrilaterals will cover the same surface area as the viewable region they are created from. The boundary points of each quadrilateral are calculated to align with the boundary points of the neighboring viewable region quadrilaterals. The morphing quadrilateral boundary points

$(P_L, P_T, P_R, P_B, P_C)$ are calculated as follows:

$$\begin{aligned}
P_L &= \left(P_1 \times \left(1 - \frac{1}{2}T_L\right)\right) + \left(P_4 \times \frac{1}{2}T_L\right),\\
P_T &= \left(P_1 \times \left(1 - \frac{1}{2}T_T\right)\right) + \left(P_2 \times \frac{1}{2}T_T\right),\\
P_R &= \left(P_2 \times \left(1 - \frac{1}{2}T_R\right)\right) + \left(P_3 \times \frac{1}{2}T_R\right),\\
P_B &= \left(P_4 \times \left(1 - \frac{1}{2}T_B\right)\right) + \left(P_3 \times \frac{1}{2}T_B\right),\\
P_C &= \left(P_1 \times \left(1 - \frac{1}{2}T_C\right)\right) + \left(P_3 \times \frac{1}{2}T_C\right).
\end{aligned}$$

Given a viewable region span θ at point P and applied offset λ, written as $\theta(P, \lambda)$, we are able to calculate a morphing factor $T(P, \lambda)$, by using the following formula:

$$T(P, \lambda) = \begin{cases} 0, & \theta(P, \lambda) \leq \frac{1}{2}\theta_{\max},\\ (\theta(P, \lambda)/\theta_{\max}) \times 2 - 1, & \frac{1}{2}\theta_{\max} < \theta(P, \lambda) < \theta_{\max},\\ 1, & \theta(P, \lambda) \geq \theta_{\max}. \end{cases}$$

We calculate each of the general morphing factors $(T_L, T_T, T_R, T_B, T_C)$ for a viewable region R with a *center position* R_C and applied offset R_λ as follows:

$$\begin{aligned}
T_L &= T(\beta_L, R_\lambda),\\
T_T &= T(\beta_T, R_\lambda),\\
T_R &= T(\beta_R, R_\lambda),\\
T_B &= T(\beta_B, R_\lambda),\\
T_C &= T(R_C, R_\lambda),
\end{aligned}$$

$$\begin{aligned}
\beta_L &= (R_{Cx} - R_\lambda, R_{Cy}),\\
\beta_T &= (R_{Cx}, R_{Cy} + R_\lambda),\\
\beta_R &= (R_{Cx} + R_\lambda, R_{Cy}),\\
\beta_B &= (R_{Cx}, R_{Cy} - R_\lambda).
\end{aligned}$$

These general morphing factors will be applied when rendering a viewable region, and assumes that the neighboring viewable regions have the same applied offset R_λ. Figure 11.6 provides a diagram of the various positions used when calculating the general morphing factors.

There are two special boundary cases we need to handle when calculating the morphing factors. These special cases arise when one viewable region is adjacent to another viewable region of a larger or smaller applied offset λ. The cases are defined as follows:

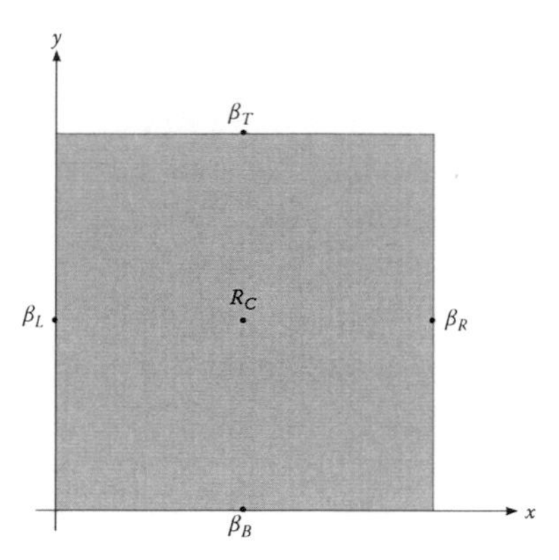

Figure 11.6. Diagram of the various positions used with each viewable region when calculating the general morphing factors.

Boundary Case 1. The viewable region R with applied offset R_λ is adjacent to a smaller viewable region with applied offset of $\frac{1}{2}R_\lambda$.

We will conditionally set the following morphing factors as follows:

- $T_L = 1$, if $\theta(\beta_{NL}, R_\lambda) \geq \theta_{\max}$;
- $T_T = 1$, if $\theta(\beta_{NT}, R_\lambda) \geq \theta_{\max}$;
- $T_R = 1$, if $\theta(\beta_{NR}, R_\lambda) \geq \theta_{\max}$;
- $T_B = 1$, if $\theta(\beta_{NB}, R_\lambda) \geq \theta_{\max}$;
- $\beta_{NL} = (R_{Cx} - (2 \times R_\lambda), R_{Cy})$;
- $\beta_{NT} = (R_{Cx}, R_{Cy} + (2 \times R_\lambda))$;
- $\beta_{NR} = (R_{Cx} + (2 \times R_\lambda), R_{Cy})$;
- $\beta_{NB} = (R_{Cx}, R_{Cy} - (2 \times R_\lambda))$.

The above set of conditionals test for the cases where an adjacent viewable region has been split into smaller viewable regions. We therefore need to lock the affected morphing factors to 1. This is to ensure that all of the overlapping quadrilateral vertices exactly match with those of the smaller adjacent viewable region. Figure 11.7 provides a diagram of the various positions used when calculating the morphing factors for Boundary Case 1.

Boundary Case 2. The viewable region with applied offset R_λ is adjacent to a larger viewable region with applied offset of $(2 \times R_\lambda)$.

In order to be able to test for this case, we need to ensure we have some additional information regarding the current viewable region we are rendering. In Stage 2 of our algorithm (explained in Section 11.2.3), we needed to store

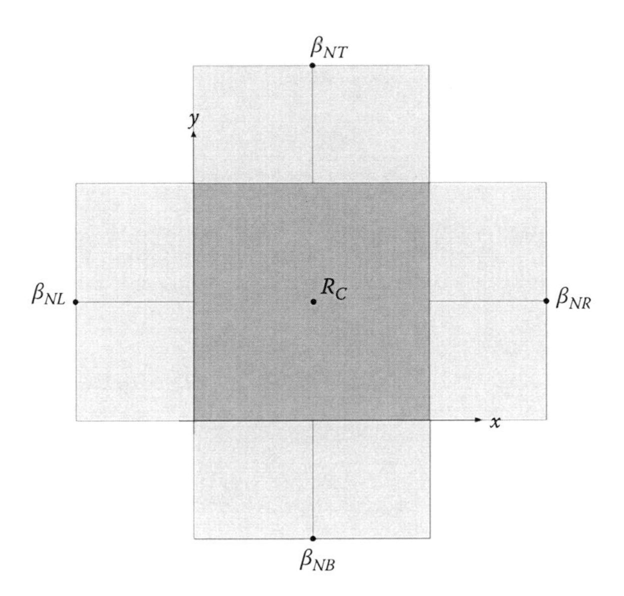

Figure 11.7. Diagram of the various positions used with each viewable region when calculating the morphing factors for Boundary Case 1.

the relative quadrant code for each viewable region. This is so we are able to correctly calculate the center positions for larger neighboring viewable regions. This also allows us to calculate the associated edge of the viewable region that may be adjacent to larger neighboring viewable regions.

We will conditionally set the following morphing factors as follows, based on the relative quadrant code for the viewable region:

$$
\begin{aligned}
R1: T_L &= 0, \text{ if } \theta(\beta_{FL}, (2 \times R_\lambda)) < \theta_{\max}, \\
T_T &= 0, \text{ if } \theta(\beta_{FT}, (2 \times R_\lambda)) < \theta_{\max}, \\
\beta_C &= (R_{Cx} + R_\lambda, R_{Cy} - R_\lambda). \\
R2: T_T &= 0, \text{ if } \theta(\beta_{FT}, (2 \times R_\lambda)) < \theta_{\max}, \\
T_R &= 0, \text{ if } \theta(\beta_{FR}, (2 \times R_\lambda)) < \theta_{\max}, \\
\beta_C &= (R_{Cx} - R_\lambda, R_{Cy} - R_\lambda). \\
R3: T_R &= 0, \text{ if } \theta(\beta_{FR}, (2 \times R_\lambda)) < \theta_{\max}, \\
T_B &= 0, \text{ if } \theta(\beta_{FB}, (2 \times R_\lambda)) < \theta_{\max}, \\
\beta_C &= (R_{Cx} - R_\lambda, R_{Cy} + R_\lambda). \\
R4: T_B &= 0, \text{ if } \theta(\beta_{FB}, (2 \times R_\lambda)) < \theta_{\max}, \\
T_L &= 0, \text{ if } \theta(\beta_{FL}, (2 \times R_\lambda)) < \theta_{\max}, \\
\beta_C &= (R_{Cx} + R_\lambda, R_{Cy} + R_\lambda),
\end{aligned}
$$

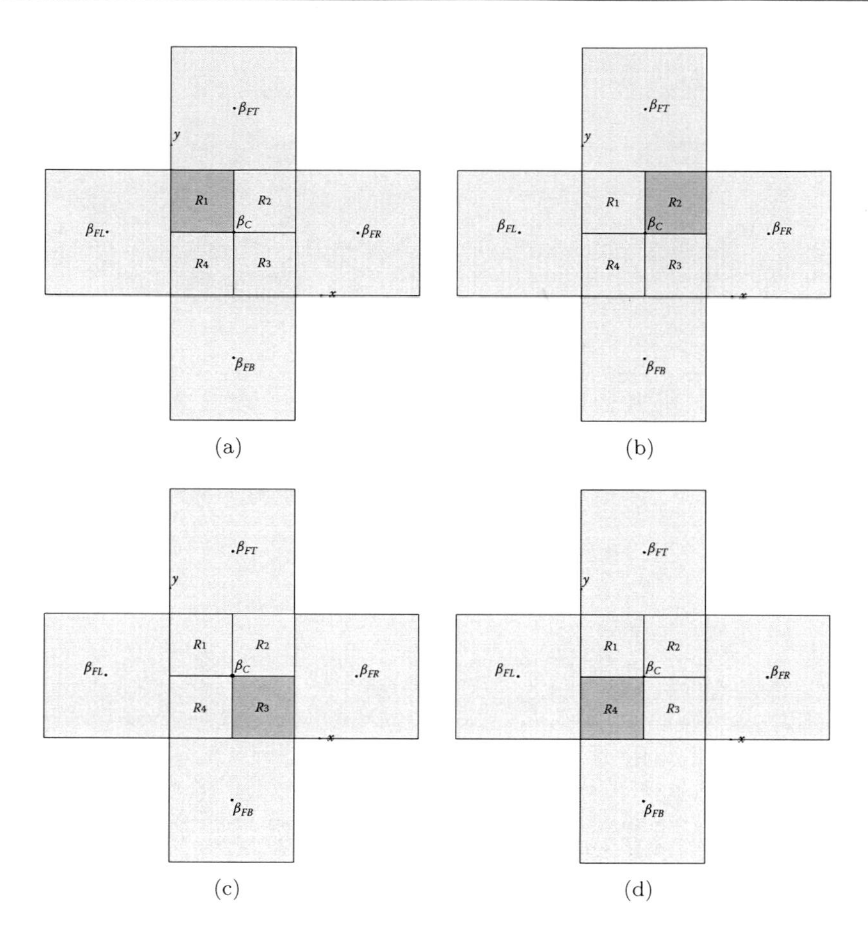

Figure 11.8. When calculating the morphing factors for Boundary Case 2, diagrams of the various positions used with each viewable region with a relative quadrant code of (a) $R1$, (b) $R2$, (c) $R3$, and (d) $R4$.

$$\begin{aligned}
\beta_{FL} &= (\beta_{Cx} - (4 \times R_\lambda), \beta_{Cy}),\\
\beta_{FT} &= (\beta_{Cx}, \beta_{Cy} + (4 \times R_\lambda)),\\
\beta_{FR} &= (\beta_{Cx} + (4 \times R_\lambda), \beta_{Cy}),\\
\beta_{FB} &= (\beta_{Cx}, \beta_{Cy} - (4 \times R_\lambda)).
\end{aligned}$$

The above set of conditionals test for the cases where an adjacent viewable region has not been split to the same size as the current viewable region and instead has an applied offset of $(2 \times R_\lambda)$. We therefore need to lock the affected morphing factors to 0. This is to ensure that all of the overlapping quadrilateral vertices exactly match with those of the larger adjacent viewable region. Figures 11.8(a)–11.8(d) provide diagrams of the various

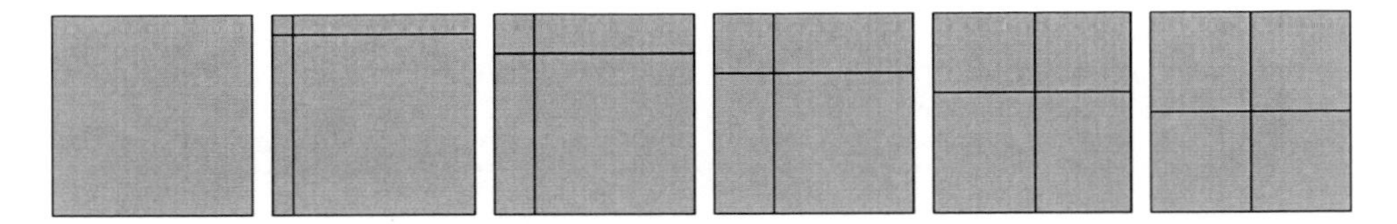

Figure 11.9. Examples of a single viewable region through various stages of the LOD transition. All of the morphing factors (T_L, T_T, T_R, T_B, T_C), for this diagram, are set to 0.0 on the left and increase up to 1.0 on the right in increments of 0.2.

positions used with each viewable region when calculating the morphing factors for Boundary Case 2, based on the relative quadrant code for the viewable region.

The various morphing stages of a single visible region are shown in Figure 11.9. As a result of how the morphing quadrilateral boundary points are calculated, neighboring visible regions share adjacent boundary points. This results in no cracks or seams when rendering, and smooth transitions when moving through a scene. Figure 11.10 shows examples of rendered viewable regions without (see Figure 11.10(a)) and with (see Figure 11.10(b)) the LOD Transition Algorithm in effect. Notice the T-junctions and visible transition boundaries that are prevalent in Figure 11.10(a).

11.2.5 Procedural Height Generation Algorithm

Any number of methods can be used in conjunction with the subdivision algorithm described in this paper. We chose to base ours on the "Ridged Multifractal Terrain Model" algorithm described by F. Kenton Musgrave [Ebert et al. 98].

The use of this procedural algorithm for us resulted in highly detailed and realistic terrain, as seen in our demo video below. The adaptive nature of the algorithm effectively eliminated high-frequency noise or aliasing and fit quite nicely with our LOD Transition Algorithm. We used a tileable noise texture when the "Ridged Multifractal Terrain Model" algorithm called for a noise value, with several optimizations to speed up the height and surface normal calculations.

The height is calculated at each of the visible region quadrilateral boundary points, resulting in displaced geometry when rendering (see Figure 11.11(c)). We calculate the surface normals on a per pixel basis to achieve even greater surface detail (see Figure 11.11(d)).

This portion of the algorithm quickly became one of the bottlenecks that reduced frame rates. The desire to attain higher levels of detail meant that we needed to come up with some optimizations to facilitate faster rendering. We were able to utilize three main optimizations that enabled us to greatly reduce the cost involved in our use of this algorithm.

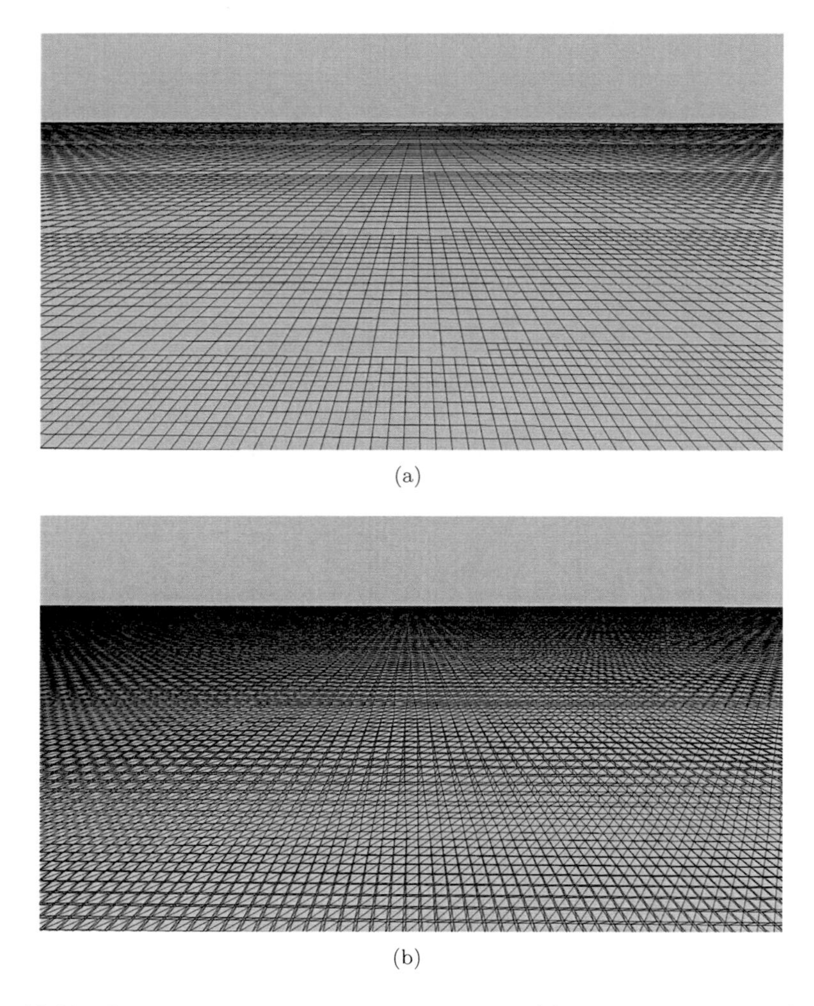

Figure 11.10. Example of rendered viewable regions (a) without using and (b) using the LOD Transition Algorithm described in this section.

The first optimization involved us separating the height calculation based on whether we were using the algorithm to displace geometry, or to calculate the surface normal. In the case of geometry displacement, we reduced the number of accumulation iterations (octaves) in the algorithm. This resulted in a less detailed displacement with the benefit of a moderate speed increase. To counteract the loss of displacement detail, we let the surface normal calculation proceed with a higher number of accumulation iterations (octaves). This resulted in greater visual detail. Each application of the algorithm will have its own balance between speed and detail, but the gains when finding the right balance can be significant.

The second optimization involved using a "layering factor" when creating the noise texture used for sampling the height. The original "Ridged Multifractal Terrain Model," in general, accumulates a number of noise octaves (layers) based in part on the distance from the viewer. As described in the algorithm, some processing is performed on each noise value to achieve the desired behavior within the terrain. Using the same noise processing calculations, we embed multiple layers of the noise within a texture. We are calling the number of added layers to the noise texture the "layering factor." The result of this was that we were able to reduce the number of accumulation iterations (octaves) by the same "layering factor," which provided a significant reduction in the total cost of the height calculation. One of the tradeoffs with this optimization was a reduction in the randomness of the resulting heights, although this was hard to detect visually. Another tradeoff was the increase in size of the noise texture in order to preserve the necessary detail of the additional layers.

The third optimization was to embed gradient information within the noise texture itself. Accumulated much like the height values, these gradient values allowed us to attain the surface normal using a variable number of iterations (octaves) without having to take additional samples to calculate a gradient. This also reduced the sharp grid-like "knife edges" that became visually problematic and reduced the overall believability of the terrain model. While this increased the memory footprint of the texture through the use of the additional color channels, the overall speed improvement was significant.

11.3 Results

In our tests, the CPU utilization averaged approximately 1%. Frame rendering times for our demo averaged between 20 to 25 ms, while running at a resolution of 1,024 × 768. The demo was run on a 2.9 GHz AMD Phenom CPU with 8 GB system memory, and a 512 MB AMD RADEON HD 645 video card.

A breakdown of the timing for each stage of the algorithm, for a typical frame of our demo, is as follows:

Algorithm Stage	Vertex Shader (ms)	Geometry Shader (ms)	Pixel Shader (ms)	Total (ms)
1	0.005	0.021	0	0.026
2	0.071	0.661	0	0.732
3	0.014	0.955	19.404	20.373

From the timing results listed above, it is interesting to note that the most significant costs involved in the use of the algorithm were not found within the geometry shader, as expected. Both the iterative subdivision in Stage 2, as well as the LOD algorithm used in Stage 3, added comparably marginal costs. The pixel shader when compositing the resulting subdivided mesh in Stage 3, dominated the incurred costs of the algorithm. Furthermore, when profiling Stage 3, 66% of the time was spent within the GPU Texture Units. This highlights the fact that our choice or implementation of the procedural height algorithm on a per pixel basis, while visually effective, could be improved upon.

Figures 11.11(a)–11.11(e) show some of the various steps taken to create the final rendered scene. In Figure 11.11(e), we added some diffuse lighting from a single directional light source, and modified the color of the terrain based on the calculated surface normal. Greater amounts of realism can be easily added through any number of techniques, such as the use of textures, detail maps, etc.

An example application, as well as the complete source code implementing this algorithm, has been included with this publication.

11.4 Conclusions

The results of our algorithm show promise, and we are continuing our research in this area in an effort to make further gains.

A lot of the speed costs involved in our usage example come from the multi-layered procedural calculation of the terrain height at the specified locations. Swapping our procedural algorithm out and using a simpler method, such as a single texture based height map, results in much faster rendering times. When we swapped out our procedural height calculation for a simpler height map lookup, we were able to achieve frame rendering times of 10 ms or even lower at the same resolution.

One problem with our approach is the visual phenomenon described as "vertex swimming." This can be noticed when moving towards large features that contain a high frequency of detail. The problem can be greatly reduced by increasing the amount of subdivisions created before rendering (i.e., by lowering the $\theta_{\max}$ value used in the Subdivision Algorithm). When we swapped out our LOD Transition Algorithm and instead added logic to correct the geometry (T-junctions) at the LOD boundaries, we were able to eliminate the "vertex swimming" phenomenon. Unfortunately, this also meant that the LOD boundaries became very noticeable, and in our opinion, were more of a visual distraction than the "vertex swimming" itself. There may be other ways to eliminate this visual distraction, and we are hopeful that this issue will be improved with continued research.

Care must be taken to ensure high precision floating-point calculations are used throughout this algorithm. Artifacts can sometimes appear if care is not taken in this regards, and floating-point errors are inadvertently allowed to compound, which can show up as sporadic pixel noise in the final rendered scene. We solved this problem by using integer based values for our visible region's center positions and applied offsets. Our initial *applied offsets* are set to large factors of two, in order to accommodate the maximum number of iterative subdivisions. By only applying a floating-point conversion factor when we required world space coordinates, we were able to completely eliminate all floating-point difference errors, while also scaling the data appropriately.

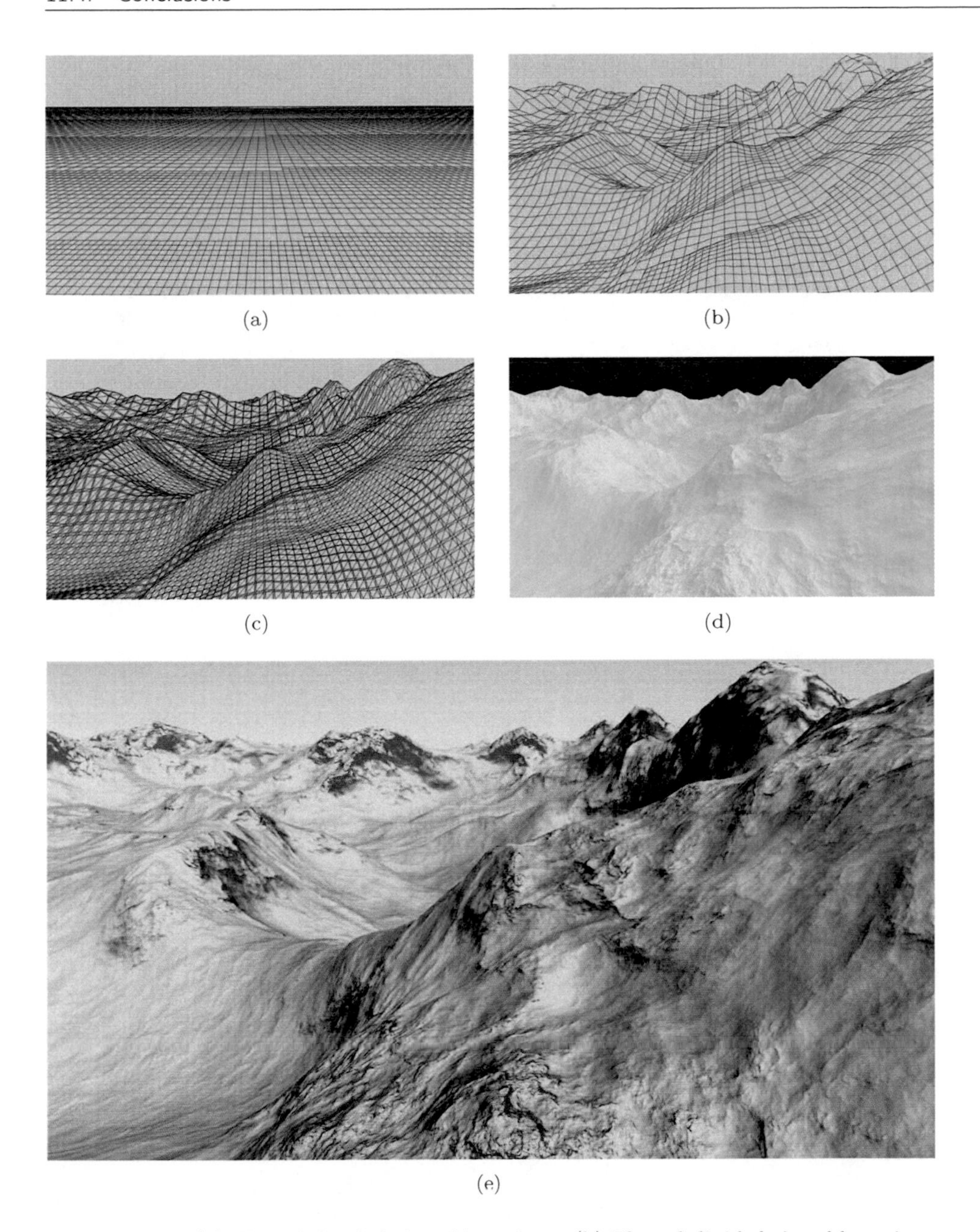

(a) (b) (c) (d) (e)

Figure 11.11. (a) The subdivided viewable regions. (b) The subdivided viewable regions shown with the heights added to the viewable region quadrilateral boundary points. (c) The subdivided viewable regions shown with the LOD Transition Algorithm applied. (d) The subdivided viewable regions shown with the per-pixel surface normals. (e) The final scene, with added directional lighting and simple procedural coloring.

Improvements to the subdivision algorithm and resulting data amplification could be made to better utilize hardware depth culling when rendering the resulting mesh. Increasing the subdivision granularity per iteration, and adding the subdivided regions to the output streams in increasing order of average depth (nearest to furthest) would enable the final rendering pass to benefit more from depth culling.

Combining this technique with the built-in hardware tessellation now available in modern graphics hardware could have some benefits, perhaps resulting in increased frame rates. Once a minimally acceptable level of subdivision has been achieved through our iterative process described in Stage 2 of Section 11.2.2, a single additional pass utilizing hardware tessellation may reduce unnecessary computations. Further research into this possibility would be needed to measure possible benefits.

Bibliography

[Ebert et al. 98] D. S. Ebert, F. K. Musgrave, D. Peachey, K. Perlin, and S. Worley. *Texturing and Modeling: A Procedural Approach*, Second Edition. San Diego: Academic Press, 1998.

[Strugar 10] Filip Strugar. "Continuous Distance-Dependent Level of Detail for Rendering Heightmaps (CDLOD)." http://www.vertexasylum.com/downloads/cdlod/cdlod_latest.pdf, 2010.

[Johanson 04] Claes Johanson. "Real-Time Water Rendering Introducing the Projected Grid Concept." Master of Science Thesis, Lund University, 2004. (Available at http://fileadmin.cs.lth.se/graphics/theses/projects/projgrid/projgrid-hq.pdf.)

Introducing the Programmable Vertex Pulling Rendering Pipeline

Christophe Riccio and Sean Lilley

12.1 Introduction

We believe that today's GPUs provide us high computing power and enough bandwidth to create scenes with much higher complexity than what is currently found in most real-time applications. Unfortunately scene complexity is mainly bound by the CPU, which limits the number of draw calls an application can submit per frame.

In this chapter we introduce what we call the *Programmable Vertex Pulling Rendering Pipeline*, which aims to remove the CPU bottleneck by moving more tasks onto the quickly evolving GPU (Figures 12.1 and 12.2). With this design, we show a way to increase the number of draws per frame leading to scenes with a higher level of complexity.

First, we describe the current limitations of existing draw submission designs. Then, we present the Programmable Vertex Pulling Rendering Pipeline. Finally, we propose new API directions we could take advantage of to create real-time scenes with yet unreached levels of complexity.

We base this study on AMD's Southern Islands architecture used for both the AMD Radeon HD 7000 series and the AMD FirePro W series. Though we discuss the hardware design and driver stack details of this architecture, we expect that these concepts would be similar on other post-OpenGL 4 GPUs.

All the performance tests in this chapter have been measured on an AMD FirePro W8000 graphics card with Catalyst 9.01 drivers running on an AMD Phenom X6 1050T and 8 GB of memory.

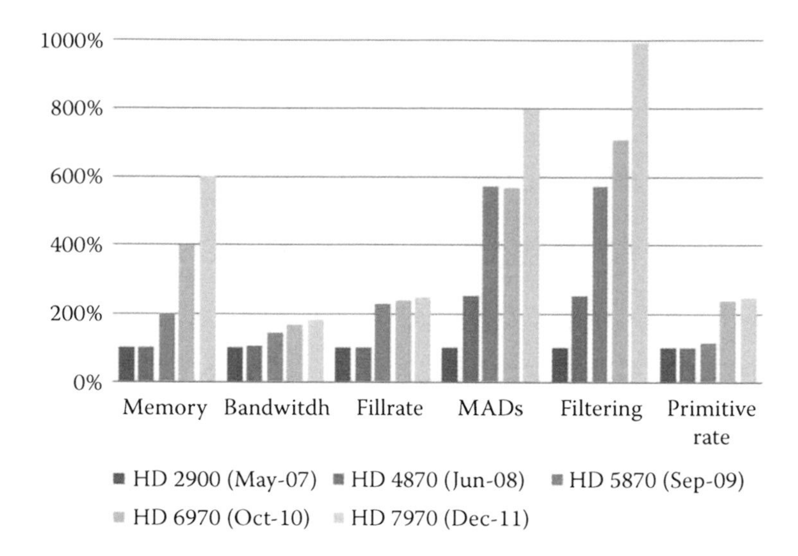

Figure 12.1. Relative evolution of AMD GPU specifications over the last six years.

	Memory MB	Bandwitdh GB/s	Fillrate Gpixels/s	MADs Gflops	Filtering Gtexels/s	Primitive rate Mtriangles/s
HD 2900	512	98.7	11.872	474.9	11.9	742
HD 4870	512	107.3	12	1,200	30	750
HD 5870	1,024	143.1	27.2	2,720	68	850
HD 6970	2,048	163.9	28.2	2,703	84.5	1,760
HD 7970	3,072	179.2	29.6	3,789	118.4	1,850

Figure 12.2. Absolute evolution of AMD GPU specifications over the last six years.

12.2 Draw Submission Limitations and Objectives

It is hard to appreciate how powerful today's GPUs can be. For example, a typical desktop GPU consumes at least two triangles per GPU clock, but if a programmer uses immediate mode to submit these two triangles, then comparatively many more CPU clocks must be expended. While the performance may seem strong, the programmer is still not fully utilizing the hardware.

Extending this discussion to draw calls using vertex buffers, the number of draws an application can submit per frame remains very limited compared to the peak ability of the GPU. Again, this is due to the CPU overhead of draw calls. We believe that this limit for real-time software is about 1,000 to 5,000 draw calls per frame, depending on the state changes and resource switching occurring between them. Another important limitation is the primitive peak rate on the

GPU, which defines the number of primitives that the GPU can render per frame without becoming the bottleneck. Because the draw call limit is so small while the peak primitive rate is proportionally high, programs must aim to render a lot of primitives per draw call.

To tackle this issue, Direct3D 11 introduced the concept of deferred contexts, where multiple contexts can record commands executed later by the main context. Unfortunately, this strategy is not particularly effective and doesn't scale linearly across the number of cores utilized. This is mainly due to the synchronization in the Direct3D 11 runtime and the fact that today's GPUs only have a single graphics ring, which can only process a single command queue at a time.

An earlier solution introduced by OpenGL 3 hardware was instancing. It provides one way to deal with a growing demand for scene complexity without adding CPU overhead. Unfortunately, instancing is nothing but duplicating a mesh multiple times, which limits the complexity we can reach. Another approach is to use batching, which aggregates multiple meshes into a single set of buffer objects and issues a single draw call. This performs well for perfectly static geometry and is relatively practical to use. However, it limits the amount and the granularity of the culling we can perform. It can also waste some memory when padding meshes into fixed-size memory chunks.

To reach a much higher scene complexity, we are looking for a solution where

- we could submit a lot more draws per frame;
- each draw would render meshes with different geometry, number of vertices, and even vertex formats;
- each draw could render a small number of primitives but still hit the GPU primitive rate;
- each draw could access different resources.

12.3 Evaluating Draw Call CPU Overhead and the GPU Draw Submission Limitation

12.3.1 The Performance Test

Due to the complex nature of real-time graphics software, it is difficult to understand the source for the cost of a single draw call: Is the 1,000 to 5,000 draw calls limit a GPU or a CPU limitation?

In this section, we follow our intuition that tells us that the more resource switching we do between draws, the higher the CPU overhead will be. Meanwhile, the GPU has a fixed cost for each draw. To build a relevant test to prove this hypothesis, we define the following criteria:

```
for(size_t i = 0; i <VertexFormat. size (); ++i)
{
  glBindVertexArray(VertexFormat[i]. Name);
  for(size_t j = 0; j <Mesh.size (); ++j)
    glDrawElementBaseVertex(
        Mesh[j].Mode, Mesh[j]. Count, Mesh[j]. Type,
        Mesh[j].Offset, Mesh[j]. BaseVertex);
}
```

Listing 12.1. Efficient draws submissions.

- The test should not be CPU bound.
- The test should not be primitive limited.
- The test should not be shading limited.
- The test should not be blending limited.

Our tests render a single quad (two triangles) per draw. The quad we render is only 16 pixels to ensure that the GPU or the drivers do not discard the draws. Rendering a slightly larger number of pixels doesn't affect the frame rate, which confirms that shading or blending is not the bottleneck.

We create three tests. The first one uses instancing so that only a single draw call is performed while multiple draws are submitted. Instancing ensures that the GPU command processor [AMD 12] does the draw submission, which guarantees that we are not CPU bound. Hence, we use this test to evaluate the constant GPU cost for a draw submission. In a second test (Listing 12.1), we render different meshes of identical vertex formats using a shared vertex array object (VAO), utilizing the base vertex parameter to avoid state changes [Romanick 08]. In this test we evaluate the constant CPU cost for a draw call. Finally in a third test (Listing 12.2), we bind a different VAO [Koch 09] for every draw call to evaluate the cost of switching a single resource.

We observe from the results of our tests (Figures 12.3 and 12.4) that the GPU is extremely efficient at processing draws. For example, Figure 12.3 shows that

```
for(size_t i = 0; i <Mesh.size (); ++i)
{
glBindVertexArray(Mesh[i]. Name);
glDrawElement(Mesh[i].Mode, Mesh[i]. Count, Mesh[i].Type, Mesh[i]. Offset);
}
```

Listing 12.2. Intensive resource switching approach to submit draws.

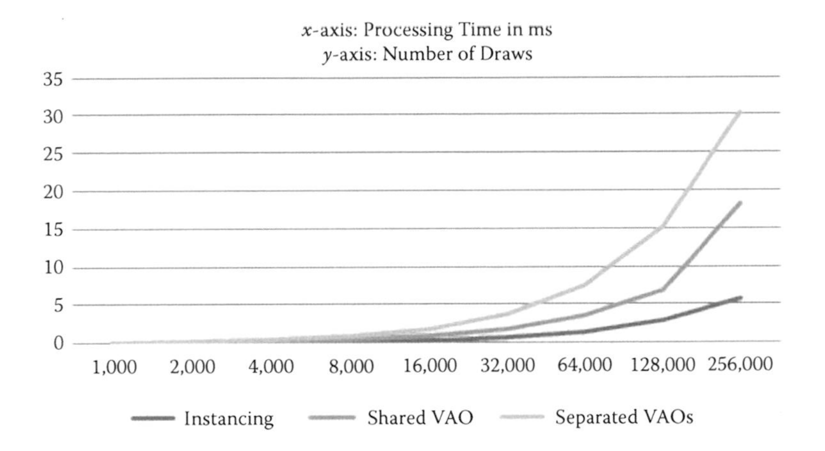

Figure 12.3. Absolute draw submission performance tests comparing instancing against a shared VAO for all draws and separated VAO per draw.

256,000 draws are processed in just above 5 ms when instancing is used in our basic scenario.

As soon as the CPU submits each individual draw (shared VAO), the performance drop is significant, but this impact is much higher with a single resource switching between draws (separated VAOs). Thus, we can expect that this performance cost will rise significantly when we increase the number of resources switching per draw, such as programs, textures, and uniforms. Switching states

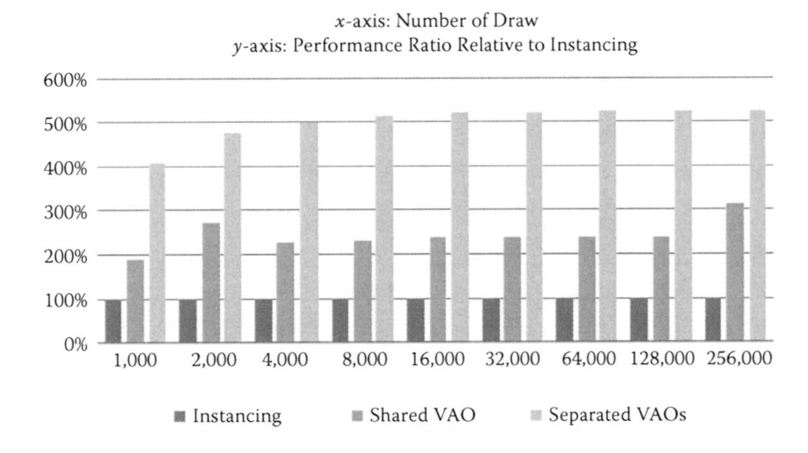

Figure 12.4. Relative draw submission performance tests comparing instancing against a shared VAO for all draws and separated VAO per draw.

and resources between draws actually consumes most of the draw submission performance. From these numbers we can confirm that the number of draw calls per frame is mainly a CPU overhead limitation and that we can reach the GPU submission limit somewhere between the instancing and the shared VAO results. This is the level of performance we are looking for.

12.3.2 Understanding the Nature of the CPU Overhead in Our Test

What is the nature of the CPU overhead in this VAO case? First, there is a validation step where the drivers check that no OpenGL error is generated by the OpenGL commands, as well as checking whether the vertex format and the bound buffers have changed. The second part concerns the vertex setup where the drivers generate a *fetch shader* devoted to building the vertices by indexing the array buffer according to the vertex format. Using a fetch shader allows reusing of the unified arithmetic logic units (ALUs) on the GPU to do the vertex fetching. To avoid increasing the number of VAO validations and vertex setups, the application should sort the rendering by vertex format and pack multiple meshes of identical vertex formats into a single VAO.

12.3.3 Avoiding CPU Overhead by Reducing Resource Switching

In the previous sections we showed that switching resources per draw has a significant impact on performance due to CPU overhead. From a software design point of view, we can avoid this cost by packing multiple resources together and using the GPU to index those resources.

For VAOs, we can pack together multiple meshes sharing the same vertex format and relying on base vertex to access the right data per draw. For textures, we can rely on texture 2D arrays to expose many textures per texture unit. For uniforms, we can pack them into large uniform buffers sorted by update rate and index the uniform blocks in the shader to access the right data per draw. The resource number limits (Figure 12.5) define how many resources we can index inside shaders, hence how much CPU side resource switching we can avoid.

Some resources like uniform blocks are extremely limited but others are generously provided. On AMD Southern Islands the maximum size for a texture 2D array is 16,384 (width) by 16,384 (height) by 8,192 (layers) by 16 (RGBA32F) bytes for a total of 32 TB for a single texture. Obviously we can't store this much memory on a graphics card but we can address it. This is one of the motivations behind the creation of the `AMD_sparse_texture` [Sellers 12a] extension enabling partially resident memory of the GPU resources.

12.3.4 Indexing Resources in Shaders, Dynamically Uniform Expressions

Indexing resources is typically performed by relying on some of the built-in variables provided by OpenGL (Figure 12.6).

Resources	OpenGL 4.3	Southern Islands	Kepler
Max texture units	16	32	32
Max texture 2D layers	2,048	8,192	2,048
Max texture 2D size	16,384	16,384	16,384
Max texture 3D size	2,048	8,192	2,048
Max texture bu er size	64	256	128
Max shader storage bu er binding	8	8	
Max combined shader storage blocks	8		
Max shader bu er size	16 MB		
Max uniform blocks per stage	14	15	14
Max combined uniform blocks	84	90	84
Max uniform block size	64 KB	64 KB	64 KB
Max vertex attributes	16	29	16
Max subroutines	256	4,096	1,024
Max texture image units per stage	16	16	32
Max combined texture image units	96	96	192

Figure 12.5. Some implementation-dependent values used for the count of available resources.

Built-in	Shader Stage	Description
gl_InstanceID	vertex	The instance number of the current draw in an instanced draw call
gl_VertexID	vertex	The integer index i implicitly passed by one of the other drawing commands
gl_PrimitiveID	control, evaluation	The number of primitives processed by the shader since the current set of rendering primitives was started
gl_PrimitiveID	fragment	The value written to the gl_PrimitiveID geometry shader output if a geometry shader is present. Otherwise, it is assigned in the same manner as with tessellation control and evaluation shaders
gl_PrimitiveIDIn	geometry	Filled with the number of primitives processed by the shader since the current set of rendering primitives was started
gl_SampleID	fragment	The sample number of the sample currently being processed
gl_InvocationID	control	The number of the output patch vertex assigned to the tessellation control shader invocation
gl_InvocationID	geometry	The invocation number assigned to the geometry shader invocation
gl_Layer	fragment	Selected framebu er layer number
gl_ViewportIndex	fragment	Selected viewport number
gl_WorkGroupID	compute	The three dimensional index of the global work group that the current invocation is executing in
gl_LocalInvocationID	compute	The three-dimensional index of the local work group within the global work group that the current invocation is executing in

Figure 12.6. GLSL built-in variables for shader indexing.

Because the constant engine of AMD OpenGL 4 GPUs can only fetch a single resource header per workgroup, indexing an array of resources must be done using what OpenGL calls *dynamically uniform expressions* [Kessenich 12, Section 3.8.3]. All work items in a work group must use the same index to access the same resource. Before OpenGL 4.3 introduced the compute shader stage, OpenGL didn't have the notion of workgroup or work item but we can consider each vertex, primitive or fragment as a work item. If an index is set per primitive, it will be a dynamically uniform expression on the fragment shader stage because all the fragments will belong to the same workgroup. Resources that must be indexed by dynamically uniform expressions are sampler arrays, image arrays, uniform block arrays, atomic counter buffer arrays, shader storage block arrays, and subroutine index arrays. Furthermore, GLSL shaders may access resources through a series of if statements. This is nothing but another embodiment of resource indexing that requires following the same constraints as other dynamically uniform expressions.

12.4 Programmable Vertex Pulling

Relying on resource batching and GPU indexing of resources can significantly reduce the CPU overhead [Hilaire 12]. However, this approach still suffers from several limitations. First, CPU overhead still exists in the form of CPU draw call submissions as shown in the shared VAO case in Figures 12.3 and 12.4. In addition, actual real-time rendering applications don't just submit draws; they need to select the draws that they expect to be visible first, performing culling. This task is not trivial as it often relies on space partitioning techniques to quickly analyze the scene typically consuming a lot of CPU time. To hide this cost, many applications use a dedicated thread for this task, introducing a frame of latency. When the scene increases in complexity, like we are imagining in this chapter, the time consumed by this thread increases until its latency can't be hidden anymore.

One idea is to move the culling and sorting from the CPU to the GPU by relying on OpenCL or the OpenGL compute shader stage so that the GPU selects and submits itself the draws. We call this approach the *Programmable Vertex Pulling Rendering Pipeline* (Figure 12.7). The initial pipeline is composed of two stages. On the one hand, the *Programmable Draw Dispatch* stage uses compute shaders with OpenGL 4.3 multi draw indirect buffers [Sellers 12b]. On the other hand, the *Programmable Vertex Fetching* stage uses the vertex shader stage to index into shader storage buffers or texture buffers to manually compose each vertex instead of using the VAO.

12.4.1 Programmable Draw Dispatch

OpenGL 4.0 introduced the draw indirect functionality that allows storing the parameters of draw commands into a buffer object. Unfortunately, a call to such

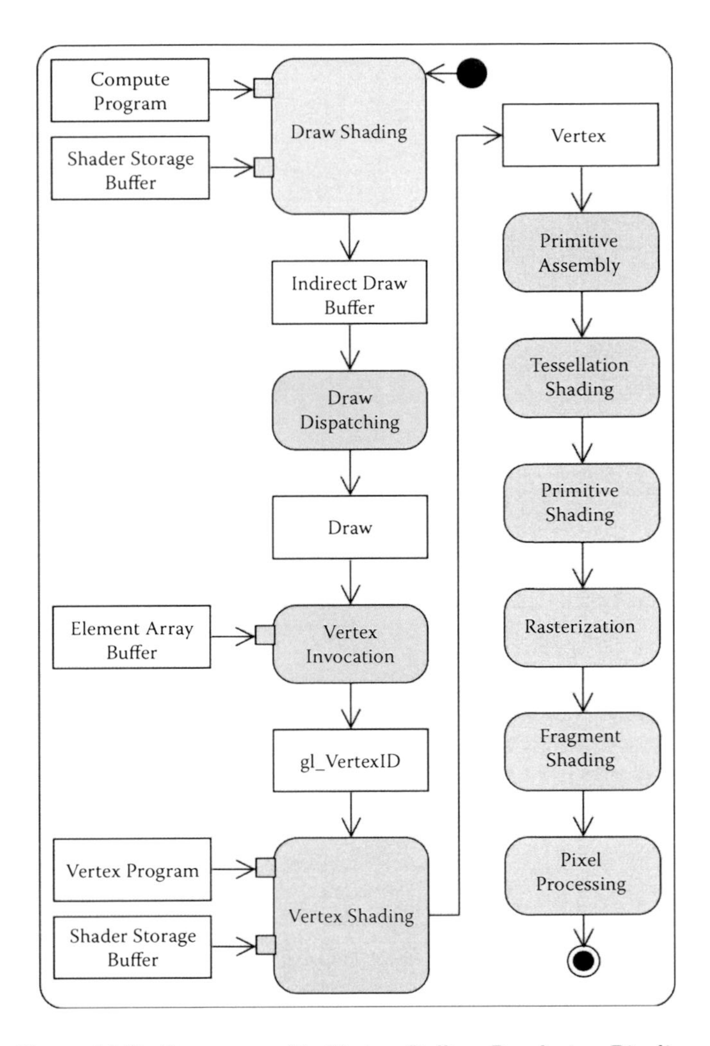

Figure 12.7. Programmable Vertex Pulling Rendering Pipeline.

a draw function is quite expensive on the CPU side, which originally decreased the interest of the draw indirect functionality.

Fortunately, `ARB_multi_draw_indirect`, a version of `AMD_multi_draw_indirect` promoted to the core specification in OpenGL 4.3, extends the draw indirect functionality by packing multiple draw indirect calls into a single call, amortising the high constant cost of the call [Rákos 12a]. With the new functions `glMultiDrawArraysIndi rect` and `glMultiDrawElementsIndirect` the command processor submits the draws in place of the CPU, thus removing nearly all CPU overhead per draw. (See Listing 12.3 and Figure 12.8.)

```
structdrawElementsIndirectCommand
{
 GLuint count;
 GLuintinstanceCount;
 GLuintfirstIndex;
 GLintbaseVertex;
 GLuintbaseInstance;
};
...
// At creation time
glBindBuffer(GL_DRAW_INDIRECT_BUFFER, BufferName);

glBufferData(GL_DRAW_INDIRECT_BUFFER,
    sizeof(drawElementsIndirectCommand) * DrawCount, NULL, GL_STATIC_COPY);
glBindBuffer(GL_DRAW_INDIRECT_BUFFER, 0);
...
// In the rendering loop
glBindBuffer(GL_DRAW_INDIRECT_BUFFER, BufferName);
glMultiDrawElementsIndirect(GL_TRIANGLES,
GL_UNSIGNED_INT, 0, DrawCount, sizeof(drawElementsIndirectCommand));
```

Listing 12.3. Creation and use of a multi draw indirect buffer.

An OpenCL kernel or an OpenGL compute shader is capable of building the multi draw indirect buffer. Just like the CPU visibility culling thread, the GPU processes a batch of object-bounding volumes. When the bounding volumes pass the visibility tests, the corresponding draw parameters for the objects fill the draw

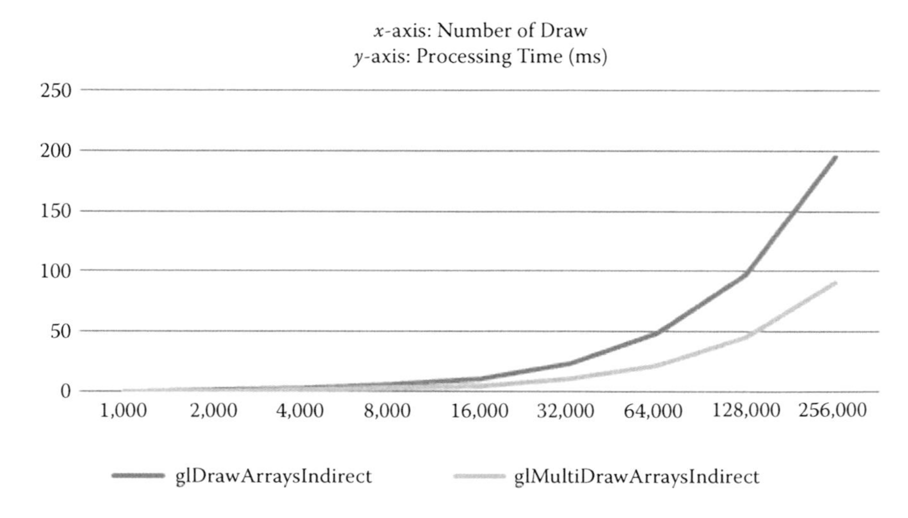

Figure 12.8. Performance comparisons between `glDrawArraysIndirect` and `glMultiDrawArraysIndirect`.

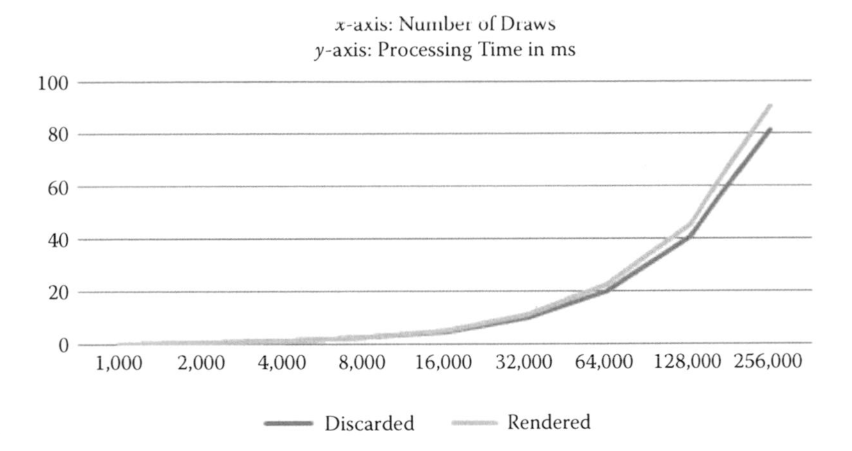

Figure 12.9. Comparing performance between discarded draws and draws rendering a single quad per draw producing 16 pixels on the framebuffer.

indirect buffer. For this chapter we used a brute force OpenCL kernel to check every object in the scene, but nothing prevents us from using space partitioning techniques to continue scaling up the scene complexity.

Taking a closer look at the new draw call functions, it is both interesting and annoying to acknowledge that the `glMultiDrawElementsIndirect` and `glMultiDrawArraysIndirect` parameter `drawcount` is not sourced from the indirect draw buffer. It would be a wrong idea to query the number of draws that the compute program effectively wrote in the indirect draw buffer and use this number for `drawcount`. Such an idea is worse than a CPU draw call overhead as it implies a synchronization point between the CPU and the GPU, preventing them from running in parallel. A better alternative is to use the maximum number of elements in the multi draw indirect buffer for `drawcount` and ask the command processor to discard the draws we don't need by writing 0 for the `primCount` parameter of each draw we want to discard. As Figure 12.9 shows, discarding draws is not free so that the number of draws needs to be carefully chosen.

12.4.2 Programmable Vertex Fetching

Using Programmable Draw Dispatching is an interesting step but it can show some limitations as soon as the scene contains meshes with multiple vertex formats. Sorting the multiple indirect draws per vertex format is not very practical but is necessary. Each vertex format requires a separate CPU call, which increases the CPU overhead. If there are many different vertex formats, then the high CPU cost of indirect draw calls may lead to an overall lower level of performance from the application.

Instead, we propose to no longer rely on the VAO and build the vertices ourselves in the vertex shader stage so that we don't need to sort draws per vertex format. We can group all the meshes into a single set of shader storage buffers, regardless of their vertex format. This allows us to rely on GPU indexing to fetch vertex data, avoiding the CPU overhead of resource switching.

Unfortunately OpenGL 4.3 doesn't provide a built-in variable `gl_DrawID` that would allow us to identify each draw. One approach that might be considered but is not successful would be to use the expression `gl_VertexID == 0` to detect that a new draw is invocated and to increment an atomic counter each time this case happens. However, the OpenGL specification doesn't specify the order of execution of the atomic operations so we can't identify which DrawID corresponds to which draw.

To generate the DrawID, the proposed solution is to store it in a 32-bit integer vertex attribute. Using `baseInstance` and an attribute equal to one, this attribute is shared with all vertex invocations of a draw. This DrawID is a bit-field where some bits are used to encode the vertex format ID and the rest of the bits are used to index the first vertex of a draw. (See Listing 12.4.)

```
layout (binding = PART_DYNAMIC) buffer partDynamicBuffer
{
    vec4 Kueken [];
} PartDynamicBuffer;

structpartStaticStorage
{
    vec2 Ovtsa;
    vec2 Varken;
};

layout (binding = PART_STATIC) buffer partStaticBuffer
{
    partStaticStorageVertex[];
} PartStaticBuffer;

struct vertex
{
    vec4 Kueken;
    vec2 Ovtsa;
    vec2 Varken;
};

layout (binding = FULL_STATIC) buffer staticBuffer
{
    vertex Vertex [];
} StaticBuffer;

// Only vertex attribute
layout (location = DRAW_OFFSET) in intDrawID;

// The fetch function gathers the vertex data from multiple buffers
vertex vertexFetchPartiallyDynamic(in intDrawOffset, in intVertexID)
{
    intBufferOffset = DrawOffset + VertexID;
```

```
    vertex Vertex;
    Vertex.Kueken = PartDynamicBuffer.Kueken [BufferOffset];
    Vertex.Ovtsa = PartStaticBuffer.Vertex [BufferOffset].Ovtsa;
    Vertex.Varken = PartStaticBuffer.Vertex [BufferOffset].Varken;
    return Vertex;
}

// The fetch function gathers the vertex data from a single static buffer
vertex vertexFetchStatic(in intDrawOffset, in intVertexID)
{
    intBufferOffset = DrawOffset + VertexID;

    return StaticBuffer.Vertex [BufferOffset];
}

// Select the right fetch function por draw according to the vertex format
vertex vertexFetch(in intDrawID, in intVertexID)
{
intDrawOffset = extractDrawOffset(DrawID);
intDrawFormatID = extractDrawFormat(DrawID);

    if(DrawFormatID== PARTIALLY_DYNAMIC)
        return vertexFetchPartiallyDynamic(DrawOffset, gl_VertexID);
    else if(DrawFormatID== FULLY_STATIC)
        return vertexFetchStatic(DrawOffset, gl_VertexID);
    else // ERROR, unknown vertex format
        return vertex ();
}

void main()
{
    vertex Vertex = vertexFetch(DrawID, gl_VertexID);
    ...
}
```

Listing 12.4. Code sample of programmable vertex fetching.

Here are a few comments regarding the usage of programmable vertex fetching.

- Many applications rely on *uber-shaders* where all the resources are declared but not necessarily used. When using programmable vertex fetching we can declare a large user defined `vertex` structure where only the vertex attributes used by the program pipeline would be filled.
- The DrawID is used to index GPU resources; hence, it must be a dynamically uniform expression.
- The more bits we use for vertex format IDs, the less vertices we can store in each shader storage buffer.
- We don't have to create a dedicated buffer for the DrawIDs. We can encode it inside a variable in a custom draw indirect structure (Listing 12.5). This way the compute shader stage can write all the draw parameters into a

```
structdrawArraysIndirectCommand
{
    GLuint Count;
    GLuint InstanceCount;
    GLuint First;
    GLuint BaseInstance;
    GLuint DrawID;
};
```

Listing 12.5. User-defined draw indirect structure with interleaved DrawID.

single buffer, but using a `stride` parameter larger than the size of the draw indirect structure will cost performance on AMD hardware.

12.5 Side Effects of the Software Design

12.5.1 Reaching the Primitive Peak Rate

Our quest for higher scene complexity involves performing more draws with less triangles per draw. However, reducing the number of triangles per object too much will have a significant performance hit because the draws wouldn't be able to reach the primitive peak rate of the GPU. To evaluate the minimum number of triangles we should submit per draw, we built a test giving us the performance chart in Figure 12.10.

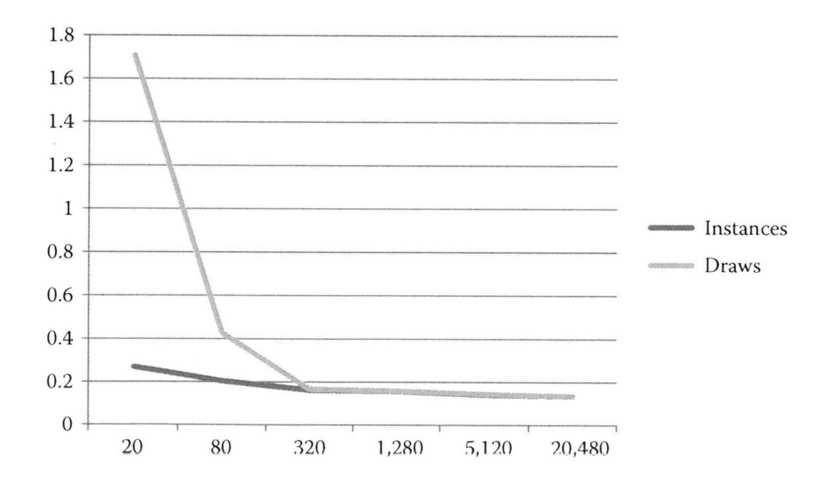

Figure 12.10. Evaluation of the minimum number of primitive to reach the GPU primitive peak rate.

We reach the primitive peak rate with about 32 primitives per draw when using multi draw indirect. This gives us the opportunity to use tessellation where triangle complexity is low, ensuring that we hit the peak primitive rate. Tessellation is not only a great tool to add geometric details; it is also a great tool to ensure that the pixel per primitive rate remains constant.

12.5.2 Memory Repacking

When rendering dynamic scenes, some objects will need to be created and deleted during the lifetime of the program execution. When using the batching approach, meshes must be added and deleted from an existing set of buffers, which typically leads to some level of memory fragmentation. There are multiple approaches to repack the memory and avoid wasting priceless graphics memory:

- The application can rely on `glCopyBufferSubData` and `glCopyImageSubData` to fill empty space in buffers and textures. If the granularity of the data is too thin, then the application would need to make more subdata CPU calls and thus create too much CPU overhead.
- The application can use an OpenCL kernel to move the data around. Such a kernel will probably underutilize the GPU's ALUs. However, because AMD Southern Islands allows us to run one graphics ring and two compute rings in parallel, such kernel execution could be hidden by other shader and kernel executions.
- The application can rely on the virtual memory capability of GPU using `AMD_sparse_buffer` and `AMD_sparse_texture` to manage the memory pages that need to be allocated or not. This relies on memory addressing to avoid moving the data while still effectively using the graphics memory.

12.6 Future Work

At the time of this chapter's writing, the Programmable Vertex Pulling Rendering Pipeline remains a work in progress considering that OpenGL drivers are suboptimized for this purpose. Many API improvements could strengthen this design for post-OpenGL 4 hardware:

- A built-in `gl_DrawID` would allow us to remove the need for vertex attributes and hence for vertex array objects. All the setup, mostly a CPU overhead and the bandwidth needs, would be avoided and replaced by a simple command processor counter.
- Currently when using programmable vertex fetching, we are basically losing the capabilities of certain draw call parameters: base vertex, base instance,

and the offset to the first element or vertex. All those parameters are used to compute the actual index of each vertex. We believe that such parameters are not necessarily useful for all scenarios. The registers used for those parameters should become user-defined variables to store the DrawID, for example.

- The series of if statements required to select the right vertex format inside the vertex fetching function is not very elegant as it introduces a level of indirection. By backing subroutines in buffers, we could select a subroutine per draw and effectively hide this indirection.
- The strategy behind the Programmable Vertex Pulling Rendering Pipeline is to replace CPU resource switching by GPU-based indexing of the resources. For this to be possible, we need to be able to access enough different resources. AMD Southern Islands architecture supports bindless buffers and textures so that an unlimited number of resources could be bound. By working with partially resident memory, we believe that both features would enable rendering of more complex scenes.

Beyond API improvements, we can also consider additional software design research:

- Generating the draw indirect buffer by itself is a complex task that we currently solve by using the brute force performance of the GPU in an OpenCL kernel. Instead, could we rely on GPU-based space partitioning techniques? Octrees, k-d trees, or bounding volume hierarchies (BVHs)? Which space partitioning techniques can be efficiently implemented on the GPU?
- Could we use programmable vertex pulling to bring deferred tile based rendering on immediate rendering GPUs? Is there an efficient GPU-based algorithm to build lists of triangles and dispatch them using separate draws per tile? Could we enable Order Independent Transparency [Knowles 12] in a single pass if we expose a portion of the Local Data Store [AMD 12] in the fragment shader stage?
- Can we use `AMD_query_buffer_object` [Rákos 12b] to build a heuristic to reorganize the memory in a memory management kernel?

12.7 Conclusion

In this chapter we presented the Programmable Vertex Pulling Rendering Pipeline, which can render more complex scenes by significantly reducing the CPU overhead caused by resource switching between draw calls. We detailed the possibilities given by GPU batching and indexing in the two main parts of this approach:

- With Programmable Draw Dispatch we use the GPU to both select the draws necessary to render a frame and to dispatch those draws, releasing the CPU from these tasks.
- With Programmable Vertex Fetching we extend the GPU indexing capability to ensure that each draw submitted by the GPU can render a different mesh with no interference by the CPU.

A special thanks to Arnaud Masserann and Dimitri Kudelski who reviewed this chapter.

Bibliography

[AMD 12] AMD. "AMD Graphics Cores Next (GCN) Architecture." Whitepaper, Radeon Graphics, June 2012.

[Hilaire 12] Sebastien Hillaire. "Improving Performance by Reducing Calls to the Driver." In *OpenGL Insights: OpenGL, Open GL ES, and WebGL Community Experiences*, edited by Patrick Cozzi and Christophe Riccio, Chapter 25. Boca Raton: CRC Press, 2012.

[Kessenich 12] John Kessenich (editor). *The OpenGL Shading Language 4.30.6.* http://www.opengl.org/registry/doc/GLSLangSpec.4.30.6.pdf, 2012.

[Knowles 12] Pyarelal Knowles, Geoff Leach, and Fabio Zambetta. "Efficient Layered Fragment Buffer Techniques." In *OpenGL Insights: OpenGL, Open GL ES, and WebGL Community Experiences*, edited by Patrick Cozzi and Christophe Riccio, Chapter 20. Boca Raton: CRC Press, 2012.

[Koch 09] Daniel Koch. "GL_ARB_draw_elements_base_vertex." http://www.opengl.org/registry/specs/ARB/draw_elements_base_vertex.txt, 2009.

[Rákos 12a] Daniel Rákos. "Programmable Vertex Pulling." In *OpenGL Insights: OpenGL, Open GL ES, and WebGL Community Experiences*, edited by Patrick Cozzi and Christophe Riccio, Chapter 21. Boca Raton: CRC Press, 2012.

[Rákos 12b] Daniel Rákos. "GL_AMD_query_buffer_object." http://www.opengl.org/registry/specs/AMD/query_buffer_object.txt, 2012.

[Romanick 08] Ian Romanick. "GL_ARB_vertex_array_object." http://www.opengl.org/registry/specs/ARB/vertex_array_object.txt, 2008.

[Sellers 12a] Graham Sellers. "GL_AMD_sparse_texture." http://www.opengl.org/registry/specs/AMD/sparse_texture.txt, 2012.

[Sellers 12b] Graham Sellers. "GL_ARB_multi_draw_indirect." http://www.opengl.org/registry/specs/ARB/multi_draw_indirect.txt, 2012.

A WebGL Globe Rendering Pipeline

Patrick Cozzi and Daniel Bagnell

13.1 Introduction

WebGL brings hardware-accelerated graphics based on OpenGL ES 2.0 to the web. Combined with other HTML5 APIs, such as gamepad, fullscreen, and web audio, the web is becoming a viable platform for hardcore game development. However, there are also other killer applications for WebGL; one we are particularly interested in is mapping. The web has a long history of 2D maps such as MapQuest, OpenStreetMap, and Google Maps. WebGL enables web mapping to move from flat 2D maps to immersive 3D globes.

In this chapter, we present a WebGL globe rendering pipeline that integrates with hierarchical levels of detail (HLOD) algorithms used to manage high resolution imagery streamed from standard map servers, such as Esri or OpenStreetMap. Our pipeline uses screen-space techniques, including filling cracks between adjacent tiles with different LODs with a masked Gaussian blur, filling holes in the north and south pole with masking and ray casting, and avoiding depth fighting with vector data overlaid on the globe by rendering a *depth plane*.

We use these techniques in Cesium, http://cesium.agi.com, our open-source WebGL globe and map engine. We found them to be pragmatic, clean, and light on the CPU.

13.2 Rendering Pipeline Overview

Imagery is commonly served from map servers using 256×256 RGB tiles. Tiles are organized hierarchically in a quadtree, where the root node covers the entire globe, $-180°$ to $180°$ longitude and $-90°$ to $90°$ latitude,[1] at a very low resolu-

[1] As we'll see in Section 13.4 the latitude bounds are actually $\approx \pm 85°$ for the most common projection.

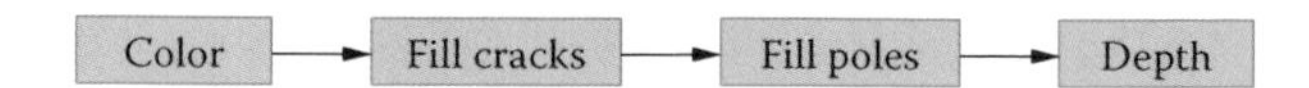

Figure 13.1. Our globe rendering pipeline.

tion. As we continue down the tree, tiles remain 256 × 256 but cover a smaller longitude-latitude extent, thus increasing their resolution. The extent of one of the root's child tiles is −180° to 0° longitude and 0° to 90°, and the extent of one of its grandchild tiles is −180° to −90° longitude and 45° to 90° latitude.

A 2D map can easily request tiles for rendering based on the visible longitude-latitude extent clipped to the viewport, and the zoom level. In 3D, when tiles are mapped onto a WGS84 ellipsoid representing the globe, more general HLOD algorithms are used to select geometry, i.e., tessellated patches of the ellipsoid at different resolutions, and imagery tiles to render based on the view parameters and a pixel error tolerance. HLOD algorithms produce a set of tiles, both geometry and texture, to be rendered for a given frame.

Our pipeline renders these tiles in the four steps shown in Figure 13.1. First, tiles are rendered to the color buffer only; the depth test is disabled. Next, to fill cracks between adjacent tiles with different geometric LODs, a screen-space Gaussian blur is performed. A fragment is only blurred if it is part of a crack; therefore, most fragments are not changed.

Next, since most map servers do not provide tiles near the poles, two viewport-aligned quads are rendered, and masked and ray-casted to fill the holes in the poles.

Finally, a depth-only pass renders a plane perpendicular to the near plane that slices the globe at the horizon. A ray is traced through each fragment and discarded if it does not intersect the globe's ellipsoid. The remaining fragment's depth values are written, allowing later passes to draw vector data on the globe without z-fighting or tessellation differences between tiles and vector data rendering later.

Let's look these steps in more detail.

13.3 Filling Cracks in Screen Space

Cracks like those in Figure 13.2 occur in HLOD algorithms when two adjacent tiles have different geometric LODs. Since the tile's tessellations are at different resolutions, together they do not form a watertight mesh, and gaps are noticeable.

There are a wide array of geometric techniques for filling cracks. A key observation is that cracking artifacts need to be removed, but adjacent tiles don't necessarily need to line up vertex-to-vertex. For example, in terrain rendering, it is common to drop flanges, ribbons, or skirts vertically or at a slight angle to minimize noticeable artifacts [Ulrich 02]. This, of course, requires creating extra geometry and determining its length and orientation. Other geometric

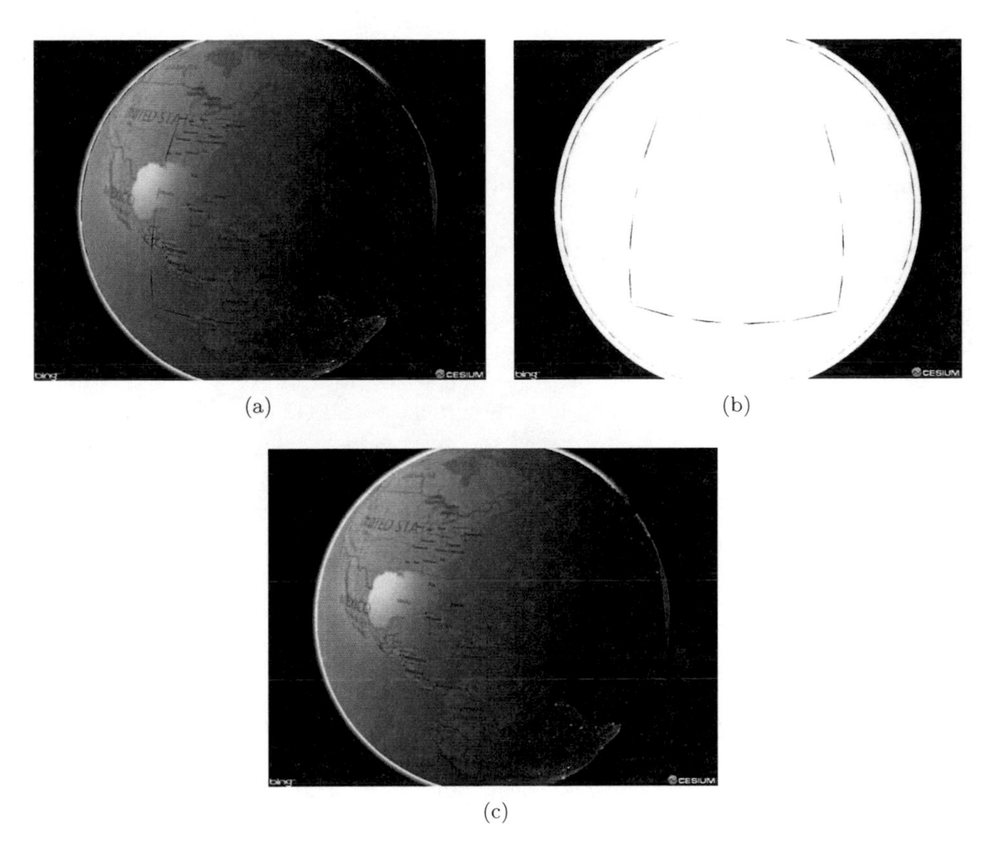

Figure 13.2. (a) Cracking between adjacent tiles with different geometric LODs (exaggerated for clarity). (b) An alpha mask that is white where tiles are rendered. (c) Cracks are detected using the alpha mask, and filled using a Gaussian blur.

techniques involve matching vertices in adjacent tiles by marking triangles that overlap a boundary as restricted during decimation [Erikson et al. 01].

Instead of filling cracks geometrically, we fill them in screen space. Cracks need to be filled with something plausible, but not necessarily extra geometry. To shade a fragment in a crack, we use a Gaussian blur that only includes samples from surrounding fragments not in the crack. The result is visually plausible, simple to implement, and well-suited to WebGL since it requires no extra geometry and is light on the CPU.

In the first rendering pass, each tile's color is rendered, and the alpha channel is set to 1.0. As shown in Figure 13.2(b), the alpha mask is 1.0 where tiles were rendered, and 0.0 in cracks and the sky.

To fill the cracks, a bounding sphere encompassing the ellipsoid is projected into screen space, and the bounding rectangle is found to reduce fragment workload. A viewport-aligned quad is rendered in two passes, one vertical and one horizontal, to perform a masked Gaussian blur. Using two passes instead of one reduces the number of texture reads from n^2 to $n + n$ for an $n \times n$ kernel [Rákos 10].

The fragment shader first reads the alpha mask. If the alpha is 1.0, the color is passed through since the fragment is part of a tile. Otherwise, two other values are read from the alpha mask. For the vertical pass, these are the topmost and bottommost texels in the kernel's column for this fragment. For example, for a 7×7 kernel, the texels are $(0, \pm 3)$ texels. If the alpha for both texels is 0.0, sky is detected, and the color is passed through. Otherwise, the blur is performed including only texels in the kernel with an alpha of `1.0`. Essentially, pixels surrounding cracks are bled into the cracks to fill them.

Kernel size selection presents an important tradeoff. An $n \times n$ kernel can only fill cracks up to $n - 2 \times n - 2$ pixels because of the sky check. We found that a 7×7 kernel works well in practice for our engine. However, we do not restrict geometry such that cracks will never exceed five pixels. It is possible cracks will not be completely filled, but we have found these cases to be quite rare.

13.4 Filling Poles in Screen Space

Most standard map servers provide tiles in the Web Mercator projection. Due to the projection, tiles are not available above 85.05112878° latitude and below −85.05112878°. Not rendering these tiles results in holes in the poles as shown in Figures 13.3(a) and 13.4(a). For many zoomed-in views, the holes are not visible. However, for global views, they are obvious.

There are several geometric solutions to fill the holes. Tiles can be created above and below the latitude bounds, essentially creating the same geometry as if image tiles were available. This is simple but can lead to over-tessellation at the poles. To avoid this, a projection tailored to the poles can be used similar to Miller and Gaskins [Miller and Gaskins 09]. However, this requires a good bit of code, and cracking between the two different tessellation methods needs to be addressed. Alternatively, a coarsely tessellated globe can be rendered after the imagery tiles, but tessellation and lighting discontinuities can be noticeable.

Instead of a geometric solution, holes can be filled in screen space, avoiding any concerns about over-tessellation or cracking. We render viewport-aligned quads covering the poles, and detect and shade the holes in a fragment shader.

For each pole, first compute a bounding sphere around the pole's longitude-latitude extent; for example, the north pole extent is −180° to 180° longitude, and 85.05112878° to 90° latitude. The sphere can be frustum and occlusion culled like any other geometry. If visible, the bounding rectangle of the sphere projected into screen space is computed, and a viewport-aligned quad is rendered as shown in Figure 13.3(b).

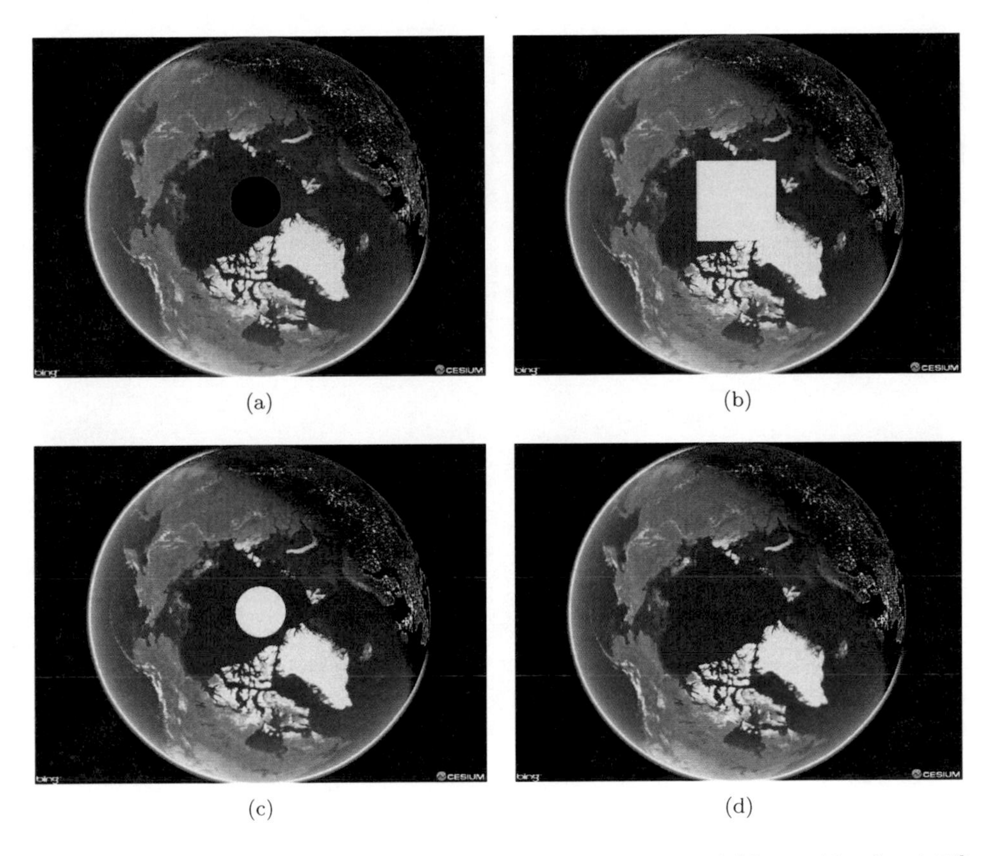

Figure 13.3. (a) A hole at the north pole since tiles are not available outside of $\approx \pm 85^\circ$ latitude. (b) A rectangle covering the hole in screen space. (c) Masking and ray casting to detect the hole. (d) Shading the hole.

The quad is larger than the hole behind it. For horizon views, it extends high above the ground. The fragment shader discards fragments that are not in front of the hole by first checking the alpha mask; if it is 1.0, a tile was rendered to this fragment, and it is not a hole. To discard fragments above ground, a ray is cast from the eye through the fragment to the ellipsoid. If the ray doesn't hit the ellipsoid, the fragment is above the ground, and discarded. At this point, if the fragment was not discarded, it covers the hole as in Figure 13.3(c)

We shade the fragment by computing the geodetic surface normal on the ellipsoid where the ray intersects, and use that for lighting with a solid diffuse color. Texture coordinates can also be computed and used for specular maps and other effects. Given that the poles are mostly uniform color, a solid diffuse color looks acceptable as shown in Figures 13.3(d) and 13.4(b).

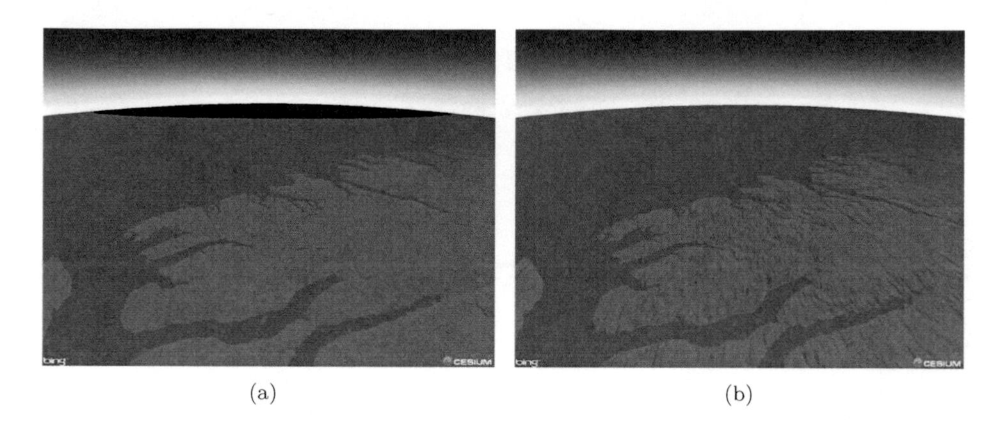

(a) (b)

Figure 13.4. Horizon views. (a) A hole at the north pole. (b) The filled hole.

Compared to geometric solutions, our screen-space approach has excellent visual quality with no tessellation or geometric cracking artifacts, uses very little memory, and is light on the CPU, requiring only the bounding rectangle computations, culling, and two draw calls. However, its fragment load can be higher given that many fragments are discarded from extreme horizon views,[2] early-z and hierarchical-z are disabled due to using `discard`, and that a ray/fragment intersection test is used. Given the speed difference between JavaScript and C++, and CPUs and GPUs, we believe this is a good tradeoff.

13.5 Overlaying Vector Data

Maps often overlay vector data, i.e., points, polylines, and polygons, on top of the globe. For example, points may represent cities, polylines may represent driving directions, and polygons may represent countries. Either raster or vector techniques can be used to render this data [Cozzi and Ring 11]. Raster techniques burn vector data into image tiles with an alpha channel. These tiles are then overlaid on top of the base map imagery. This is widely used; it keeps the rendering code simple. However, as the viewer zooms in, aliasing can become apparent, it is slow for dynamic data, and does not support points as viewport-aligned labels and billboards.

To overcome these limitations, we render vector data using point, line, and triangle primitives, which requires subdividing polylines and polygons to approximate the curvature of the globe. Given that these polygonal representations only represent the true globe surface when infinitely subdivided, the primitives are

[2]This could be reduced by using a screen-space rectangle that is the intersection of the pole's projected rectangle and the ellipsoid's projected rectangle.

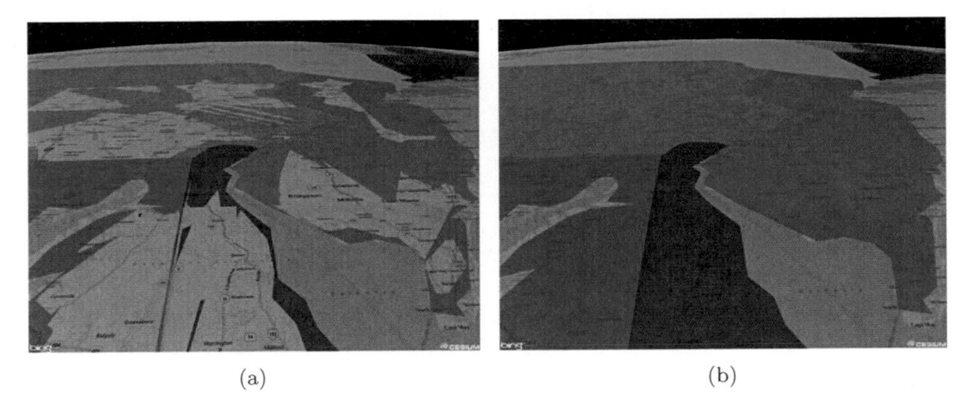

(a) (b)

Figure 13.5. (a) Vector data drawn without the depth plane result in artifacts. (b) With the depth plane.

actually under the true globe, which itself is approximated by triangles. With standard depth testing, parts of the vector data will fail the depth test and z-fight with the globe as shown in Figure 13.5.

We solve this using a method that is well-suited for JavaScript and WebGL; it uses very little CPU and does not rely on being able to write `gl_FragDepth`.[3] The key observation is that objects on or above the backside of the globe should fail the depth test, while objects on—but actually under—the front side of the globe should pass. We achieve this by rendering a *depth plane*, shown in Figure 13.6(a),

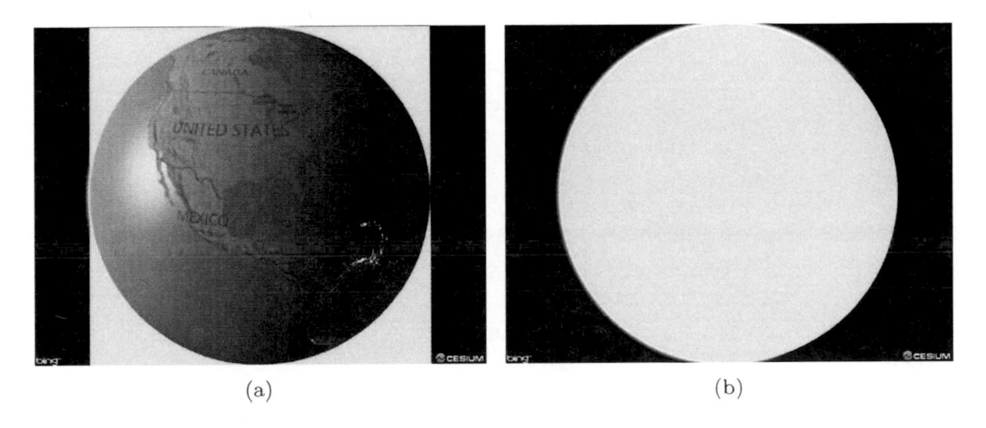

(a) (b)

Figure 13.6. (a) The depth plane intersects the globe at the horizon. (b) A ray is sent through each fragment in the plane to determine which fragments intersect the globe, and therefore, need to write depth.

[3]We expect an extension for writing depth in the future.

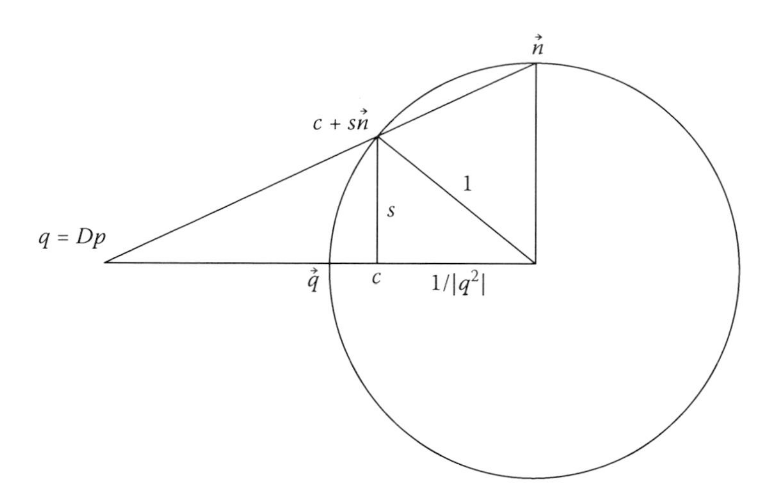

Figure 13.7. Computing the depth plane: a cross section of the ellipsoid scaled to a sphere as viewed from east of the camera.

in a depth-only pass that is perpendicular to the near plane and intersects the globe at the horizon. This plane is computed by determining the visible longitude and latitude extents given the camera position.

Following Figure 13.7, for an ellipsoid, $\frac{x^2}{a^2} + \frac{y^2}{b^2} + \frac{z^2}{c^2} = 1$, and camera position in WGS84 coordinates, p, we compute

$$D = \begin{pmatrix} \frac{1}{a} & 0 & 0 \\ 0 & \frac{1}{b} & 0 \\ 0 & 0 & \frac{1}{c} \end{pmatrix},$$
$$q = Dp,$$

where D is a scale matrix that transforms from the ellipsoid to a unit sphere and q is the camera position in the scaled space. Next, we compute the east vector, $\vec{e}$; the north vector, $\vec{n}$; the center of the circle where the depth plane intersects the unit sphere, c; and the radius of that circle, s:

$$\vec{e} = (0, 0, 1) \times \vec{q},$$
$$\vec{n} = \vec{q} \times \vec{e},$$
$$c = \frac{q}{|q|^2},$$
$$s = \sin \arccos \frac{1}{|q|^2}.$$

Finally, we compute the corners of the visible extent in WGS84 coordinates.

$$\begin{aligned}\text{upper left} &= D^{-1}(c + s(\vec{n} - \vec{e}))\\ \text{upper right} &= D^{-1}(c + s(\vec{n} + \vec{e}))\\ \text{lower left} &= D^{-1}(c + s(-\vec{n} - \vec{e}))\\ \text{lower right} &= D^{-1}(c + s(-\vec{n} + \vec{e}))\end{aligned}$$

In the fragment shader, a ray is cast from the eye through each fragment in the depth plane. If the fragment does not intersect the ellipsoid, the fragment is discarded as shown in Figure 13.6(b). The result is the globe's depth is replaced with the depth plane's depth, which allows objects on the front side of the globe to pass the depth test and those on the backside to fail without z-fighting or tessellation differences between image tiles and vector data. This is like backface culling, except it doesn't require the primitives to be backfacing; for example, a model of a satellite works with the depth plane but does not work with backface culling alone.

We originally used a technique based on backface culling. First, we rendered the tiles without depth. Next, we rendered polygons and polylines on the ellipsoid's surface without depth, and with backface culling implemented by discarding in the fragment shader based on the ellipsoid's geodetic surface normal. Finally, we rendered the tile's depth. Like the depth plane, this did not require writing `gl_FragDepth`; however, it had created a shortcoming in our API. Users needed to specify if a polygon or polyline was on the surface or in space. The depth plane works in both cases except for the rare exception of polylines normal to and intersecting the ellipsoid. The backface-culling technique also relies on two passes over the tiles, which increases the number of draw calls. This is a major WebGL bottleneck.

13.6 Conclusion

As long-time C++ and desktop OpenGL developers, we have found JavaScript and WebGL to be a viable platform for serious graphics development. We hope this chapter provided both inspiration for what is possible with WebGL, and concrete techniques for globe rendering that are well-suited to WebGL. To see these techniques in action, see our live demos at http://cesium.agi.com.

13.7 Acknowledgments

We thank Matt Amato, Norm Badler, Wolfgang Engel, Scott Hunter, and Kevin Ring for reviewing this chapter. We especially thank Frank Stoner for deriving the equations for the depth plane.

Bibliography

[Cozzi and Ring 11] Patrick Cozzi and Kevin Ring. *3D Engine Design for Virtual Globes*. Boca Raton: CRC Press, 2011. (Information at http://www.virtualglobebook.com.)

[Erikson et al. 01] Carl Erikson, Dinesh Manocha, and William V. Baxter III. "HLODs for Faster Display of Large Static and Dynamic Environments." In *Proceedings of the 2001 Symposium on Interactive 3D Graphics*, pp. 111–120. New York: ACM, 2001. (Available at http://gamma.cs.unc.edu/POWERPLANT/papers/erikson2001.pdf.)

[Miller and Gaskins 09] James R. Miller and Tom Gaskins. "Computations on an Ellipsoid for GIS." *Computer-Aided Design* 6:4 (2009), 575–583. (Available at http://people.eecs.ku.edu/~miller/Papers/CAD_6_4_575-583.pdf.)

[Rákos 10] Daniel Rákos. "Efficient Gaussian Blur with Linear Sampling." *RasterGrid Blogosphere*, http://rastergrid.com/blog/2010/09/efficient-gaussian-blur-with-linear-sampling/, 2010.

[Ulrich 02] Thatcher Ulrich. "Rendering Massive Terrains Using Chunked Level of Detail Control." *SIGGRAPH 2002 Super-Size It! Scaling Up to Massive Virtual Worlds Course Notes*. http://tulrich.com/geekstuff/sig-notes.pdf, 2002.

Dynamic GPU Terrain

David Pangerl

14.1 Introduction

Rendering terrain is crucial for any outdoor scene. However, it can be a hard task to efficiently render a highly detailed terrain in real time owing to huge amounts of data and the complex data segmentation it requires. Another universe of complexity arises if we need to dynamically modify terrain topology and synchronize it with physics simulation. (See Figure 14.1.)

This article presents a new high-performance algorithm for real-time terrain rendering. Additionally, it presents a novel idea for GPU-based terrain modification and dynamics synchronization.

Figure 14.1. Dynamic terrain simulation in action with max (0.1 m) resolution rendered with 81,000 tris in two batches.

14.2 Overview

The basic goal behind the rendering technique is to create a render-friendly mesh with topology that can smoothly handle lowering resolution with distance with minimal render calls.

14.3 Terrain Data

Because rendering and manipulation of all the data is performed on a GPU, we need to conserve the amount of data (i.e., reduce the number of rendering and simulation parameters to a minimum) and prepare the data in a GPU-compatible form. Terrain data are saved in a R16G16B16A16 texture format.

Terrain data attributes include

- terrain height—a normalized terrain height,
- texture blend—a texture index and blend parameters,
- flowability—a measure used to simulate condensed cliffs produced by a plow modification,
- compression—a measure used to simulate wheel compression.

Flowability. Terrain flowability is used to simulate terrain particles' ability to spread to neighboring particles. A flowability parameter is fundamental in dynamic erosion modification for cliff creation.

14.4 Rendering

Rendering terrain was one of the most important parts of the algorithm development. We needed a technique that would require as few batches as possible with as little offscreen mesh draw as possible.

We ended up with a novel technique that would render the whole terrain in three or fewer batches for a field of view less than 180 degrees and in five batches for a 360-degree field of view.

This technique is also very flexible and adjustable for various fields of view and game scenarios.

14.4.1 Algorithm

It all starts with the render mesh topology and vertex attributes. A render mesh is designed to discretely move on per level resolution grid with the camera field of view in a way that most of the mesh details are right in front of the camera view. A GPU then transforms the render mesh with the terrain height data.

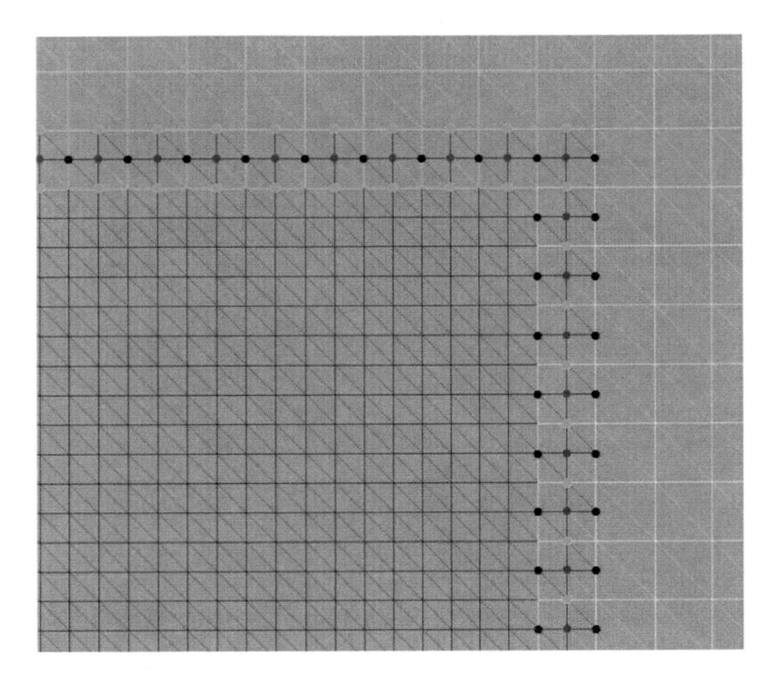

Figure 14.2. The two neighboring levels showing the intersection and geomorphing attributes.

Render mesh topology. As mentioned before, the terrain mesh topology is the most important part of the algorithm.

Terrain render mesh topology is defined by quad resolution R, level size S, level count L, and center mesh level count L_c:

- R, the quad resolution, is the edge width of the lowest level ($Level_0$) and defines the tessellation when close to the terrain.

- S, the level size, defines the number of edge quads. Level 0 is a square made of $S \times S$ quads, each of size $R \times R$.

- L, the level count, defines the number of resolution levels.

- L_c, the center mesh level count, is the number of levels (from 0 to L_c) used for the center mesh.

Each resolution level R is doubled, which quadruples the level area size. Levels above 0 have cut out the part of the intersection with lower levels except the innermost quad edge, where level quads overlap by one tile to enable smooth geomorphing transition and per-level snap movement. (See Figure 14.2.)

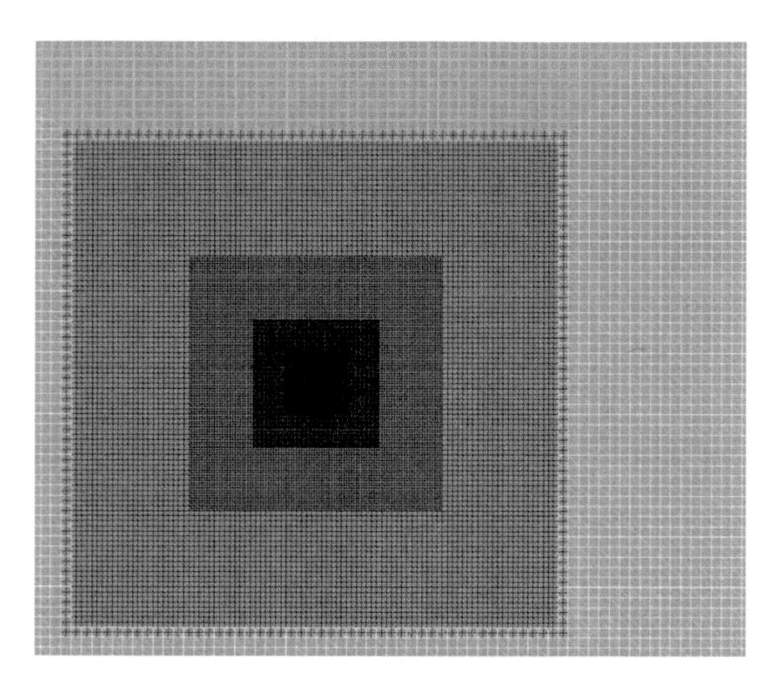

Figure 14.3. A blue center mesh (Mesh 0); green top side mesh (Mesh 1); and white left, bottom, and right side meshes (Mesh 2, Mesh 3, and Mesh 4, respectively). On the intersection of Mesh 0 and Mesh 1, the mesh tri overlap is visible. It is also very important that all mesh rectangles are cut into triangles in the same way (which is why we cannot use the same mesh for Mesh 0 and Mesh 1).

All vertices have a level index encoded in the vertex color G channel. The vertex color channels R and B are used to flag geomorphing X and Z blending factors.

With this method, we get a large tri-count mesh that would, if rendered, have most of the triangles out of the rendering view. To minimize the number of offscreen triangles, we split the render mesh into five parts: the center mesh with L_c levels (Mesh 0) and four-sided meshes with levels from $L_c + 1$ to L (Mesh 1–Mesh 4).

The center mesh is always visible, whereas side meshes are tested for visibility before rendering.

With this optimization, we gain two additional render batches; however, the rendering is reduced by 76% when a field of view is less than 180 degrees. (See Figure 14.3.)

For low field of view angles (60 degrees or less), we could optimize it further by creating more side meshes. For example, if we set L_c to 2 and create eight side meshes, we would reduce the render load by an additional 55% (we would render

58,000 triangles). However, if we looked at the terrain from straight above, we would end up rendering the entire render mesh because the center mesh is so small that it would not fill the screen.

Choosing terrain parameters. Render mesh topology parameters play a very important role in performance, so they should be chosen according to each project's requirements.

Consider a project where we need a landfill with a rather detailed modification resolution and neither big rendering size ($\sim$ 200 $\times$ 200 m) nor view distance ($\sim$500 m).

And now a bit of mathematics to get render mesh numbers:

- View extend (how far will the terrain be visible?)—$V = \frac{R \times S \times 2^{L-1}}{2}$.
- Max level quad resolution—$Q = R \times 2^{L-1}$.
- Level 0 tri count—$T_{L_0} = 2 \times S^2$.
- Level n tri count—$T_{L_n} = 2(S^2 - (\frac{S}{2} - 2)^2)$.
- Total tri count—$T = T_{L_0} + L \times T_{L_n}$.
- Mesh 0 tri count—$T_{M_0} = T_{L_0} + (L_c - 1) \times T_{L_n}$.
- Mesh n tri count—$T_{M_n} = \frac{(L - L_c) \times T_{L_n}}{4}$.

Because we had lots of scenarios where the camera was looking down on the terrain from above, we used a reasonably high center mesh level count (L_c 4), which allowed us to render the terrain in many cases in a single batch (when we were rendering the center mesh only).

We ended up with the quad resolution R 0.1 m, the level size S 100, the level count L 8, and the center mesh level count L_c 4. We used a 2048 $\times$ 2048 texture for the terrain data. With these settings we got a 10 cm resolution, a render view extend of $\sim$1 km, and a full tri count of 127,744 triangles. Because we used a field of view with 65 degrees, we only rendered $\sim$81,000 triangles in three batches.

As mentioned previously, these parameters must be correctly chosen to suit the nature of the application. (See Figures 14.4 and 14.5.)

CPU calculations. We calculate per-resolution-level snapping on a CPU. Each resolution level snap value is its edge size. This is the only terrain calculation made directly on the CPU.

A terrain render position is snapped to a double level Q size so that each level is aligned with a higher level. A vertex shader snaps all vertices at $\vec{x}$ and $\vec{z}$ position to a vertex level snap position.

Figure 14.4. Color coded mesh topology for $L = 8$ and $L_c = 4$.

Figure 14.5. Wire frame showing different levels of terrain mesh detail.

The following level shift is used to skip resolution levels that are too small for the camera at ground height:

```
int shift=(int)floor( log( 1 + cameragroundheight / 5 ) );
```

The CPU code is

```
    float snapvalue=Q;
    float snapmax=2 * snapvalue;
    possnap0.x=floor( camerapos.x / snapmax + 0.01f ) * snapmax;
```

```
    possnap0.z=floor( camerapos.z / snapmax + 0.01f ) * snapmax;
    float levelsnap=snapvalue;

    TTerrainRendererParams[0].z=possnap0.x - camerapos.x;
    TTerrainRendererParams[0].w=possnap0.z - camerapos.z;

    for(int a=1; a<levels; a++)
    {
        levelsnap=levelsnap * 2;
        float l=levelsnap * 2;

        TVector lsnap;
        lsnap.x=floor( possnap0.x / l + 0.01f ) * l;
        lsnap.z=floor( possnap0.z / l + 0.01f ) * l;
        TTerrainRendererParams[a].x=lsnap.x - possnap0.x;
        TTerrainRendererParams[a].y=lsnap.z - possnap0.z;
        TTerrainRendererParams[a].z=lsnap.x - camerapos.x;
        TTerrainRendererParams[a].w=lsnap.z - camerapos.z;
    }
```

Vertex shader. All other terrain-rendering algorithm calculations are done in the vertex shader:

- perform vertex shader texture fetch,
- calculate world-space position,
- calculate level resolution shift,
- calculate geomorphing parameters and blending factors.

Geomorphing is performed on the inner-level edge where lower-level points lie on edges of a higher level. These points are smoothly shifted onto the edge position while they are closing the distance to where they are hidden and the higher level is shown.

```
float4 pos0=TTerrainRendererParams[16];
float4 siz0=TTerrainRendererParams[17];
//
float4 posWS=input.pos;
//
int level=input.tex1.g;
posWS.xz+=TTerrainRendererParams[ level ].xy;
//
int xmid=input.tex1.r;
int zmid=input.tex1.b;
float geomorph=input.tex1.a;
//
float levelsize =input.tex2.x;
float levelsize2 =input.tex2.y;
//
output.color0=1;
//
float4 posterrain=posWS;
//
posterrain=(posterrain - pos0) / siz0;
```

```
//
output.tex0.xy=posterrain.xz;
//
float4 geo0=posWS;
float4 geox=posWS;
float4 geo1=posWS;
//
geox=(geox - pos0) / siz0;

//////////////////////////////////
// output center geo as tex0
//////////////////////////////////
output.tex0.xy=geox.xz;

//////////////////////////////////
// sample center height
//////////////////////////////////
float heix =tex2Dlod( User7SamplerClamp , float4( geox.x , geox.z,
0 , 0 ) ).r;
//
heix=heix * siz0.y + pos0.y;

//////////////////////////////////
// geomorphing
//////////////////////////////////
if( geomorph > 0 )
{
    float geosnap=levelsize;
    //
    if( xmid )
    {
    geo0.x-=geosnap;
    geo1.x+=geosnap;
    }
    //
    if( zmid )
    {
        geo0.z-=geosnap;
        geo1.z+=geosnap;
    }
    //
    geo0=(geo0 - pos0) / siz0;
    geo1=(geo1 - pos0) / siz0;
    //
    float hei0 =tex2Dlod( User7SamplerClamp ,
                          float4( geo0.x , geo0.z , 0 , 0 ) ).r;
    float hei1 =tex2Dlod( User7SamplerClamp ,
                          float4( geo1.x , geo1.z , 0 , 0 ) ).r;

    // geomorph
    float heigeo=(hei0+hei1) * 0.5 * siz0.y + pos0.y;
    //
    posWS.y=lerp( heix , heigeo , geomorph );
}
else
{
    posWS.y=heix;
}
//
posWS.w=1;
output.pos =mul( posWS , TFinalMatrix );
```

Figure 14.6. A sample of a static render mesh for a dynamic terrain on a small area.

14.4.2 Rendering Terrain for Small Areas

For a small contained dynamic area (e.g., a dump truck cargo area or a dump hole), we use a standard rendering technique with a static mesh (with the level of details) that covers the area. (See Figure 14.6.)

Topology of the mesh in this case is not important because it is small and always rendered as a whole.

14.5 Dynamic Modification

Dynamic terrain modification was the second important aspect of the new algorithm. Previously, we developed several techniques that used CPU terrain modification; however, it was difficult to optimize these techniques and therefore the main target of the new modification algorithm was the one executed on the GPU.

14.5.1 Algorithm Overview

The following is a high-level overview of the algorithm.

As shown in Figure 14.7, we took advantage of the fact that all modifications (red rectangle) in the large main terrain texture (blue rectangle) are mostly done in a very small area (a few meters at the most).

Initially, we created a small temporary modification render texture (black rectangle) that we use as a ping-pong data buffer. While processing, we first selected this temporary modification texture as a render target and the main texture as a source and copied the modified location of the main mesh into the temporary modification texture with a plain data copy shader to maintain the texture pixel size.

Next, we swapped the roles and selected the main texture as a render target and the small temporary modification texture as the texture source. Then we rendered the rectangle only on the modified part of the main texture with the

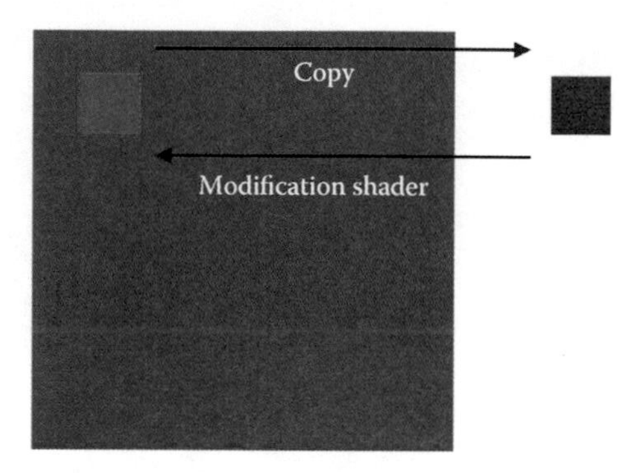

Figure 14.7. Modifications (red rectangle) in the large main terrain texture (blue rectangle) are done in a very small area.

modification shader. Modification shaders can have additional masks to perform the desired modification (e.g., a plow mask, cylinder mask for wheels, or sphere mask).

The temporary texture is sampled in many effects several times around the target pixel to get the final result (e.g., an erosion shader or plow shader).

We use a 128×128 temporary modification texture (covering 12.8×12.8 m changes).

14.5.2 Plow

A plow modification shader is the most complex terrain modification that we do. The idea is to displace the volume moved by the plow in front of the plow while simulating the compression, terrain displacement, and volume preservation.

We use the texture query to measure how much volume the plow would remove (the volume displaced from the last plow location). Then we use the special plow distribution mask and add the displaced volume in front of the plow.

Finally, the erosion simulation creates a nice terrain shape.

14.5.3 Erosion

Erosion is the most important terrain modification. It is performed for a few seconds everywhere a modification is done to smooth the terrain and apply a more natural look.

Erosion is a simple function that sums target pixel height difference for neighboring pixels, performs a height adjustment according to the pixel flowability parameter, and adds a bit of a randomization for a natural look.

Unfortunately, we have not yet found a way to link the erosion simulation with the volume preservation.

14.5.4 Wheels

Wheel modification is a simulation of a cylindrical shape moving over a terrain. It uses a terrain data compression factor to prevent oversinking and to create a wheel side supplant.

We tried to link this parameter with the terrain data flowability parameter (to reduce the texture data), but it led to many problems related to the erosion effect because it also changes the flowability value.

14.6 Physics Synchronization

One drawback of GPU-only processing is that sometimes data needs to be synchronized with the physics, which is in the CPU domain. To do that, we need to transfer data from the GPU memory to the CPU memory to perform synchronization.

14.6.1 Collision Update

Because upon downloading a full terrain data texture (2000×2000 in our case) every frame would be a performance killer, we have to collect and localize eventual terrain changes.

These changes are copied from the main texture into a smaller one for every few frames and downloaded into the main memory and used to update collision mesh information.

We found out that using a 64×64 texture (capturing 6.4×6.4 m) was totally adequate for our needs. Preparation, downloading, and synchronizing in this manner takes less than 0.1 ms.

14.7 Problems

14.7.1 Normals on Cliffs

Normals are calculated per pixel with the original data and with a fixed offset (position offset to calculate slope). This gives a very detailed visual terrain shape even from a distance, where vertex detail is very low. (See Figure 14.8.)

The problem occurs where flowability is very low and the terrain forms cliffs. What happens is that the triangle topology is very different between high and low details, and normals, which are calculated from the high-detailed mesh, appear detached. (See Figure 14.9.)

One way of mitigating this would be to adjust normal calculation offset with the edge size, where flowability is low, but with this we could lose other normal details.

Figure 14.8. Normal on cliffs problem from up close. High-detail topology and normals are the same, and this result is a perfect match.

14.7.2 Physics Simulation on Changing Terrain

Physics simulation (currently we are using Physx 3.3) is very temperamental about changing the cached contact point collision, which we are constantly doing by changing the terrain topology below wheels. If the ground penetrates a collision too deeply, it usually causes a dynamic object to be launched into orbit.

To remedy this behavior we have to adjust the physics solver to limit the maximum penetration depth.

14.7.3 Inconsistent Texture Copy Pixel Offset

When we are performing a dynamic terrain modification, we need to copy from the main texture into the smaller temporary modification texture and back again. With the bilinear texture filtering, this can cause a minor texture shift that is very noticeable when performed repeatedly. Somehow, the per-pixel texture offset is linked to the device resolution even if the texture size is the same.

We have to make an initialization calibration to find an appropriate pixel offset whenever the resolution is changed.

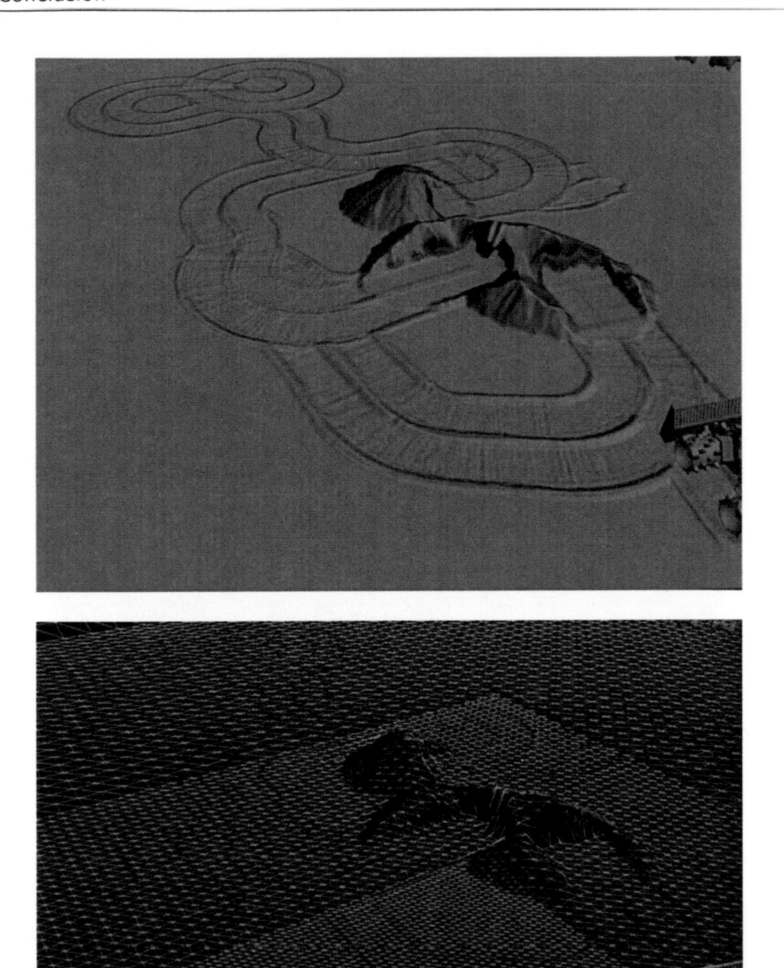

Figure 14.9. Normal on cliffs problem from a distance. Low-detail topology (clearly visible in the bottom wire frame image) and per-pixel normals are not the same.

14.8 Conclusion

14.8.1 Future Work

At the moment, the algorithm described here uses a single texture for the whole terrain and as such is limited by either the extend or the resolution. By adding a

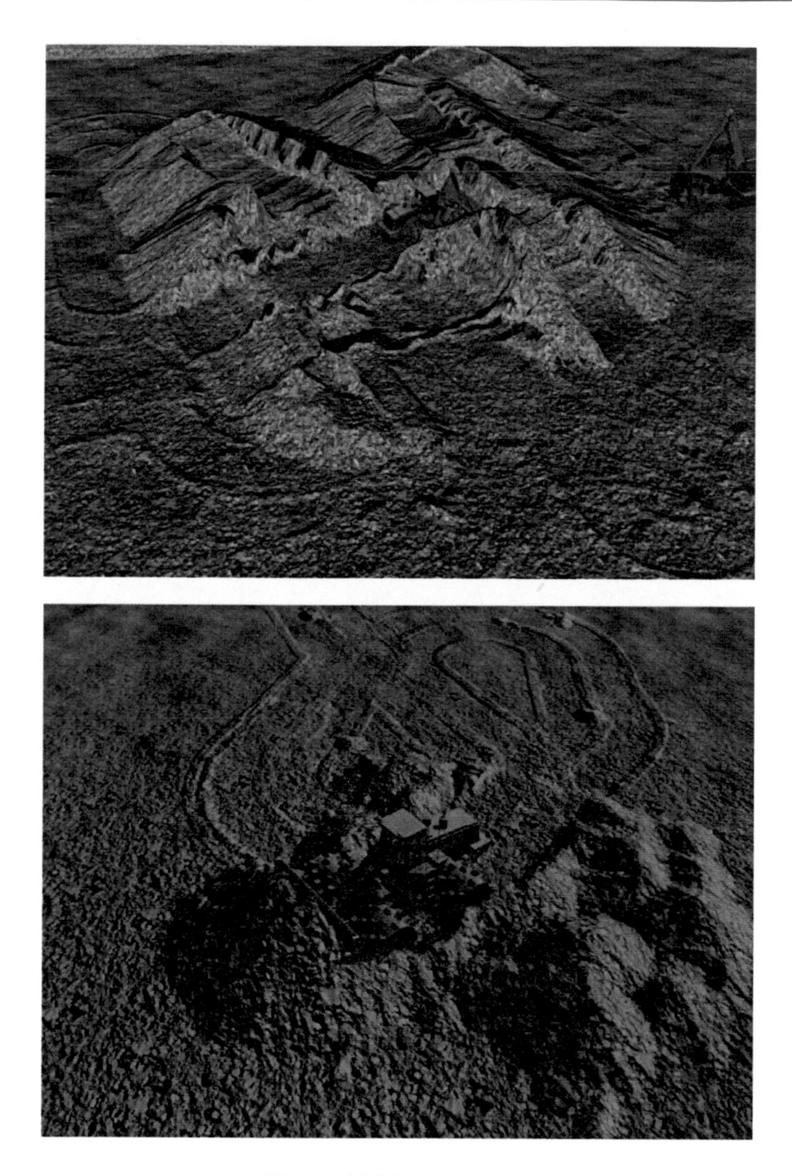

Figure 14.10. Examples.

texture pyramid for coarser terrain detail levels, we could efficiently increase the render extend and not sacrifice the detail.

Mesh 0 and Mesh 2 (as well as Mesh 1 and Mesh 2) are theoretically the same, so we could reuse them to optimize their memory requirements.

Only one level quad edge makes a noticeable transition to a higher level (a lower-lever detail) at a close distance. By adding more overlapping quad edges on lower levels, we would be able to reduce the effect and make smoother geomorphing.

Currently, we have not yet found a way to maintain the terrain volume, so the simulation can go into very strange places (e.g., magically increasing volume).

Because we have already downloaded change parts for collision synchronization, we could also use this data to calculate the volume change and adjust simulation accordingly.

14.8.2 Summary

This paper presents a novel algorithm for terrain rendering and manipulation on a GPU.

In Section 14.4, "Rendering," we showed in detail how to create and efficiently render a very detailed terrain in two or three render batches.

In Section 14.5, "Dynamic Modification," we demonstrated how the terrain can be modified in real time and be synchronized with the CPU base collision.

Figure 14.10 provides an example of the algorithm at work.

Bandwidth-Efficient Procedural Meshes in the GPU via Tessellation

Gustavo Bastos Nunes and João Lucas Guberman Raza

15.1 Introduction

Memory bandwidth is still a major bottleneck in current off-the-shelf graphics pipelines. To address that, one of the common mechanisms is to replace bus consumption for arithmetic logic unit (ALU) instructions in the GPU. For example, procedural textures on the GPU mitigate this limitation because there is little overhead in the communication between CPU and GPU. With the inception of DirectX 11 and OpenGL 4 tessellator stage, we are now capable of expanding procedural scenarios into a new one: procedural meshes in the GPU via parametric equations, whose analysis and implementation is the aim of this article.

By leveraging the tessellator stage for generating procedural meshes, one is capable of constructing a highly detailed set of meshes with almost no overhead in the CPU to GPU bus. As a consequence, this allows numerous scenarios such as constructing planets, particles, terrain, and any other object one is capable of parameterizing. As a side effect of the topology of how the tessellator works with dynamic meshes, one can also integrate the procedural mesh with a geomorphic-enabled level-of-detail (LOD) schema, further optimizing their shader instruction set.

15.2 Procedural Mesh and the Graphics Pipeline

To generate a procedural mesh in the GPU via the tessellator, this article proposes leveraging parametric meshes. The points of a parametric mesh are generated via a function that may take one or more parameters. For 3D space, the mathematical function in this article shall be referenced as a parametric equation of $g(u, v)$, where u and v are in the $[0, 1]$ range. There are mechanisms other than parametric surface equations, such as implicit functions, that may be used to generate procedural meshes. However, implicit functions don't map well to tessellator use, because its results imply if a point is in or out of a surfaces mesh, which is best used in the geometry shader stage via the marching cubes algorithm [Tatarchuk et al. 07]. Performance-wise, the geometry shader, unlike the tessellator, was not designed to have a massive throughput of primitives.

Although the tessellator stage is performant for generating triangle primitives, it contains a limit on the maximum number of triangle primitives it can generate. As of D3D11, that number is 8192 per patch. For some scenarios, such as simple procedural meshes like spheres, that number may be sufficient. However, to circumvent this restriction so one may be able to have an arbitrary number of triangles in the procedural mesh, the GPU must construct a patch grid. This is for scenarios such as terrains and planets, which require a high poly count. Each patch in the grid refers to a range of values within the $[0, 1]$ domain, used as a source for u and v function parameters. Those ranges dissect the surface area of values into adjacent subareas. Hence, each one of those subareas that the patches define serve as a set of triangles that the tessellator produces, which themselves are a subset of geometry from the whole procedural mesh.

To calculate the patch range p we utilize the following equation:

$$p = \frac{1}{\sqrt{\alpha}},$$

where α is the number of patches leveraged by the GPU. Because each patch compromises a square area range, p may then serve for both the u and the v range for each produced patch. The CPU must then send to the GPU, for each patch, a collection of metadata, which is the patches u range, referenced in this article as $[p_{u_{\min}}, p_{u_{\max}}]$, and the patches v range, referenced in this article as $[p_{v_{\min}}, p_{v_{\max}}]$. Because the tessellator will construct the entire geometry of the mesh procedurally, there's no need to send geometry data to the GPU other than the patch metadata previously described. Hence, this article proposes to leverage the point primitive topology as the mechanism to send metadata to the GPU, because it is the most bandwidth-efficient primitive topology due to its small memory footprint. Once the metadata is sent to the GPU, the next step is to set the tessellation factors in the hull shader.

15.3 Hull Shader

The hull shader's purpose is to receive geometry data, which in this article would be one control point per patch. With that geometry data, the hull shader may then set the tessellation factor per domain edge as well as the primitive's interior. The tessellation factor determines the number of triangle primitives that are generated per patch. The higher the tessellation factor set in the hull shader for each patch, the higher the number of triangle primitives constructed. The hull shader's requirement for this article is to produce a pool of triangle primitives, which the tessellator shader then leverages to construct the mesh's geometry procedurally. Hence, the required tessellation factor must be set uniformly to each patch edges and interior factors, as exemplified in the code below:

```
HS_CONSTANT_DATA_OUTPUT BezierConstantHS(InputPatch<VS_
CONTROL_POINT_OUTPUT,
INPUT_PATCH_SIZE> ip, uint PatchID : SV_PrimitiveID)
{
        HS_CONSTANT_DATA_OUTPUT Output;
        Output.Edges[0] = g_fTessellationFactor;
        Output.Edges[1] = g_fTessellationFactor;
        Output.Edges[2] = g_fTessellationFactor;
        Output.Edges[3] = g_fTessellationFactor;
        Output.Inside[0] = Output.Inside[1]= g_fTessellationFactor;
        return Output;
}
```

Because the patch grid will have primitives that must end up adjacent to each other, the edges of each patch must have the same tessellation factor, otherwise a patch with a higher order set of tessellation might leave cracks in the geometry. However, the interior of the primitive might have different tessellation factors per patch because those primitives are not meant to connect with primitives from other patches. A scenario where altering the tessellation factor may be leveraged is for geomorphic LOD, where the interior tessellation factor is based from the distance of the camera to the procedural mesh. The hull shader informs the tessellator how to constructs triangle primitives, which the domain shader then leverages. This LOD technique is exemplified in the high poly count procedural mesh shown in Figures 15.1 and 15.2, with its subsequent low poly count procedural mesh in Figures 15.3 and 15.4.

15.4 Domain Shader

The domain shader is called for each vertex generated by the tessellator. It also receives a pair (u, v) of parametric coordinates for each generated vertex. For this article, we shall reference that coordinate pair as d_u and d_v. Because these parametric coordinates are in domain space, the domain shader must then map them into patch grid space. To do so, we do a linear interpolation:

$$p_u = p_{u_{\min}} + d_u * p_{u_{\max}},$$
$$p_v = p_{v_{\min}} + d_v * p_{v_{\max}},$$

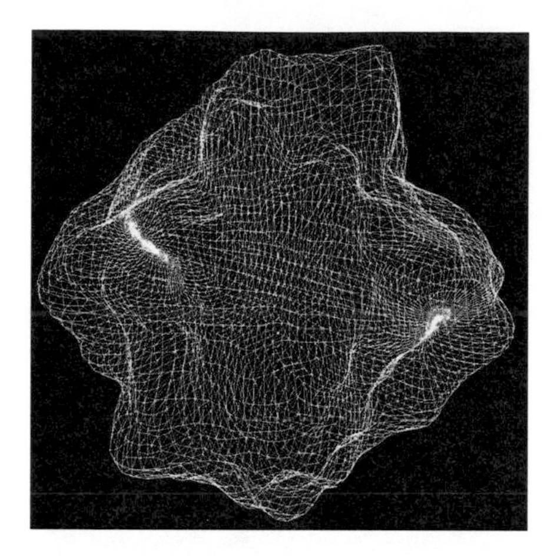

Figure 15.1. A high poly count mesh with noise.

Figure 15.2. The same mesh in Figure 15.1, but shaded.

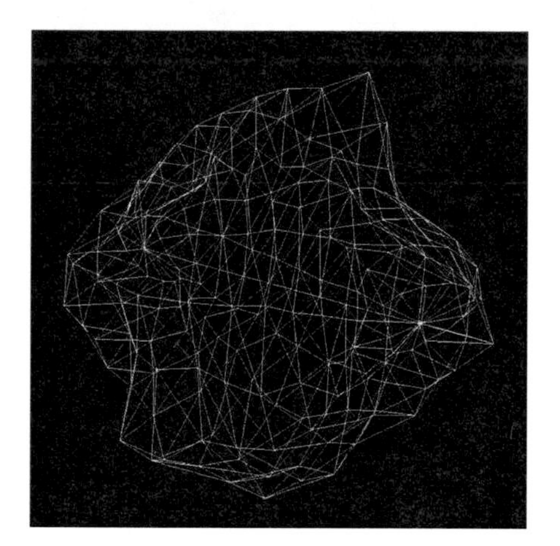

Figure 15.3. The same mesh in Figure 15.1, but with a lower tessellation factor.

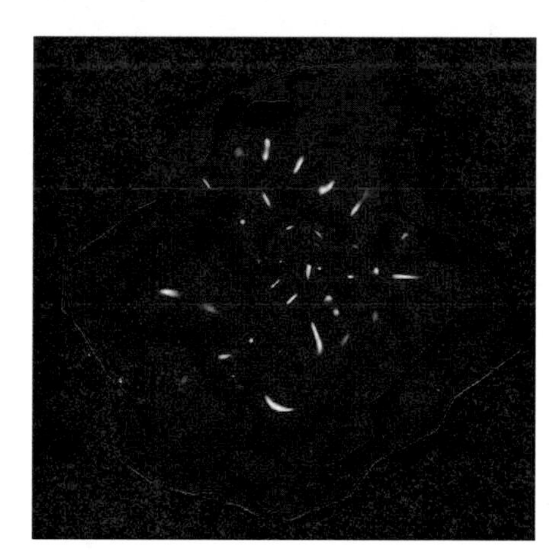

Figure 15.4. The same mesh in Figure 15.3, but shaded.

where p_u and p_v are the parameters to be leveraged in a parametric equation of the implementer's choice. The example in Figure 15.5 uses the following code snippet:

```
float3 heart(float u, float v)
{
  float pi2 = 2 * PI;
  float pi = PI;
  float x, y, z;
  float s = u;
```

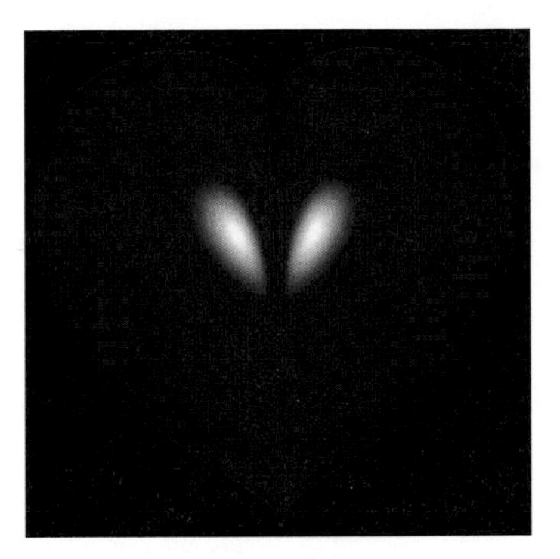

Figure 15.5. Parametric heart generated in the GPU.

Figure 15.6. Deformed cylinder generated in the GPU.

```
    float t = v;
    x = cos(s * pi) * sin(t * 2 * pi) -
      pow(abs(sin(s * pi) * sin(t * 2 * pi)), 0.5f) * 0.5f;
     y = cos(t * 2 * pi) * 0.5f;
     z = sin(s * pi) * sin(t * 2 * pi);
     float3 heart = float3(x, y, z);
     return heart;
}
```

15.5 Noise in Procedural Meshes

Noise with procedural meshes allows applications to generate a myriad of different mesh outputs based on a common factor. For example, with the algorithm proposed in this article for a sphere, noise may allow an application to construct several different types of planets, asteroids, rocks, etc., by altering the generated vertices of the mesh. A possible effect is exemplified in Figure 15.6, as it displays the deformation of a cylinder with Perlin noise as described in [Green 05].

15.6 Performance Optimizations

Because all the primitives are being generated in the GPU, primitives won't be subject to frustum culling. To optimize this aspect, clients should determine if the triangles generated from a patch will be in the frustum or not. This can be done by a heuristic that verifies if the points of the given patch are within a volume that intersects with the frustum. Once that's done, the application can adjust the patch's parametric range or cull the patch altogether from being sent to further states in the GPU. Depending on the optimization, the former can be done on the client or in the GPU hull shader via setting the tessellation factor to 0.

Another set of optimizations relates to normal calculations when using noise functions. Calculating the normal of the produced vertices might be one of the main bottlenecks, because one needs to obtain the nearby positions for each pixel (on per-pixel lighting) or per vertex (on per-vertex lighting). This circumstance becomes even further problematic when leveraging a computationally demanding noise implementation. Take the example in the proposal by [Perlin 04]. It calculates the new normal ($\overrightarrow{N_n}$) by doing four evaluations of the noise function while leveraging the original noiseless normal ($\overrightarrow{N_o}$):

$$\begin{aligned}
F_0 &= F(x, y, z), \\
F_x &= F(x + \epsilon, y, z), \\
F_y &= F(x, y + \epsilon, z), \\
F_z &= F(x, y, z + \epsilon), \\
\overrightarrow{dF} &= \left| \frac{F_x - F_0}{\epsilon}, \frac{F_y - F_0}{\epsilon}, \frac{F_z - F_0}{\epsilon} \right|, \\
\overrightarrow{N_n} &= \text{normalize}(\overrightarrow{N_o} + \overrightarrow{dF}).
\end{aligned}$$

However, given that the domain shader for each vertex passes its coordinates (u, v) in tangent space, in relation to the primitive that each vertex belongs to, one might be able to optimize calculating the normal vector ($\overrightarrow{N}$) by the cross product of the tangent ($\overrightarrow{T}$) and binormal ($\overrightarrow{B}$) vectors (which themselves will also be in tangent space) produced by the vertices in the primitive:

$$\begin{aligned}
F_0 &= g(u, v) + \text{normalize}(\overrightarrow{N_o}) \times F(g(u, v)), \\
F_x &= g(u + \epsilon, v) + \text{normalize}(\overrightarrow{N_o}) \times F(g(u + \epsilon, v)), \\
F_y &= g(u, v + \epsilon) + \text{normalize}(\overrightarrow{N_o}) \times F(g(u, v + \epsilon)), \\
\overrightarrow{T} &= F_x - F_0, \\
\overrightarrow{B} &= F_y - F_0, \\
\overrightarrow{N} &= T \times B,
\end{aligned}$$

where the parametric function is $g(u, v)$, the noise function that leverages the original point is $F(g(u, v))$, and the original normal is $\overrightarrow{N_o}$. This way, one only does three fetches, as opposed to four, which is an optimization in itself because noise fetches are computationally more expensive than doing the cross product.

Lastly, in another realm of optimization mechanisms, the proposed algorithm produces a high quantity of triangles, of which the application might not be able to have a predefined understanding of its output topology. Deferred shading could then be used to reduce the number of operations done in the resulting fragments.

15.7 Conclusion

For the proposed algorithm, the number of calculations linearly increases with the number of vertices and patches, thus making it scalable into a wide range of scenarios, such as procedural terrains and planets. An example of such a case would be in an algorithm that also leverages the tessellation stages, such as in [Dunn 15], which focuses on producing volumetric explosions. Other domains of research might also be used to extend the concepts discussed herein, due to their procedural mathematical nature, such as dynamic texture and sounds. Lastly, as memory access continues to be a performance bottleneck, especially in hardware-constrained environments such as mobile devices, inherently mathematical processes that result in satisfactory visual outputs could be leveraged to overcome such limitations.

15.8 Acknowledgments

João Raza would like to thank his family and wife for all the support they've provided him. Gustavo Nunes would like to thank his wife and family for all their help. A special thanks goes to their friend F. F. Marmot.

Bibliography

[Green 05] Simon Green. "Implementing Improved Perlin Noise." In *GPU Gems 2*, edited by Matt Farr, pp. 409–416. Reading, MA: Addison-Wesley Professional, 2005.

[Owens et al. 08] J. Owens, M. Houston, D. Luebke, S. Green, J. Stone, and J. Phillips. "GPU Computing." *Proceedings of the IEEE* 96:5 (2008), 96.

[Perlin 04] Ken Perlin. "Implementing Improved Perlin Noise." In *GPU Gems*, edited by Randima Fernando, pp. 73–85. Reading, MA: Addison-Wesley Professional, 2004.

[Tatarchuk et al. 07] N. Tatarchuk, J. Shopf, and C. Decoro. "Real-Time Isosurface Extraction Using the GPU Programmable Geometry Pipeline." In *Proceedings of SIGGRAPH 2007*, p. 137. New York: ACM, 2007.

[Dunn 15] Alex Dunn. "Realistic Volumetric Explosions in Games." In *GPU Pro 6: Advanced Rendering Techniques*, edited by Wolfgang Engel, pp. 271–282. Boca Raton, FL: CRC Press, 2015.

Real-Time Deformation of Subdivision Surfaces on Object Collisions

Henry Schäfer, Matthias Nießner, Benjamin Keinert, and Marc Stamminger

16.1 Introduction

Scene environments in modern games include a wealth of moving and animated objects, which are key to creating vivid virtual worlds. An essential aspect in dynamic scenes is the interaction between scene objects. Unfortunately, many real-time applications only support rigid body collisions due to tight time budgets. In order to facilitate visual feedback of collisions, residuals such as scratches or impacts with soft materials like snow or sand are realized by dynamic decal texture placements. However, decals are not able to modify the underlying surface geometry, which would be highly desirable to improve upon realism. In this chapter, we present a novel real-time technique to overcome this limitation by enabling fully automated fine-scale surface deformations resulting from object collisions. That is, we propose an efficient method to incorporate high-frequency deformations upon physical contact into dynamic displacement maps directly on the GPU. Overall, we can handle large dynamic scene environments with many objects (see Figure 16.1) at minimal runtime overhead.

An immersive gaming experience requires animated and dynamic objects. Such dynamics are computed by a physics engine, which typically only considers a simplified version of the scene in order to facilitate immediate visual feedback. Less attention is usually paid to interactions of dynamic objects with deformable scene geometry—for example, footprints, skidmarks on sandy grounds, and bullet impacts. These high-detail deformations require a much higher mesh resolution, their generation is very expensive, and they involve significant memory I/O. In

Figure 16.1. Our method allows computation and application of fine-scale surface deformations on object collisions in real time. In this example, tracks of the car and barrels are generated on the fly as the user controls the car.

most real-time applications, it is thus too costly to compute fine-scale deformations on the fly directly on the mesh. Instead, good cost-efficient approximations are deformations on a template as decal color textures, bump maps, or displacements.

Recently, we have introduced a more flexible approach to this problem [Schäfer et al. 14]. We dynamically generate and store displacements using tile-based displacement maps—that is, deformations are computed, stored, and applied individually on a per-patch level. While low-frequency dynamics are still handled by the CPU physics engine, fine-detail deformations are computed on the fly directly on the GPU. Every frame, we determine colliding objects, compute a voxelization of the overlap region, and modify displacements according to the resulting deformations. Deformed patches are then rendered efficiently using the hardware tessellator. As our algorithm runs entirely on the GPU, we can avoid costly CPU–GPU data transfer, thus enabling fine-scale deformations at minimal runtime overhead.

In this chapter, we describe the implementation of our system, which is available on GitHub as part of this publication[1]. The input to our deformation framework are large scenes composed of quadrilateral subdivision meshes.

[1] https://github.com/hsdk/DeformationGPU

More specifically, we process Catmull-Clark subdivision surfaces, which we render using feature-adaptive subdivision and the GPU hardware tessellator [Nießner et al. 12]. Deformations are realized by analytic displacements, which can be efficiently updated at runtime without a costly normal map re-computation [Nießner and Loop 13]. In order to keep GPU storage requirements at a minimum, we use dynamic memory management, thus only allocating space for displacements of surface patches affected by deformations.

16.2 Deformable Surface Representation

We represent deformable objects as displaced subdivision surfaces, where the base mesh is a Catmull-Clark mesh [Catmull and Clark 78], and high-frequency detail is stored in displacement offsets. On modern hardware, these high-quality surface representations are efficiently evaluated and rendered using the hardware tessellation unit [Nießner et al. 12]. Catmull-Clark surfaces are defined by a coarse set of control points, which are refined at render time and converge to a limit surface that is C^2 continuous everywhere except at non–valence-four vertices, where it is C^1. On top of this surface, displacements are applied along the analytic surface normal of the subdivided mesh.

If we detect object collisions causing deformations, we update the displacement data accordingly (see Section 16.4). When rendering this surface, the normals resulting from these displacements are required. If displacements are static, these normals can be precomputed and stored in a high-resolution normal map, yet in our case we have to update these normals on the fly, which is costly. We thus employ analytic displacements [Nießner and Loop 13], where normals are analytically obtained from scalar surface offsets, allowing for efficient displacement updates without costly normal re-computations (see Section 16.4.3).

16.2.1 Analytic Displacements

The key idea of analytic displacements is the combination of a C^2 base surface $s(u,v)$ with a C^1 offset function $D(u,v)$. As a result, the displaced surface $f(u,v) = s(u,v) + N_s(u,v)D(u,v)$ is C^1 everywhere and provides a continuous normal field $\frac{\partial}{\partial u}f(u,v) \times \frac{\partial}{\partial v}f(u,v)$; $s(u,v) \in \mathbb{R}^3$, $D(u,v) \in \mathbb{R}^1$, and $f(u,v) \in \mathbb{R}^3$. $D(u,v)$ is a scalar-valued, biquadratic B-spline with Doo-Sabin connectivity and special treatment at extraordinary vertices. The connectivity is dual with respect to the base Catmull-Clark surface, which provides a one-to-one mapping between base patches and the subpatches of the displacement function.

Tile-based texture format. We store scalar-valued displacement offsets (i.e., control points of the biquadratic B-spline) in a tile-based texture format similar to PTex [Burley and Lacewell 08] (see Figure 16.2). The key advantage of a tile-based format is the elimination of seam artifacts at (u,v)-boundaries because

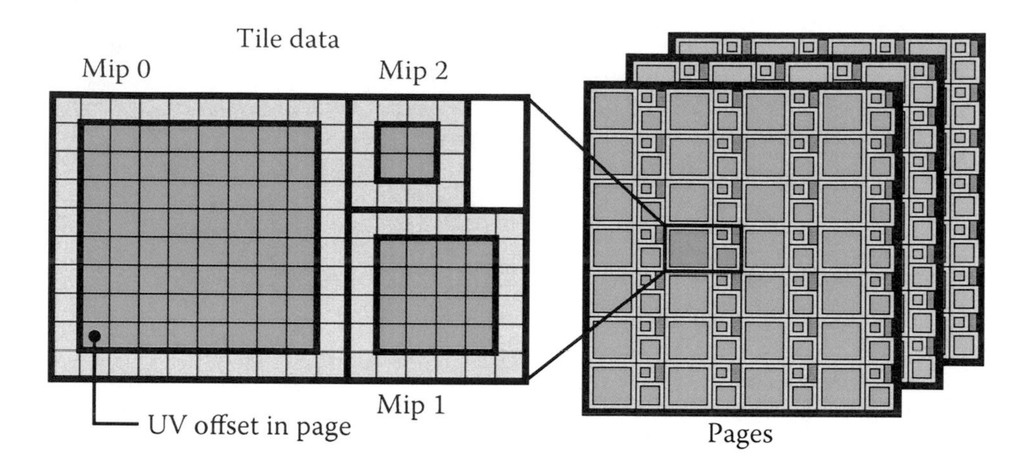

Figure 16.2. The tile-based texture format for analytic displacements: each tile stores a one-texel overlap to avoid the requirement of adjacency pointers. In addition, a mipmap pyramid, computed at a tile level, allows for continuous level-of-detail rendering. All tiles are efficiently packed in a large texture array.

texels are aligned in parameter space; that is, the parametric domain of the tiles matches with the Catmull-Clark patches, thus providing consistent (u, v)-parameters for $s(u, v)$ and $D(u, v)$. In order to evaluate the biquadratic function $D(u, v)$, 3×3 scalar control points (i.e., subpatch; see Figure 16.3) need to be accessed (see Section 16.2.1). At base patch boundaries, this requires access

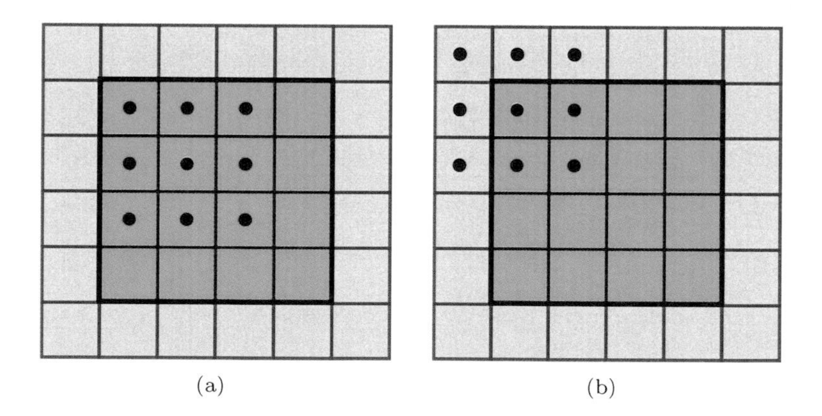

Figure 16.3. (a) A 3×3 control point array of a scalar-valued biquadratic B-spline subpatch of a texture tile storing displacement data. (b) Another set of control points of the same tile where the one-texel boundary overlap is used; overlap data is redundant with adjacent tiles. Each tile corresponds to a Catmull-Clark base patch.

```
struct TileDescriptor
{
  int page; // texture slice
  int uOffset;  // tile start u
  int vOffset;  // tile start v
  uint size;    // tile width, height
  uint nMipmap; // number of mipmaps
};

TileDescriptor GetTile(Buffer<uint> descSRV, uint patchID)
{
  TileDescriptor desc;
  uint offset     = patchID * 4;
  desc.page      = descSRV[offset];
  desc.uOffset    = descSRV[offset + 1];
  desc.vOffset    = descSRV[offset + 2];
  uint sizeMip    = descSRV[offset + 3];
  desc.size       = 1 << (sizeMip >> 8);
  desc.nMipmap    = (sizeMip & 0xff);
  return desc;
}
```

Listing 16.1. Tile descriptor: each tile corresponds to a Catmull-Clark base face and is indexed by the face ID.

to neighboring tiles. This access could be done using adjacency pointers, yet pointer traversal is inefficient on modern GPUs. So we store for each tile a one-texel overlap, making tiles self-contained and such pointers unnecessary. While this involves a slightly larger memory footprint, it is very beneficial from a rendering perspective because all texture access is coherent. In addition, a mipmap pyramid is stored for every tile, allowing for continuous level of detail. Note that boundary overlap is included at all levels.

All tiles—we assume a fixed tile size—are efficiently packed into a large texture array (see Figure 16.2). We need to split up tiles into multiple pages because the texture resolution is limited to 16,000 × 16,000 on current hardware. Each page corresponds to a slice of the global texture array. In order to access a tile, we maintain a buffer, which stores a page ID and the (u, v) offset (within the page) for every tile (see Listing 16.1). Entries of this buffer are indexed by corresponding face IDs of base patches.

Efficient evaluation. In order to efficiently render the displaced objects, we use the GPU hardware tessellation unit. The Catmull-Clark base surface $s(u, v)$ and the corresponding normal $N_s(u, v)$ are exactly evaluated using feature-adaptive subdivision [Nießner and Loop 13]. Because there is a one-to-one mapping between the Catmull-Clark base patches and texture tiles, displacement data can be retrieved by the patch ID, $u, v \in [1, 0] \times [1, 0]$ triple. That is, we obtain the 3×3 array of scalar-valued biquadratic displacement coefficients $d_{i,j}$ of the subpatch corresponding to the base patch u, v. For each base patch, the scalar displacement function $D(u,v)$ is then evaluated using the B-spline basis functions B_i^2:

$$D(u,v) = \sum_{i=0}^{2}\sum_{j=0}^{2} B_i^2(T(u))B_j^2(T(v))d_{i,j},$$

where the subpatch domain parameters $\hat{u}, \hat{v}$ are given by the linear transformation T,

$$\hat{u} = T(u) = u - \lfloor u \rfloor + \frac{1}{2} \quad \text{and} \quad \hat{v} = T(v) = v - \lfloor v \rfloor + \frac{1}{2}.$$

In order to obtain the displaced surface normal $N_f(u,v)$, the partial derivatives of $f(u,v)$ are required:

$$\frac{\partial}{\partial u} f(u,v) = \frac{\partial}{\partial u} s(u,v) + \frac{\partial}{\partial u} N_s(u,v)D(u,v) + N_s(u,v)\frac{\partial}{\partial u} D(u,v).$$

In this case, $\frac{\partial}{\partial u} N_s(u,v)$ would involve the computation of the Weingarten equation, which is costly. Therefore, we approximate the partial derivatives of $f(u,v)$ (assuming small displacements) by

$$\frac{\partial}{\partial u} f(u,v) \approx \frac{\partial}{\partial u} s(u,v) + N_s(u,v)\frac{\partial}{\partial u} D(u,v),$$

which is much faster to compute. The computation of $\frac{\partial}{\partial v} f(u,v)$ is analogous.

Rendering implementation. The rendering of subdivision surfaces with analytic displacements can be efficiently mapped to the modern graphics pipeline with the hardware tessellation unit. The Catmull-Clark base surface $s(u,v)$ is converted into a set of regular bicubic B-spline patches using DirectX Compute Shaders [Nießner et al. 12]—that is, all regular patches, which have only valence-four vertices, are directly sent to the tessellation unit as they are defined as bicubic B-splines. Irregular patches, which have at least one non–valence-four vertex, are adaptively subdivided by a compute kernel. Each refinement step turns an irregular patch into a set of smaller regular patches and an irregular patch next to the extraordinary vertex. After only a few adaptive subdivision steps, the size of irregular patches is reduced to just a few pixels and can be rendered as final patch filling quads; no further tessellation is required.

All generated regular patches are sent to the hardware tessellation unit, where the domain shader takes the 16 patch control points to evaluate $s(u,v)$ and $N_s(u,v)$ using the bicubic B-spline basis functions. In addition, the domain shader evaluates a displacement function $D(u,v)$ and computes the displaced vertices $f(u,v) = s(u,v) + N_s(u,v)D(u,v)$. Generated vertices are then passed to the rasterization stage. In the pixel shader, the shading normals $N_f(u,v)$ are computed based on the partial derivatives of the displacement function; i.e., $\frac{\partial}{\partial u} D(u,v)$ and $\frac{\partial}{\partial v} D(u,v)$ are evaluated. Code for the evaluation of analytic displacements is shown in Listing 16.2, the domain shader part in Listing 16.3, and

```
Texture2DArray<float> g_displacementData  : register(t6);
Buffer<uint>          g_tileDescriptors   : register(t7);

float AnalyticDisplacement(in uint patchID, in float2 uv,
                           inout float du, inout float dv)
{
  TileDescriptor tile = GetTile(g_tileDescriptors, patchID);

  float2 coords = float2( uv.x * tile.size + tile.uOffset,
                          uv.y * tile.size + tile.vOffset);

  coords -= float2(0.5, 0.5);
  int2 c = int2(rount(coords));

  float d[9];
  d[0] = g_displacementData[int3(c.x-1, c.y-1, tile.page)].x;
  d[1] = g_displacementData[int3(c.x-1, c.y-0, tile.page)].x;
  d[2] = g_displacementData[int3(c.x-1, c.y+1, tile.page)].x;
  d[3] = g_displacementData[int3(c.x-0, c.y-1, tile.page)].x;
  d[4] = g_displacementData[int3(c.x-0, c.y-0, tile.page)].x;
  d[5] = g_displacementData[int3(c.x-0, c.y+1, tile.page)].x;
  d[6] = g_displacementData[int3(c.x+1, c.y-1, tile.page)].x;
  d[7] = g_displacementData[int3(c.x+1, c.y-0, tile.page)].x;
  d[8] = g_displacementData[int3(c.x+1, c.y+1, tile.page)].x;

  float evalCoord = 0.5 - (float(c) - coords);
  float displacement = EvalQuadricBSpline(evalCoord, d, du, dv);

  du *= tile.size;
  dv *= tile.size;
  return displacement;
}
```

Listing 16.2. Analytic displacement lookup and evaluation.

the pixel shader computation in Listing 16.4. Note that shading normals are obtained on a per-pixel basis, leading to high-quality rendering even when the tessellation budget is low.

Evaluating $f(u, v)$ and $N_f(u, v)$ for regular patches of the Catmull-Clark patch is trivial because tiles correspond to surface patches. Regular patches generated by feature adaptive subdivision, however, only correspond to a subdomain of a specific tile. Fortunately, the feature-adaptive subdivision framework [Nießner et al. 12] provides local parameter offsets in the domain shader to remap the subdomain accordingly.

Irregular patches only remain at the finest adaptive subdivision level and cover only a few pixels. They require a separate rendering pass because they are not processed by the tessellation stage; patch filling quads are rendered instead. To overcome the singularity of irregular patches, we enforce the partial derivatives of the displacement function $\frac{\partial}{\partial u}D(u, v)$ and $\frac{\partial}{\partial v}D(u, v)$ to be 0 at extraordinary vertices; i.e., all adjacent displacement texels at tile corners corresponding to a non–valence-four vertex are restricted to be equal. Thus,

```
void ds_main_patches(in HS_CONSTANT_FUNC_OUT input,
                     in OutputPatch<HullVertex, 16> patch,
                     in float2 domainCoord : SV_DomainLocation,
                     out OutputVertex output )
{
  // eval the base surface s(u,v)
  float3 worldPos = 0, tangent = 0, bitangent = 0;
  EvalSurface(patch, domainCoord, worldPos, tangent, bitangent);
  float3 normal = normalize(cross(Tangent,BiTangent));

  float du = 0, dv = 0;
  float displacement = AnalyticDisplacement(patch[0].patchID,
              domainCoord, du, dv);
  worldPos += displacement * normal;

  output.pos      = mul(ProjectionMatrix, float4(worldPos, 1.0));
  output.tangent    = tangent;
  output.bitangent  = bitangent;
  output.patchCoord = domainCoord;
}
```

Listing 16.3. Analytic displacement mapping evaluation in the domain shader.

```
float4 ps_main(in OutputVertex input) : SV_TARGET
{
  // compute partial derivatives of D(u,v)
  float du = 0, dv = 0;
  float displacement = AnalyticDisplacement(input.patchID,
                       input.patchCoord, du, dv);
  // compute base surface normal N_s(u,v)
  float3 surfNormal = normalize(cross(input.tangent,
                                      input.bitangent));
  float3 tangent   = input.tangent   + surfNormal * du;
  float3 bitangent = input.bitangent + surfNormal * dv;
  // compute analytic displacement shading normal N_f(u,v)
  float3 normal    = normalize(cross(tangent, bitangent));

  // shading
  ...
}
```

Listing 16.4. Analytic displacement mapping evaluation in the pixel shader.

$N_f(u, v) = N_s(u, v)\ \forall(u, v)_{\text{extraordinary}}$. A linear blend between this special treatment at extraordinary vertices and the regular $N_f(u, v)$ ensures a consistent C^1 surface everywhere.

16.3 Algorithm Overview

Our aim is to provide highly detailed deformations caused by object-object collisions. To achieve instant visual feedback, we approximate collisions and apply

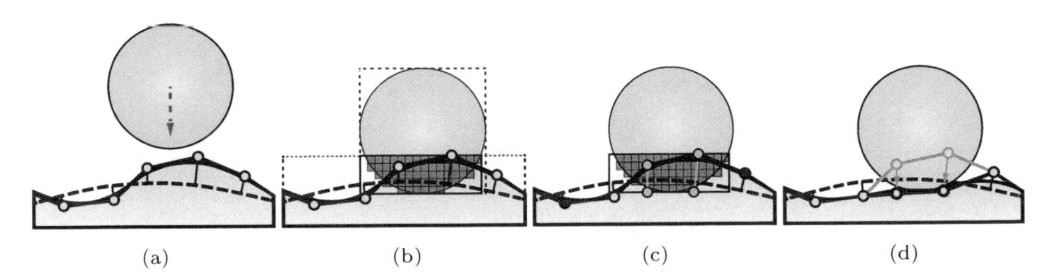

Figure 16.4. Algorithm overview: (a) Subdivision surfaces with quadratic B-spline displacements are used as deformable object representation. (b) The voxelization of the overlapping region is generated for an object penetrating the deformable surface. (c) The displacement control points are pushed out of the voxelization, (d) creating a surface capturing the impact.

deformations by updating displacement data. This is much more cost efficient than a physically correct soft body simulation and also allows for visually plausible results. In this section, we provide an overview of our algorithm as outlined in Figure 16.4. A detailed description of our implementation can be found in Section 16.4.

For simplicity, we first assume collisions only between a rigid penetrating object and a deformable one. We represent deformable objects as displaced subdivision surfaces (see Section 16.2). The penetrating object can be either a subdivision surface or a regular triangle mesh (see Figure 16.4(a)). For all colliding *deformable-penetrating* object pairs, we compute the deformation using the following algorithm:

- Approximate the penetrating object by computing a solid voxelization using an improved variant of the real-time binary voxelization approach by Schwarz [Schwarz 12] (see Figure 16.4(b)).

- From the voxelization, determine displacement offsets of deformable objects to match the shape of the impact object (Figure 16.4(c) and (d)). This is achieved by casting rays from the deformable object's surface and modifying the displacements accordingly.

In the case that both objects are deformable, we form two collision pairs, with each deformable acting as a rigid penetrating object for the other deformable and only applying a fraction of the computed deformations in the first pass.

16.4 Pipeline

In this section, we describe the implementation of the core algorithm and highlight important details on achieving high-performance deformation updates.

16.4.1 Physics Simulation

Our algorithm is designed to provide immediate visual feedback on collisions with deformable objects made of soft material such as sand or snow. Because a full soft body simulation would be too expensive in large scene environments, we interpret deformable objects as rigid bodies with fine-scale dynamic surface detail. We handle rigid body dynamics using the Bullet physics engine [Coumans et al. 06]. All dynamic objects are managed on the CPU and resulting transformation matrices are updated every frame. In theory, we could also process low-frequency deformations on base meshes if allowed by the time budget; however, we have not explored this direction.

After updating rigid bodies, we search for colliding objects and send pairs that hit a deformable to our deformation pipeline on the GPU.

16.4.2 Voxelization

Once we have identified all potential deformable object collisions (see above), we approximate the shape of penetrating objects using a variant of the binary solid voxelization of Schwarz [Schwarz 12]. The voxelization is generated by a rasterization pass where an orthogonal camera is set up corresponding to the overlap region of the objects' bounding volumes. In our implementation, we use a budget of 2^{24} voxels, requiring about 2 MB of GPU memory. Note that it is essential that the voxelization matches the shape as closely as possible to achieve accurate deformations. We thus determine tight bounds of the overlap regions and scale the voxelization anisotropically to maximize the effective voxel resolution.

Intersecting volume. In order to determine the voxelization space, we intersect the oriented bounding boxes (OBBs) of a collision pair. The resulting intersecting volume is extended such that it forms a new OBB that conservatively bounds the overlapping region (see Figure 16.5). We precompute all OBBs in model space during loading and transform OBBs at runtime using the physics rigid transformations. Exceptions are skinned animations, for which we apply skinning in a compute shader and recompute the OBB each frame.

Efficient GPU implementation. The voxelization of objects is generated by performing a simple rasterization pass using an orthographic camera. The voxel grid is filled in a pixel shader program using scattered `write` operations (see Listing 16.5). We determine the voxelization direction—i.e., the camera direction—according to the major axis of the intersection volume. Clipping an object against the intersecting volume results in nonclosed surfaces, which cannot be handled in all cases by the original voxelization approach by Schwarz. However, the voxelization will be correct if we guarantee that front-faces are always hit first—i.e., they are not being clipped. Therefore, we construct intersecting OBBs such that

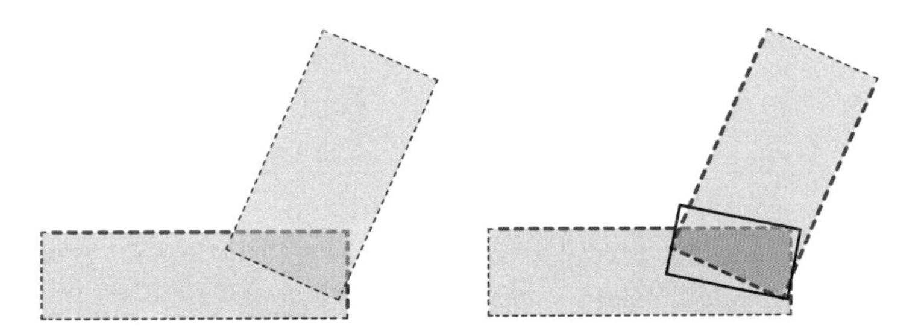

Figure 16.5. Generation of the OBB for voxelization: a new OBB is derived from the intersecting OBBs of the deformable and the penetrating object.

at least one of the faces is completely outside of the penetrating object OBB. The voxelization is then performed toward the opposite direction of the face, which is on the outside. We use either of two kernels to perform the voxelization and fill the adaptively scaled voxel grid forward or backward, respectively, as shown in Listing 16.5.

```
RWByteAddressBuffer g_voxels : register(u1);

float4 PS_VoxelizeSolid(in OutputVertex input) : SV_TARGET
{
   // transform fragment position to voxel grid
   float3 fGridPos = input.posOut.xyz / input.posOut.w;
   fGridPos.z *= g_gridSize.z;
   int3 p = int3(fGridPos.x, fGridPos.y, fGridPos.z + 0.5);

   if (p.z > int(g_gridSize.z))
      discard;

   // apply adaptive voxel grid scale
   uint address =  p.x * g_gridStride.x
                 + p.y * g_gridStride.y
                 + (p.z >> 5) * 4;

#ifdef VOXELIZE_BACKWARD
   g_voxels.InterlockedXor(address,
                           ~(0xffffffffu << (p.z & 31)));
   // flip all voxels below
   for (p.z = (p.z & (~31)); p.z > 0; p.z -= 32) {
      address -= 4;
      g_voxels.InterlockedXor(address, 0xffffffffu);
   }
#else
   g_voxels.InterlockedXor(address, 0xffffffffu << (p.z & 31));
   // flip all voxels below
   for(p.z = (p.z | 31) + 1; p.z < g_gridSize.z; p.z += 32) {
      address += 4;
      g_voxels.InterlockedXor(address, 0xffffffffu);
   }
#endif
}
```

Listing 16.5. Pixel shader implementation of the binary voxelization using atomic operations.

16.4.3 Ray Casting

In the previous stage, we generated a voxelization of the penetrating object into the space of the intersecting volume. Now, all patches of the deformable surface within the intersecting volume are to be displaced such that they no longer intersect with the (voxelization of the) penetrating object. Our implementation of this process is shown in Listing 16.6. First, we evaluate the surface at all control points of the displacement B-spline (patch parameter-space position of the displacement map texels) and compute their corresponding world-space positions. More precisely, these are Catmull-Clark surface points, evaluated at the knot points of the displacement B-spline, with applied displacement (see Figure 16.4(b)). Hence, we account for the previous surface offset in case the surface at this position is already displaced.

If such a control point lies within the penetrating object, we move it in the negative base surface normal direction until it leaves the penetrating object (the red control points in Figure 16.4). Therefore, we cast a ray that originates at the control points' corresponding world-space position pointing along the negative base surface normal. This step involves evaluating the Catmull-Clark surface. Fortunately, this evaluation is very fast using the regular B-spline patches obtained by adaptive subdivision.

We now traverse the rays through the binary voxelization using a 3D digital differential analyzer (DDA) (see Listing 16.7). We also make sure that the ray can actually hit the voxel volume by first intersecting the ray with the voxel grid's OBB. Thereby, control points outside the voxelization (e.g., the yellow one in Figure 16.4) and outside the overlap region (the red ones) are left unchanged. In case the ray can hit the voxel volume but the surface position and thus the ray originates outside the voxel volume, the initial ray distance is updated such that it lies on the intersection point with the voxel grid. Then, we trace the ray through the voxel grid and on each set voxel encountered we update the displacement distance until we leave the voxel volume. It is important to trace the ray until it leaves the voxelization, otherwise concave objects would result in an incorrect deformation, as depicted in Figure 16.6.

After ray traversal, we update the displacement by applying the negative traveled distance to account for the deformation and write the result into the displacement map.

16.4.4 Overlap Update

Once surface deformations are computed, we need to update the analytic displacement map tile overlap in order to enforce fast, watertight, and consistent evaluation during rendering. This requires copying B-spline coefficients (displacement values) from the boundary to the overlap region of the adjacent patches. To this end, we precompute patch adjacency in a preprocessing step. More precisely, per patch we store the indices of the four neighboring patches and the

```
#define NUM_BLOCKS (TILE_SIZE/DISPLACEMENT_DISPATCH_TILE_SIZE)
[numthreads(DISPLACEMENT_DISPATCH_TILE_SIZE,
            DISPLACEMENT_DISPATCH_TILE_SIZE, 1)]
void ComputeDisplacementCS( uint3 blockIdx  : SV_GroupID,
                            uint3 threadIdx : SV_GroupThreadID)
{
  uint patchID = blockIdx.x + g_PrimitiveIdBase;

  TileDescriptor tile = GetTile(g_TileInfo, patchID);
  if(threadIdx.x >= tile.size || threadIdx.y >= tile.size)
    return;

  int patchLevel = GetPatchSubDLevel(patchID);

  // threadIdx to tile coord
  float2 tileUV  = ComputeTileCoord( patchID, tile,
                                     blockIdx, threadIdx);
  // threadIdx to (sub-) patch coord
  float2 patchUV = ComputePatchCoord(patchID, patchLevel,
                                     blockIdx, threadIdx);

  int3 coord = int3(tile.uvOffset + tileUV.xy * tile.size,
                    tile.page);

  float disp = g_displacementUAV[coord];

  // eval surface and apply displacement
  float3 worldPos = 0;
  float3 normal   = 0;

  float3 controlPoints[16] = ReadControlPoints(patchID);
  EvalPatch(controlPoints, patchUV, worldPos, normal);
  worldPos += disp * normal;

  // traverse ray until leaving solid
  float3 rayOrigin = mul((g_matWorldToVoxel),
                         float4(worldPos,1.0)).xyz;
  float3 rayDir = normalize(mul((float3x3)g_matWorldToVoxel,
                                -normal.xyz));

  float distOut = 0;
  if(! VoxelDDA(rayOrigin, rayDir, dist));
    return;

  float3 p = rayDir * distOut;
  p = mul((float3x3)(((g_matNormal))), p);
  disp = disp - length(p);

  g_displacementUAV[coord] = disp;
}
```

Listing 16.6. Implementation of displacement update using voxel ray casting.

```
bool IsOutsideVolume(int3 voxel) {
   return (any(voxel < 0) || any(voxel > int3(g_gridSize)))
}
bool VoxelDDA(in float3 origin, in float3 dir, out float dist)
{
  PreventDivZero(dir);
  float3 dt = abs(1.0 / dir);

  float tEnter,tExit;
  if (!intersectRayVoxelGrid(origin, dir, tEnter, tExit))
    return false;

  // start on grid boundary unless origin is inside grid
  tEnter = max( 0.0, tEnter - 0.5 * min3(dt.xyz);

  float3 p = origin + tEnter * dir;
  int3 gridPos = floor(p);

  // check if ray is starting in voxel volume
  if(IsOutsideVolume(gridPos))
    return false;

  float3 tMin = INFINITY;
  // update step, dir components are != 0 (PreventDivZero)
  tMin.x = (dir.y < 0.0) ? (p.x-gridPos.x) : (gridPos.x-p.x+1);
  tMin.y = (dir.y < 0.0) ? (p.y-gridPos.y) : (gridPos.y-p.y+1);
  tMin.z = (dir.z < 0.0) ? (p.z-gridPos.z) : (gridPos.z-p.z+1);
  tMin *= dt;

  int3 step = 1;
  if (dir.x <= 0.0) step.x = -1;
  if (dir.y <= 0.0) step.y = -1;
  if (dir.z <= 0.0) step.z = -1;

  uint maxSteps =  g_gridSize.x + g_gridSize.y + g_gridSize.z;
  [allow_uav_condition]
  for(uint i = 0; i < maxSteps; i++) {
    t = min(tMin.x, min(tMin.y, tMin.z));

    if(tEnter + t >= tExit) break;

    if (IsVoxelSet(gridPos))
        dist = t;

    if(tMin.x <= t) { tMin.x += dt.x; gridPos.x += step.x; }
    if(tMin.y <= t) { tMin.y += dt.y; gridPos.y += step.y; }
    if(tMin.z <= t) { tMin.z += dt.z; gridPos.z += step.z; }

    if(IsOutsideVolume(gridPos)) break;
 }
 return (dist > 0);
}
```

Listing 16.7. Implementation of the voxel digital differential analyzer (DDA) algorithm for ray casting.

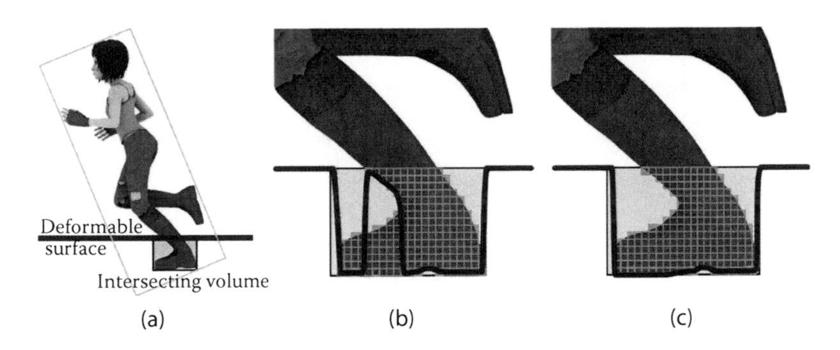

Figure 16.6. (a) Illustration of the ray casting behavior when tracing from the surface of the deformable through the voxelized volume. (b) The incorrect deformations that occur when using the distance of the first exit of the ray. (c) Tracing the ray throughout the complete volume yields correct deformations.

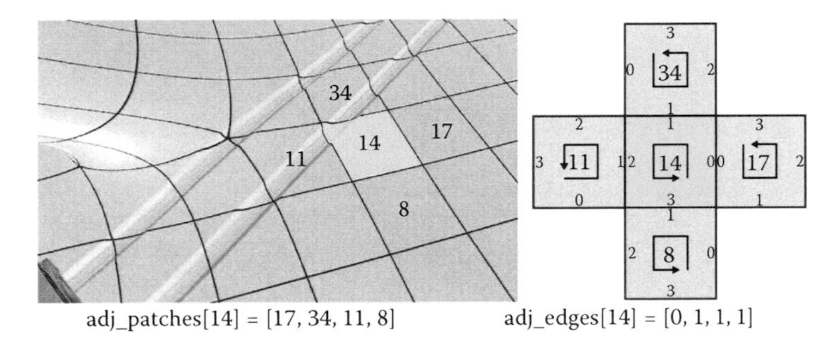

Figure 16.7. Example of the adjacency information storage scheme: for the green patch, we store all neighboring patch indices and indices of the shared edges in the neighboring patches in counterclockwise order.

shared edges' indices oriented with respect to the respective neighboring patch as depicted in Figure 16.7.

In our implementation, we handle the edge overlap separately from the corner overlap region, as the corners require special treatment depending on whether the patch is regular or connected to an irregular vertex.

Edge overlap. Using the precomputed adjacency information, we first update the edge overlap by scattering the boundary displacements coefficients to the adjacent neighbors overlap region. This process is depicted in Figure 16.8 for a single patch.

Corner overlap. Finally, we have to update the corner values of the overlap region to provide a consistent evaluation of the analytic displacement maps during rendering. The treatment of the corner values depends on the patch type. For

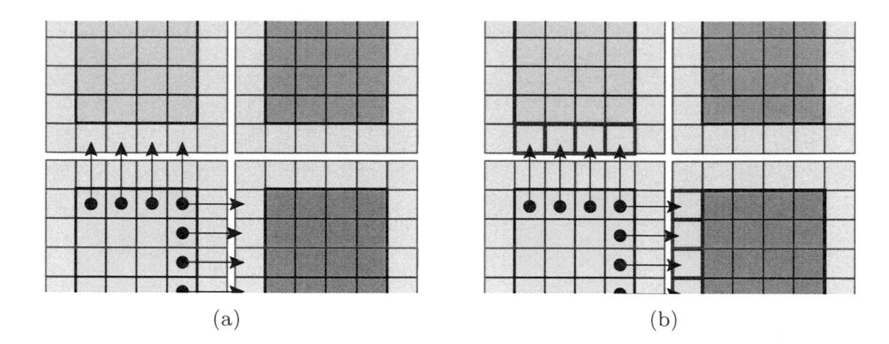

Figure 16.8. Edge overlap update for a single patch: (a) The direction of the overlap updates originates from the blue patch. (b) The two adjacent patches (red and yellow) receive their resulting overlap data by gathering the information from the blue patch.

Catmull-Clark subdivision surfaces two patch types are possible:

- Regular patches: All vertices of a patch have exactly four incoming edges.
- Irregular patches: At least one vertex of the patch has a valence different from four.

In the regular case, the corner values of a patch can simply be copied to the diagonal patch's boundary corner (see Figure 16.9(a)). In our implementation, we do not store the patch index of the diagonal adjacent patch. However, after the edges are updated, we can achieve exactly the same result by copying the correct

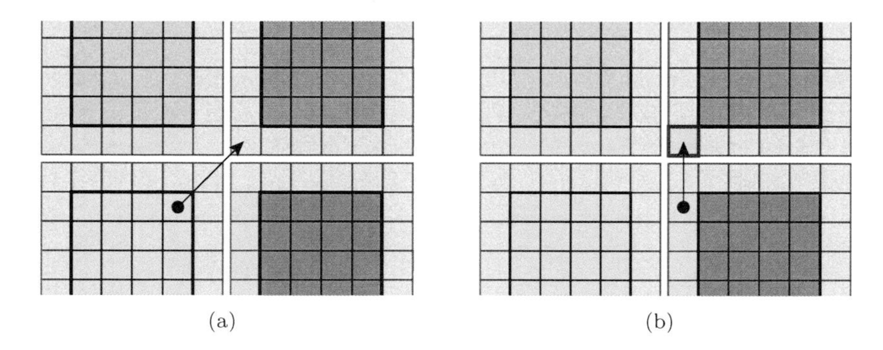

Figure 16.9. Corner overlap update at regular vertices. (a) The direction of the corner overlap update originates from the blue patch. The required information is also stored in the direct neighbors of the green patch after the edge overlap update pass. (b) The resulting corner overlap update is gathered from the overlap of the adjacent yellow patch.

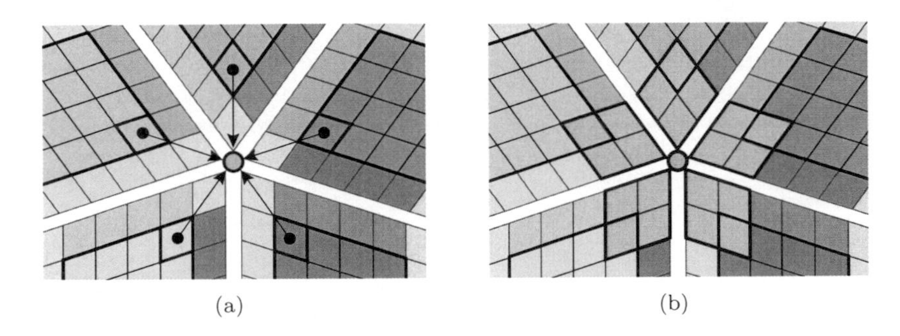

Figure 16.10. Corner overlap update at irregular vertices: (a) texels to be gathered and (b) the result of scattering the resulting average value to the adjacent tiles.

coefficient from the adjacent patch's edge to the boundary corner as depicted in Figure 16.9(b).

In order to provide a watertight and consistent evaluation in the irregular case, all four corner coefficients must contain the same value. Therefore, we run a kernel per irregular vertex and average the interior corner coefficients of the connected patches (see Figure 16.10(a)). Then, we scatter the average to the four corner texels of the analytic displacement map in the same kernel.

In the end, the overlap is updated and the deformed mesh is prepared for rendering using displacement mapping.

16.5 Optimizations

16.5.1 Penetrated Patch Detection

The approach described in the previous sections casts rays for each texel of each patch of a deformed object: a compute shader thread is dispatched for each texel to perform the ray casting in parallel. This strategy is obviously inefficient since only a fraction of the patches of a deformed object will be affected. This can be prevented by culling patches that are outside the overlap regions. To this end, we compute whether the OBB of the penetrating object and the OBB of each patch of the object to be deformed do overlap. For this test, we extend the OBBs of the patches by the maximum encountered displacement to handle already displaced patches' surfaces properly. In case an overlap is detected, the patch (likely to be intersected by the penetrating object) is marked, and its patch index is enqueued for further processing. Also, the update of tile overlaps is only necessary for these marked patches.

Intersection. The patch intersection detection stage is implemented entirely on the GPU using a compute shader. One dispatch detects the collision of a single penetrating object's OBB with all patches of the scene. For each patch, a thread is dispatched. In the compute shader the OBB of the patch is computed on the fly from the patches' control points and overlap tested against the OBB of the penetrating object. If an overlap is found, the patch index is appended to the list to be handled for further processing.

Intersection batching. The previously depicted intersection stage implementation dispatches threads for each penetrating object sequentially, thus causing unnecessary and redundant memory accesses because the same patch control points have to be read over multiple kernel dispatches. This memory I/O overhead can significantly be reduced by batching multiple penetrating objects into a single dispatch. Batching the intersection testing of multiple penetrating objects into a single dispatch additionally reduces the number of total compute shader dispatches required. For patches requiring memory allocation, their index is appended to a global append buffer. In addition, we use further append buffers for each penetrating object. If a patch is possibly affected by a penetrating object, the patch index is appended to the penetrating object's append buffer for ray casting. Finally, the overlap is updated only once per deformable object after all penetrating collisions are processed.

16.5.2 Memory Management

Because we want to support deformation on scenes with a large number of patches at high tile resolutions, statically preallocating tile memory for each possibly deformed patch would require unreasonably large amounts of GPU memory. Therefore, we preallocate a predefined number of tiles and manage a table of tile descriptors pointing to these unused tiles. In addition, we use an atomic index i for memory allocation, which points to the end of the free memory table when no tiles are in use.

If memory allocation is required for a patch, this is implemented using an atomic decrement operation on i and fetching the tile descriptor of the tile it pointed to before decrementation. (See Listing 16.8.)

16.5.3 Optimized Pipeline

Our final deformation pipeline, including the proposed optimizations, is depicted in Figure 16.11.

16.6 Results

In this section, we provide several screenshots showing the qualitative results of our real-time deformations pipeline. The screenshots in Figure 16.12 are taken from our example scene (see Figure 16.1) consisting of a snowy deformable terrain

```
Buffer<uint>     g_isctResults      : register(t0);
Buffer<uint3>    g_memoryTable      : register(t1);
RWBuffer<uint4> g_tileDescriptors : register(u0);
RWBuffer<int>    g_atomicIndex      : register(u1);

void allocTile(uint tileID) {
   int i = 0;
   InterlockedAdd(g_atomicIndex[0], -1, i);

   // copy page ID and start offsets (u, v)
   g_tileDescriptors[tileID].xyz = g_memoryTable[i].xyz;
}

[numthreads(ALLOCATOR_BLOCKSIZE, 1, 1)]
void AllocateTilesCS(uint3 DTid: SV_DispatchThreadID) {
   uint tileID = DTid.x;
   if (tileID >= g_NumTiles) return;

   if (IsTileIntersected(tileID) && IsNotAllocated(tileID))
      allocTile(tileID);
}
```

Listing 16.8. Compute shader for tile memory allocation with atomic operations.

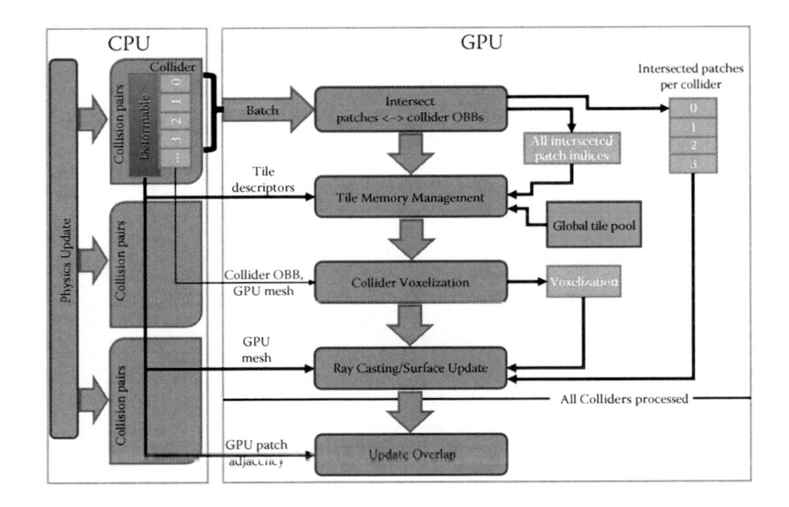

Figure 16.11. Overview of deformation pipeline with optimizations.

subdivision surface, dynamic objects like the car and barrels, and static objects such as trees and houses.

We start by presenting the parameters that impact the overall quality of the deformations.

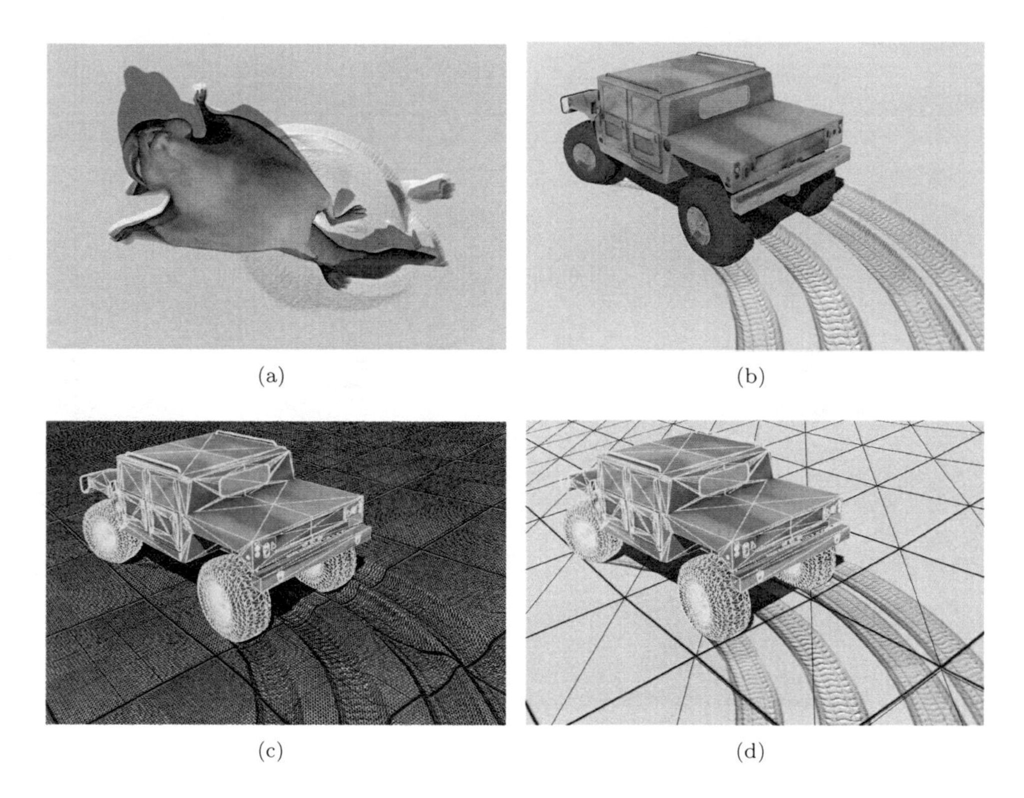

(a) (b) (c) (d)

Figure 16.12. Results of the proposed deformation pipeline on snowy surface: (a) an example of animated character deforming the surface; (b) high-quality geometric detail including shadows and occlusion; and (c) wireframe visualization of a deformed surface. (d) Even at low tessellation densities, the deformation stored in the displacement map can provide visual feedback in shading.

16.6.1 Influence of Tile Resolution

The first parameter is the tile resolution for storing the per-patch displacement coefficients. Figure 16.13 shows a comparison of the deformation quality for different tile resolutions ((a) 32×32 and (b) 128×128). Obviously, the lower resolution tiles (a) cannot represent the impact of the collider on the surface well. Employing our tile memory allocation scheme (Section 16.5.2) enables very high tile resolution (b) on the deformed patches and thus highly detailed deformations. These tile resolutions would not be possible using static preallocation for all patches without memory management due to the limited GPU memory.

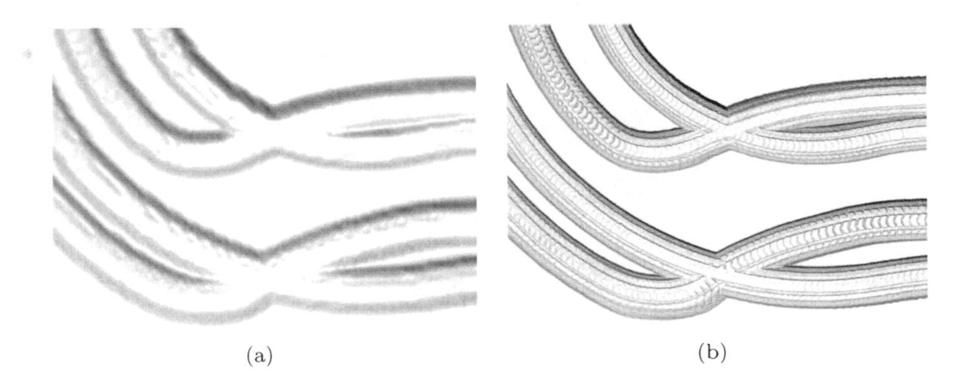

(a) (b)

Figure 16.13. Comparison of deformation quality using different tile resolutions per patch: (a) The higher resolution (128×128) captures high-frequency detailwhile (b) the lower (32×32) does not.

16.6.2 Influence of Voxelization

The second parameter that influences the overall quality most is the approximation of the penetrating object shape. The quality of the voxelization is limited by the chosen voxel grid size. Choosing too low a voxel grid resolution results in a low-quality deformation as shown in the example in Figure 16.14. In our implementation, we use a single voxel grid per penetrating object. Voxelizing the object only in the overlapping region as described in Section 16.4.2 results in a much better utilization of the available voxel space in the region of interest compared to voxelizing the entire object.

Figure 16.14. Choosing too coarse a voxel grid cannot capture the shape of the penetrating object (wheel) well and results in a low-quality deformation.

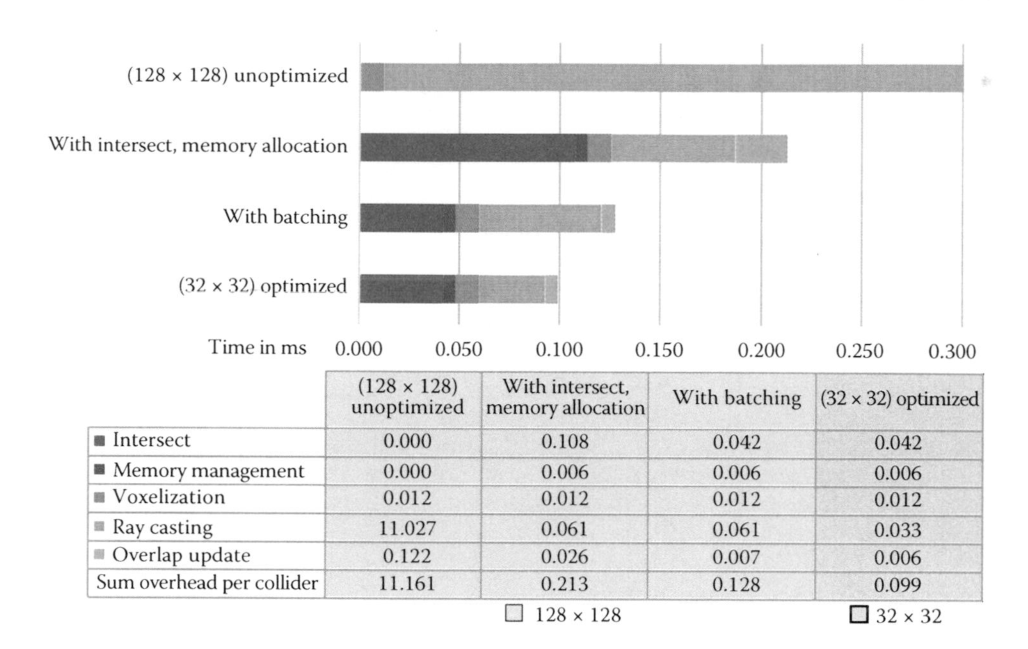

	(128 × 128) unoptimized	With intersect, memory allocation	With batching	(32 × 32) optimized
Intersect	0.000	0.108	0.042	0.042
Memory management	0.000	0.006	0.006	0.006
Voxelization	0.012	0.012	0.012	0.012
Ray casting	11.027	0.061	0.061	0.033
Overlap update	0.122	0.026	0.007	0.006
Sum overhead per collider	11.161	0.213	0.128	0.099

Figure 16.15. Timings in milliseconds on an NVIDIA GTX 780 for the different optimizations. The first three bars (from top to bottom) show the effects of our optimizations for a tile resolution of 128 × 18 texels, while the last bar shows the timings for a 32 × 32 tile resolution with all optimizations enabled.

16.6.3 Performance

In this section we provide detail timings of our deformation pipeline, including the benefits of the optimizations presented in Section 16.5. While we use the standard graphics pipeline for rendering and the voxelization of the models, including hardware tessellation for the subdivision surfaces, we employ compute shaders for patch-OBB intersection, memory management, ray casting (DDA), and updating the tile overlap regions.

Figure 16.15 summarizes the performance of the different pipeline stages and the overall overhead per deformable-penetrator collision pair measured on an NVIDIA GTX 780 using a default per-patch tile size of 128 × 128.

The measurements in Figure 16.15 show that ray casting is the most expensive stage of our algorithm. With a simple patch–voxel volume intersection test we can greatly improve the overall performance by starting ray casting and overlap updates only for the affected patches. This comes at the cost of spending additional time on the intersection test, which requires reading the control points of each patch. Because fetches from global memory are expensive, we optimize the intersection stage by computing the intersection with multiple penetrating objects after reading the control points, which further improves overall performance.

The chosen displacement tile size—as expected—only influences the ray casting and overlap stage. Because the computational overhead for the higher tile resolution is marginal, the benefits in deformation quality easily pay off.

16.7 Conclusion

In this chapter, we described a method for real-time visual feedback of surface deformations on collisions with dynamic and animated objects. To the best of our knowledge, our system is the first to employ a real-time voxelization of the penetrating object to update a displacement map for real-time deformation. Our GPU deformation pipeline achieves deformations in far below a millisecond for a single collision and scales with the number of deforming objects since only objects close to each other need to be tested. We believe that this approach is ideally suited for complex scene environments with many dynamic objects, such as in future video game generations. However, we emphasize that the deformations aim at a more detailed and dynamic visual appearance in real-time applications but cannot be considered as a physical simulation. Therefore, we do not support elasticity, volume preservation, or topological changes such as fractures.

16.8 Acknowledgments

This work is co-funded by the German Research Foundation (DFG), grant GRK-1773 Heterogeneous Image Systems.

Bibliography

[Burley and Lacewell 08] Brent Burley and Dylan Lacewell. "Ptex: Per-Face Texture Mapping for Production Rendering." In *Proceedings of the Nineteenth Eurographics Conference on Rendering*, pp. 1155–1164. Aire-la-Ville, Switzerland: Eurographics Association, 2008.

[Catmull and Clark 78] E. Catmull and J. Clark. "Recursively Generated B-Spline Surfaces on Arbitrary Topological Meshes." *Computer-Aided Design* 10:6 (1978), 350–355.

[Coumans et al. 06] Erwin Coumans et al. "Bullet Physics Library: Real-Time Physics Simulation." http://bulletphysics.org/, 2006.

[Nießner and Loop 13] Matthias Nießner and Charles Loop. "Analytic Displacement Mapping Using Hardware Tessellation." *ACM Transactions on Graphics* 32:3 (2013), article 26.

[Nießner et al. 12] Matthias Nießner, Charles Loop, Mark Meyer, and Tony DeRose. "Feature-Adaptive GPU Rendering of Catmull-Clark Subdivision Surfaces." *ACM Transactions on Graphics* 31:1 (2012), article 6.

[Schäfer et al. 14] Henry Schäfer, Benjamin Keinert, Matthias Nießner, Christoph Buchenau, Michael Guthe, and Marc Stamminger. "Real-Time Deformation of Subdivision Surfaces from Object Collisions." In *Proceedings of HPG'14*, pp. 89–96. New York: ACM, 2014.

[Schwarz 12] Michael Schwarz. "Practical Binary Surface and Solid Voxelization with Direct3D 11." In *GPU Pro 3: Advanced Rendering Techniques*, edited by Wolfgang Engel, pp. 337–352. Boca Raton, FL: A K Peters/CRC Press, 2012.

Realistic Volumetric Explosions in Games

Alex Dunn

17.1 Introduction

In games, explosions can provide some of the most visually astounding effects. This article presents an extension of well-known ray-marching [Green 05] techniques for volume rendering fit for modern GPUs, in an attempt to modernize the emulation of explosions in games

Realism massively affects the user's level of immersion within a game, and previous methods for rendering explosions have always lagged behind that of production quality [Wrennige and Zafar 11]. Traditionally, explosions in games are rendered using mass amounts of particles, and while this method can look good from a static perspective, the effect starts to break down in dynamic scenes with free-roaming cameras. Particles are camera-facing billboards and, by nature, always face the screen; there is no real concept of rotation or multiple view angles, just the same texture projected onto the screen with no regard for view direction. By switching to a volumetric system, explosions look good from all view angles as they no longer depend on camera-facing billboards. Furthermore, a single volumetric explosion can have the same visual quality as thousands of individual particles, thus, removing the strain of updating, sorting, and rendering them all—as is the case with particle systems.

By harnessing the power of the GPU and the DirectX 11 tessellation pipeline, I will show you that single-pass, fully volumetric, production-quality explosions are now possible in the current generation of video games. We will be exploring volumetric rendering techniques such as ray marching and sphere tracing, as well as utilizing the tessellation pipeline to optimize these techniques.

There are certain drawbacks to the technique, such as it not being as generic a system as particles. It's more of a bespoke explosion system and like particles, the effect is generally quite pixel heavy from a computational perspective.

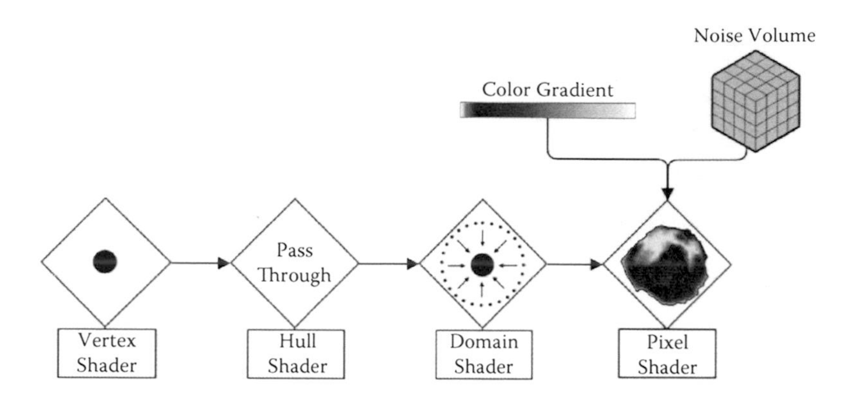

Figure 17.1. Pipeline flow overview of the technique.

17.2 Rendering Pipeline Overview

Explosions are represented by a single volumetric sphere with detail layered on top. The explosion is rendered by first generating a hemisphere mesh of a radius equal to the maximum radius of the explosion. (The explosion won't have a uniform radius, so instead we define a *maximum radius*, which is the distance from the explosion at its most extended point, to its core.) Then, shrink the hemisphere around the explosion to form a tight-fitting semi-hull. This is done in order to decrease the amount of degenerate fragments when later performing ray marching in the pixel shader. An overview of the technique is shown in Figure 17.1.

17.3 Offline/Preprocessing

First, we must create a 3D volume of noise, which we can use later to create some nice noise patterns. We can do this offline to save precious cycles later in the pixel shader. This noise is what's going to give the explosions their recognizable cloud-like look. In the implemention described here, simplex noise was used—however, it should be noted that it isn't a requirement to use simplex noise; in fact, in your own implementation you are free to use whatever type of noise you want, so long as it tiles correctly within our volume. In order to conserve bandwidth and fully utilize the cache, size, and format of this texture is detrimental to the performance of the technique. The implementation demonstrated here uses a $32 \times 32 \times 32$ sized volume with a 16-bit floating point format, `DXGI_FORMAT_R16_FLOAT`. The noise is calculated for each voxel of the volume using its UVW coordinate as the position parameter for the noise function.

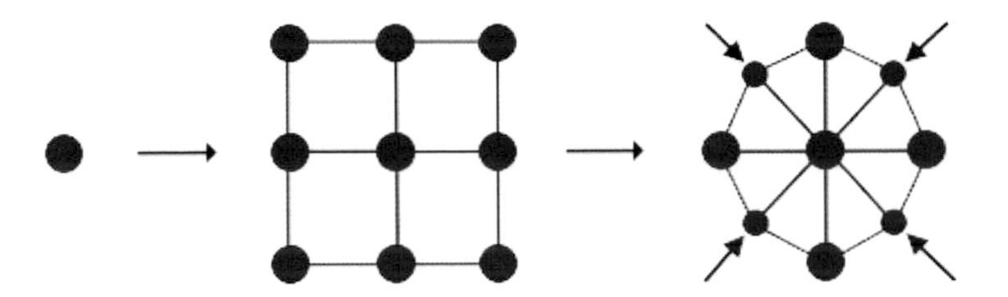

Figure 17.2. The life of a vertex. In an actual implementation, the level of subdivision should be much higher than shown in the diagram.

17.4 Runtime

As the effect will be utilizing the tessellation pipeline of the graphics card for rendering, it is required to submit a draw call using one of the various patch primitive types available in Direct X. For this technique, the `D3D11_PRIMITIVE_TOPOLOGY1_CONTROL_POINT_PATCHLIST` primitive type should be used as we only need to submit a draw call that emits a single vertex. This is because the GPU will be doing the work of expanding this vertex into a semi-hull primitive. The life of a vertex emitted from this draw call throughout this technique is shown in Figure 17.2.

17.4.1 Semi-Hull Generation

The GPU starts rendering our explosion with the vertex shader. This is run once per explosion. Its job is to read values from some data buffer, which stores position, radius, and "time lived" in seconds, and passes them down to the next shader in the pipeline, the hull shader. For your own implementation, it's entirely up to you how this information is stored, so long as it's accessible by the GPU.

The next stage of the pipeline is the hull shader. It runs once for each point of the input primitive and outputs control points for the next stage. This hull shader will be using the quad domain. When using this domain, the shader will load in a single vertex patch and the data associated with it (loaded in by the vertex shader previously) and output the four corner vertices of a quad (control points), each with its own copy of the data.

The tessellator stage is fixed function in DirectX 11. Its main purpose is to accept the control points generated by the hull shader as inputs and subdivide them. While programming the subdivision of control points is out of our reach in DirectX 11, we do have control over the level of subdivision. For the purposes of this article, we will just use a constant tessellation level. However, there is scope to adaptively tessellate your primitives based on the onscreen size of the explosions. A higher tessellation level can provide a tighter fitting hull around the explosion and thus decrease the amount of fragments rendered, which can make a big difference for high-quality, close-up explosions. The performance gains vary from case to case though, so I'd suggest profiling to find the best fit for your own implementations.

Once the patch has been subdivided, the next stage of the tessellation pipeline takes over. With the domain shader, we first transform the vertices into a screen-aligned hemisphere shape, with the inside of the sphere facing the camera and the radius set to that of the explosion. Then we perform a technique called *sphere tracing* [Hart 94] to shrink wrap the hemisphere around the explosion to form a tight-fitting hull. Sphere tracing is a technique not unlike ray marching, where starting at an originating point (a vertex on the hemisphere hull), we move along a ray toward the center of the hemisphere. Normally, when ray marching, we traverse the ray at fixed size intervals, but when sphere tracing, we traverse the ray at irregular intervals, where the size of each interval is determined by a distance field function evaluated at each step. A distance field function represents the signed distance to the closest point on an implicit surface from any point in space. (You can see an example of a signed distance function for an explosion in Listing 17.1).

17.4.2 Pixel Shading

The last programmable stage required for the effect is the pixel shader. This shader is invoked for each visible pixel of the explosion on screen; it is here where the bulk of the work will be done. For each pixel, it is required to step, or "march," through our explosion and evaluate the color at each step. The stepping will take place along a per-pixel ray, calculated using the world-space position of the pixel. The ray is then marched from front to back. At each step along the ray, a distance field function is evaluated. This function is the distance function for a sphere, perturbed by the value stored in the noise texture for the current position. A source code snippet is provided in Listing 17.1. The function `DrawExplosion`, will return the distance to the explosion from a point in world space and provide the amount of displacement caused by noise at that point. If the distance returned from this function is less than some epsilon, then this step is inside the explosion and contributes to the final color of this pixel. (For a great primer on this technique, see [Green 05].)

The method `FractalNoise`, used in Listing 17.1, calculates how perturbed the surface of the explosion will be at a given point in world space. The inner mechanics of this function can be seen in Listing 17.2.

The `FractalNoise` function uses the noise volume we created offline earlier. The volume is sampled multiple times. The location of each sample is calculated from the original sampling position by applying a constant frequency factor. Each sample read in this fashion is known as an *octave*. Once we have completed reading all the samples, the values from each are summed to give the final noise value. We found that four octaves provided a fairly reasonable visual experience.

```
// Returns the distance to the surface of a sphere.
float SphereDistance(float3 pos, float3 spherePos, float radius)
{
  float3 relPos = pos - spherePos;
  return length( relPos ) - radius
}

// Returns the distance to the surface of an explosion.
float DrawExplosion
  (
    float3 posWS,
    float3 spherePosWS,
    float radiusWS,
    float displacementWS,
    out float displacementOut
  )
{
  displacementOut = FractalNoise( posWS );

  float dist = SphereDistance(posWS, spherePosWS, radiusWS);

  return dist - displacementOut * displacementWS;
}
```

Listing 17.1. HLSL source: Distance field function for explosion.

```
// How many octaves to use when calculating the fractal noise.
static const uint kNumberOfNoiseOctaves = 4;

// Returns a noise value by texture lookup.
float Noise( const float3 uvw )
{
  return _NoiseTexRO.Sample( g_noiseSam, uvw, 0 );
}

// Calculates a fractal noise value from a world-space position.
float FractalNoise( const float3 posWS )
{
  const float3 animation = g_AnimationSpeed * g_time;

  float3 uvw = posWS * g_NoiseScale + animation;
  float amplitude = 0.5f;
  float noiseValue = 0;

  [unroll]
  for( uint i=0 ; i<kNumberOfNoiseOctaves ; i++ )
  {
    noiseValue += amplitude * Noise( uvw );
    amplitude *= g_NoiseAmplitudeFactor;
    uvw *= g_NoiseFrequencyFactor;
  }

  return noiseValue;
}
```

Listing 17.2. HLSL source: the noise function.

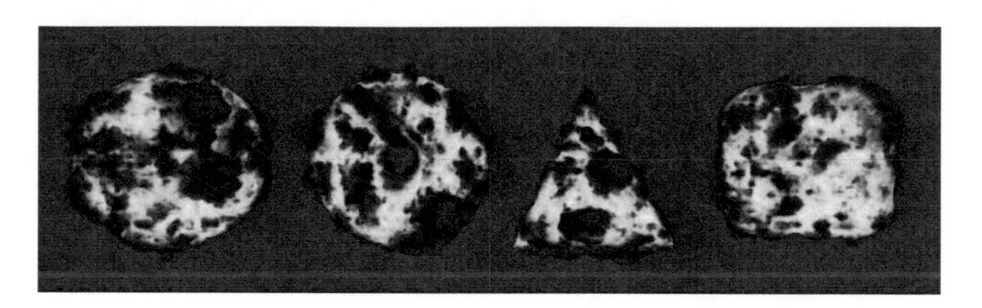

Figure 17.3. A collection of explosions rendered using different primitive types.

The color of the explosion at each step is calculated by performing a lookup into a gradient texture using the `displacementOut` parameter (from Listing 17.1) as a texture coordinate. This color is then blended with the other samples gathered from previous steps before the next step of the ray.

This process is repeated either until we hit the exit point of our ray or, as a further optimization, until the output color has reached full opacity and no further steps would contribute to the final color of the pixel. (It is important to note that this optimization will only work when marching through the explosion from front to back.)

17.5 Visual Improvements

17.5.1 Primitives

While this article has so far only demonstrated how to create an explosion based on the sphere primitive shape, it's possible to extend the technique to handle a variety of shapes, as you can see in Figure 17.3.

In Listing 17.1, there is a method called `SphereDistance`, which calculates the distance to the closest point on a sphere from some point in world space. In order to render an explosion with a different underlying primitive type, this method can be swapped for one that calculates the distance to another primitive, or even a collection of primitives. See Listing 17.3 for a list of basic primitive functions written in HLSL.

Rendering an explosion with one of the customized primitives above can be useful in situations where explosions are formed from explosive containers. For example, the cylinder primitive type would be useful when modeling an explosion that's to be associated with an old game developer favorite, the exploding barrel.

17.5.2 Extra Step

When the number of steps used to render an explosion is low, and when the explosion intersects some of the scene geometry, certain view angles relative to

```
float Sphere(float3 pos, float3 spherePos, float radius)
{
  float3 relPos = pos - spherePos;
  return length( relPos ) - radius
}

float Cone( float3 pos, float3 conePos, float radius )
{
  float3 relPos = pos - conePos;

  float d = length( relPos.xz );
  d -= lerp( radius * 0.5f, 0, 1 + relPos.y/radius );

  d = max( d,-relPos.y - radius );
  d = max( d, relPos.y - radius );

  return d;
}

float Cylinder( float3 pos, float3 cylinderPos, float radius )
{
  float3 relPos = pos - cylinderPos;

  float2 h = radius.xx * float2( 1.0f, 1.5f ); // Width, Radius
  float2 d = abs( float2( length( relPos.xz ), relPos.y ) ) - h;

  return min( max( d.x, d.y ), 0.0f) + length( max( d, 0.0f) );
}

float Box( float3 pos, float3 boxPos, float3 b )
{
  float3 relPos = pos - boxPos;
  float3 d = abs( relPos ) - b;

  return min( max( d.x, max( d.y, d.z ) ), 0.0f )
         + length( max( d, 0.0f ) );
}

float Torus( float3 pos, float3 torusPos, float radius )
{
  float3 relPos = pos - boxPos;

  float2 t = radius.xx * float2( 1, 0.01f );
  float2 q = float2( length( relPos.xz ) - t.x , relPos.y );

  return length( q ) - t.y;
}

// Rendering a collection of primitives can be achieved by
// using multiple primitive distance functions, combined
// with the 'min' function.
float Cluster( float3 pos )
{
  float3 spherePosA = float3(-1, 0, 0);
  float3 spherePosB = float3( 1, 0, 0);
  float sphereRadius = 0.75f;

  return min( Sphere(pos, spherePosA, sphereRadius),
              Sphere(pos, spherePosB, sphereRadius) );
}
```

Listing 17.3. HLSL source: a collection of distance functions for various primitives.

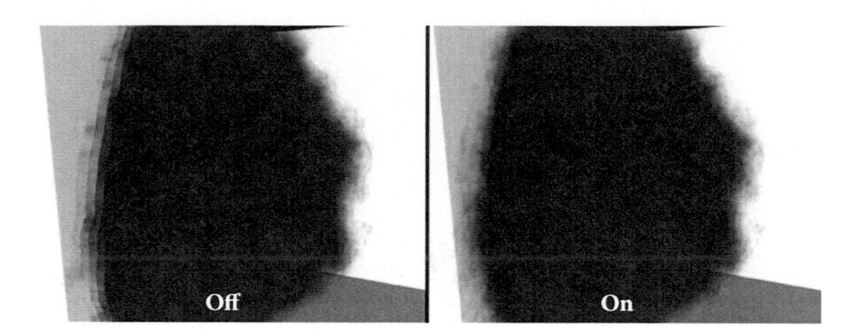

Figure 17.4. The extra step trick in action.

the intersection geometry can produce an ugly banding artifact in which the slices of the volume are completely visible.

The extra step trick [Crane et al. 07] attempts to minimize this artifact by adding one final step at the end of the ray marching, passing in the world-space position of the pixel instead of the next step position. Calculating the world-space position of the pixel can be done any way you see fit; in the approach demonstrated here, we have reconstructed world-space position from the depth buffer. (See Figure 17.4.)

17.5.3 Lighting

Lighting the explosion puffs from a directional light is possible by performing a similar ray-marching technique to the one seen earlier while rendering the explosion [Ikits et al. 03]. Let's go back to when we were rendering via ray marching. In order to accurately calculate lighting, we need to evaluate how much light has reached each step along the ray. This is done at each rendering step by ray marching from the world-space position of the step toward the light source, accumulating the density (in the case of the explosion this could be the opacity of the step) until either the edge of the volume has been reached or the density has reached some maximum value (i.e., the pixel is fully shadowed).

Because this is rather expensive, as an optimization you don't need to calculate the lighting at every step. Depending on the amount of steps through the volume and the density of those steps, you can just calculate the lighting value for one in every x steps and reuse this value for the next steps. Use best judgement and check for visual artifacts while adjusting the x variable.

17.6 Results

The screenshots in Figures 17.5–17.7 were rendered using 100 steps (but only a few rays will actually use this much) with the shrink wrapping optimization enabled.

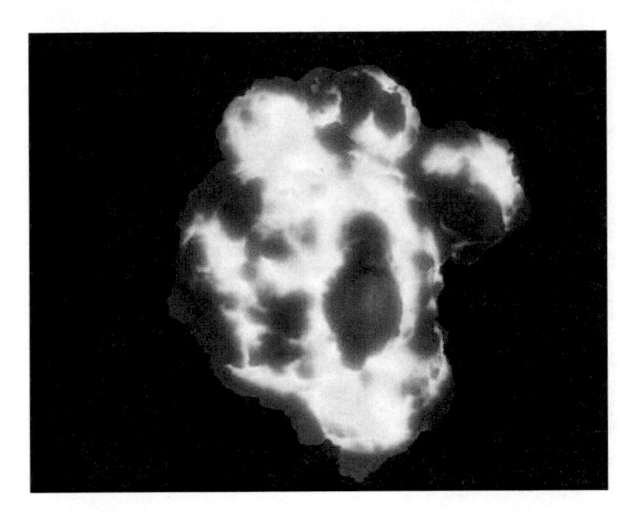

Figure 17.5. A shot of a clustered volumetric explosion. Here, a collection of spheres has been used to break up the obvious shape of a singular sphere.

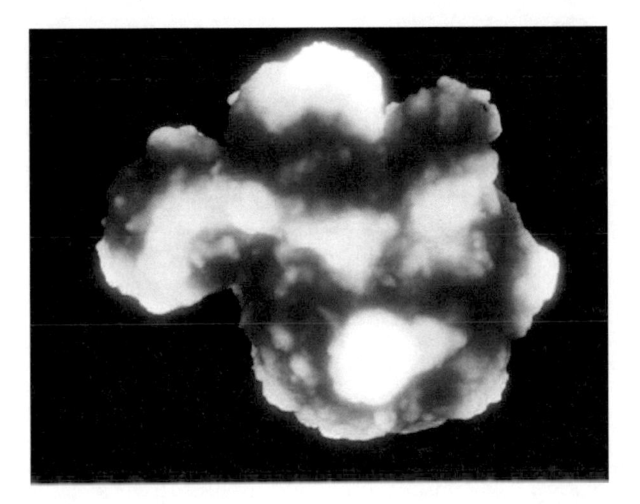

Figure 17.6. Varying the displacement to color gradient over time can provide a powerful fourth dimension to the effect.

17.7 Performance

The performance of this explosion technique is certainly comparable to that of a particle-based explosion. With the shrink wrapping optimization, rendering times can be significantly reduced under the right circumstances. In Figure 17.8, you'll see a visual comparison of the shrink wrapping technique and the effect it

Figure 17.7. As the life of an explosion comes to an end, the entire effect turns more smoke than flame.

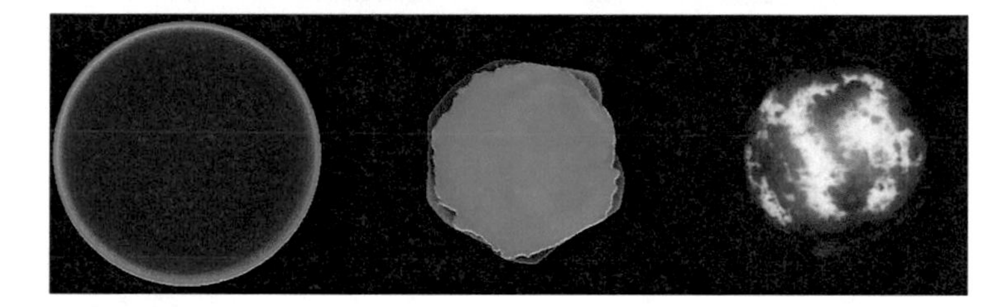

Figure 17.8. Here you can see the amount of rays required to step through the explosion (more red means more steps are required): with no shrink wrapping optimizations (left), with shrink wrapping and early out for fully opaque pixels (middle), and the final rendering (right).

has on the number of steps required to render a volumetric explosion. Following this, Figure 17.9 shows a graph detailing the amount of time taken to render the same explosion with shrink wrapping on and off across a range of rendering steps. For both datasets, the exact same GPU and driver were used (NVIDIA GTX980 with driver v344.11) and the time taken has been calculated by recording an average over 200 frames.

17.7.1 Further Optimizations

Half-resolution up-sampling. With a slight modification to [Cantlay 07], we can significantly reduce the pixel workload by reducing the size of the rendering buffer. This technique works by binding a low-resolution render target in place of the

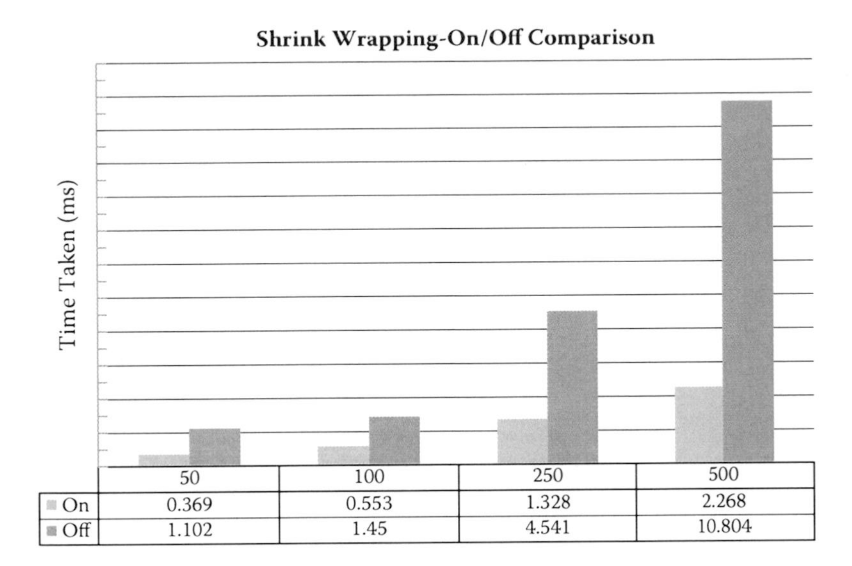

	50	100	250	500
On	0.369	0.553	1.328	2.268
Off	1.102	1.45	4.541	10.804

Figure 17.9. See how the shrink wrapping optimization improves the render time. All numbers were captured using a GTX980 (driver v344.11). Timings were averaged over 200 frames.

back buffer just before rendering an explosion, then, once the explosions have been rendered, up-sampling the texture associated with the low-resolution render target by rendering it to the full-resolution back buffer.

There are several corner cases to be aware of—depth testing and edge intersections to name a couple—that are out of the scope of this article. I recommend reading [Cantlay 07], in which these are thoroughly explained.

Depth testing. Currently, when rendering an explosion, we do so by rendering the back faces (front-face culling) of the sphere geometry. The upside of this is that we can still see the back faces while inside the explosion, which allows us to keep rendering. The downside is that we can't perform hardware depth testing for early exiting pixels, which are occluded by nearer geometry but have to resort to performing a texture read-dependent branch in the pixel shader.

It's possible, with very little tweaking, to render the explosion by using the front faces (back face culling) which will allow the correct use of hardware depth testing on a read-only depth buffer (we still need to read from the depth buffer in the shader to figure out the depth of the scene, so the DSV bound must be readonly, that is, created with 'flags = 0'). The only downside is that once the camera moves inside the explosion, the front faces are no longer visible, and the explosion disappears. The solution to this is to switch to the back faces once the camera enters the volume.

17.8 Conclusion

Volumetric explosions undoubtedly provide a much richer visual experience over particle-based techniques, and as I've shown, it's possible to use them now in the current generation of games. This article has demonstrated how to best utilize the modern graphics pipeline and DirectX, taking full advantage of the tessellation pipeline. The optimization methods described allow for implementing this effect with a minimal impact on frame times.

17.9 Acknowledgments

The techniques described in this article are an extension of the previous works of Simon Green in the area of real-time volume rendering.

Bibliography

[Cantlay 07] Iain Cantlay. "High-Speed, Off-Screen Particles." In *GPU Gems 3*, edited by Hubert Nguyen, pp. 535–549. Reading, MA: Addison-Wesley Professional, 2007.

[Crane et al. 07] Keenan Crane, Ignacio Llamas, and Sarah Tariq. "Real-Time Simulation and Rendering of 3D Fluids." In *GPU Gems 3*, edited by Hubert Nguyen, pp. 653–694. Reading, MA: Addison-Wesley Professional, 2007.

[Green 05] Simon Green. "Volume Rendering For Games." Presented at Game Developer Conference, San Francisco, CA, March, 2005.

[Hart 94] John C. Hart. "Sphere Tracing: A Geometric Method for the Antialiased Ray Tracing of Implicit Surfaces." *The Visual Computer* 12 (1994), 527–545.

[Ikits et al. 03] Milan Ikits, Joe Kniss, Aaron Lefohn, and Charles Hansen. "Volume Rendering Techniques." In *GPU Gems*, edited by Randima Fernando, pp. 667–690. Reading, MA: Addison-Wesley Professional, 2003.

[Wrennige and Zafar 11] Magnus Wrennige and Nafees Bin Zafar. "Production Volume Rendering." SIGGRAPH Course, Vancouver, Canada, August 7–11, 2011.

Deferred Snow Deformation in *Rise of the Tomb Raider*

Anton Kai Michels and Peter Sikachev

18.1 Introduction

Procedural snow deformation is one of the defining graphical features in the *Rise of the Tomb Raider*. (See Figure 18.1.) It creates a discernable connection between Lara and her environment while serving as a key gameplay element, allowing the titular Tomb Raider to track her targets while remaining hidden

Figure 18.1. Deformable snow in *Rise of the Tomb Raider*. [Image courtesy of Square Enix Ltd.]

from sight. At the core of this technology is a novel technique called *deferred deformation*, which decouples the deformation logic from the geometry it affects. This approach can scale with dozens of NPCs, has a low memory footprint of 4 MB, and can be easily altered to handle a vast variety of procedural interactions with the environment. This chapter aims to provide the reader with sufficient theoretical and practical knowledge to implement deferred deformation in a real-time 3D application.

Procedural terrain deformation has remained an open problem in real-time rendering applications, with past solutions failing to provide a convincing level of detail or doing so with a very rigid set of constraints. Deferred deformation delivers a scalable, low-memory, centimeter-accurate solution that works on any type of terrain and with any number of deformable meshes. It renders not only the depression in the trail center but also elevation on the trail edges and allows for gradual refilling of snow tracks to emulate blizzard-like conditions.

Some terminology used in this publication with regards to snow trails and deformation will be outlined in Section 18.2. Section 18.3 then takes a look at past approaches, where they succeeded and why they were ultimately not suitable for *Rise of the Tomb Raider*. Section 18.4 outlines a simple, straightforward algorithm for rendering snow deformation that will serve as a prelude to deferred deformation in Section 18.5, which elaborates on the core ideas behind the technique and the use of compute shaders to achieve it. Section 18.6 details the deformation heightmap used in our algorithm and how it behaves like a sliding window around the player. Section 18.7 explains how the snow tracks fill over time to emulate blizzard-like conditions. Section 18.8 covers the use of adaptive hardware tessellation and the performance benefits gained from it. Finally, Section 18.9 discusses alternate applications of this technique and its future potential.

18.2 Terminology

Here is a collection of terms used throughout the article (see Figure 18.2):

- Snow height: The vertical coordinate (in our case `vertex.z`) of the snow prior to any deformation.
- Deformation points: 3D points estimating Lara's feet and other objects that cause deformation.
- Foot height: The vertical height of a deformation point (`point.z`).
- Trail depression: The part of the trail that is stomped down and is lower than the original snow height.
- Trail elevation: The small bump along the edges of the trail caused from pushing snow out of the way.

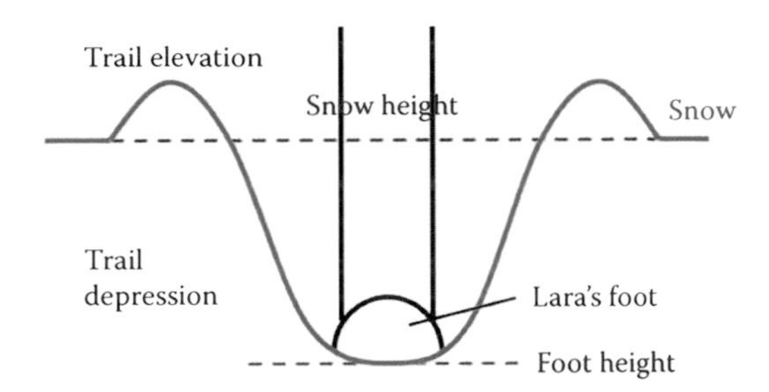

Figure 18.2. The various components of a snow trail.

- Deformation heightmap: Single 32-bit texture, 1024 × 1024 pixels, that stores the deformation.
- Depression depth: abs(snow height − foot height).
- Deformation shader: Compute shader used to output the deformation to the heightmap.
- Fill shader: Compute shader used to fill snow tracks during a blizzard.
- Snow shader: The shader used for objects with deformable snow material.

18.3 Related Work

Deformation from dynamic objects is a key component in making snow look believable in 3D applications. The following two titles took very different approaches in this regard, each with its own advantages and disadvantages.

18.3.1 Assassin's Creed III

As a Playstation 3 and Xbox 360 title, *Assassin's Creed III* could not make use of Shader Model 5.0 features like compute shaders and hardware tessellation for the console version of the game. Instead, a render-to-vertex-buffer trick was used to create tessellated triangles at runtime using the GPU, with the limitation that all triangles created this way must have the same tessellation factor. These tessellated triangles are then pushed down using a geometrical approximation of the character's movement [St-Amour 13].

Advantages of this technique include the creation of persistent tracks on a large scale and support of various terrain (forests, slopes, etc.). The disadvantages are a lack of support for filling the trails in snowy conditions and not producing an elevation along the trail edges. This technique also requires encoding the maximum possible deformation in snow mesh vertices to avoid pushing the snow below the terrain, a further drawback.

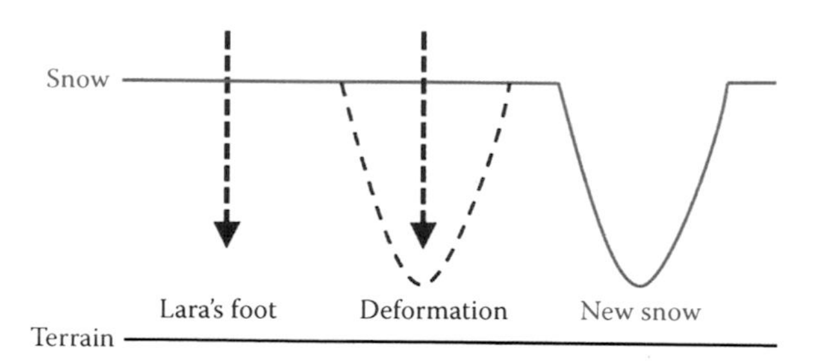

Figure 18.3. Basic snow deformation uses the snow and the terrain height to clamp the deformation height when it is rendered.

18.3.2 Batman: Arkham Origins

The most recent AAA title using procedural snow deformation, *Batman: Arkham Origins*, takes place in an urban environment devoid of slopes and terrain and thus uses rectangular rooftop meshes for its deformable snow. These rectangular boxes form orthogonal view frustums into which dynamic objects affecting the snow are rendered. The resulting render target is used as a heightmap to displace the vertices of the snow mesh [Barré-Brisebois 14].

This technique leverages the rendering pipeline to create very accurate snow deformation, which is enhanced by GPU tessellation on DirectX 11–compatible hardware. Filling tracks during a snowstorm is also supported. The disadvantage is that this technique is unusable for anything other than flat rectangular surfaces. And like *Assassin's Creed III*, it does not produce elevation along the trail edges.

18.4 Snow Deformation: The Basic Approach

Consider a terrain mesh, a snow mesh on top of it, and a number of dynamic objects deforming the snow. One approach to rendering the snow deformation is to first render the terrain and snow meshes from a bird's-eye view into two separate heightmaps, then render the dynamic objects or some approximation of these objects into a deformation heightmap and clamp the rendered values between the terrain and snow height. Finally, the deformation heightmap is sampled when rendering the snow to displace vertices and calculate normals (Figure 18.3).

The simplicity of this approach has several drawbacks. First is the need to gather all necessary terrain and snow meshes and render them from a bird's-eye view. Second is that each dynamic object affecting the snow requires its own draw call. Both of these problems are solved with deferred deformation, as shown in the next section.

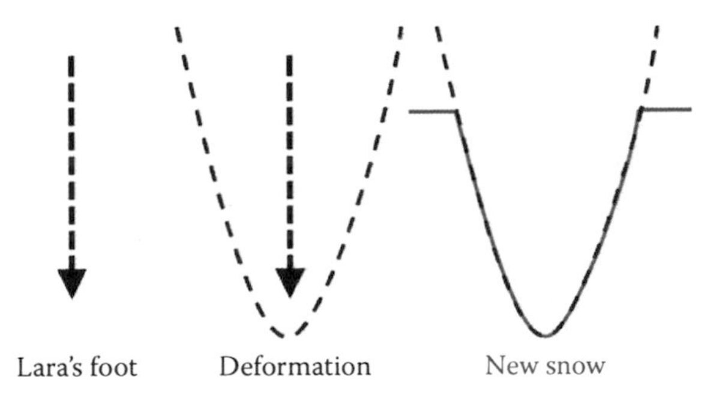

Figure 18.4. Deferred deformation forgoes the initial use of the snow and terrain height and instead clamps the deformation height during the snow rendering.

18.5 Deferred Deformation

The idea behind deferred deformation is as follows: during the snow render pass, the snow height is already provided by the vertices of the snow mesh. There is therefore no need to pre-render the snow mesh into a heightmap, and the deformation height can be clamped when it is sampled instead of when it is rendered. This allows the heightmap to be rendered with an approximate deformation using the dynamic objects only. The exact deformation is calculated later during the actual rendering of the snow, hence the term *deferred deformation* (Figure 18.4). Note that it is important to pass the original snow height from the snow vertex shader to the snow pixel shader for per-pixel normals. (See Listing 18.1 for an overview of the deferred deformation algorithm.)

```
Deformation Shader (compute shader)
  affected_pixels = calculate_deformation(dynamic_object)
Fill Shader (compute shader)
  all_pixels += snow_fill_rate
Snow Shader
  Snow Vertex Shader
    snow_height = vertex.Z
    deformation_height = sample_deformation_heightmap()
    vertex.Z = min(snow_height, deformation_heightmap)
    pixel_input.snow_height = snow_height
  Snow Pixel Shader
    snow_height = pixel_input.snow_height
    deformation_height = sample_deformation_heightmap()
    calculate_deformed_normal()
```

Listing 18.1. Deferred deformation algorithm overview.

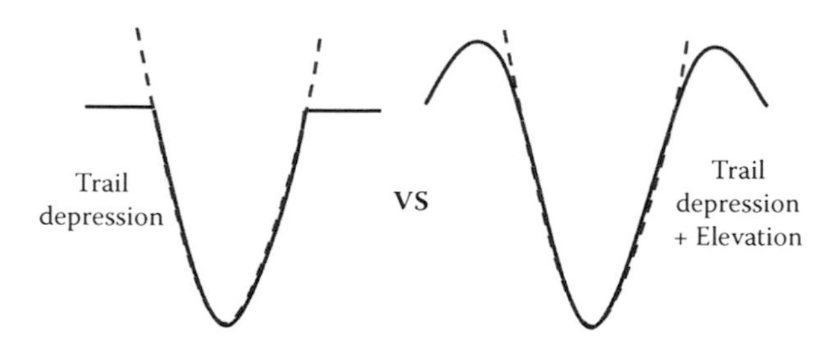

Figure 18.5. It is desirable to add an elevation along the edges of the trail to enhance the overall look.

18.5.1 Rendering the Deformation Heightmap

A key insight during the development of the deferred deformation algorithm was observing that the desired trail shape closely resembles a quadratic curve. By approximating dynamic objects with points, the deformation height around these points can be calculated as follows:

$$\text{deformation height} = \text{point height} + (\text{distance to point})^2 \times \text{artist's scale}.$$

These deformation points are accumulated into a global buffer, and the deformation shader is dispatched with one group for each point. The groups write in a 32^2 pixel area (1.64 m^2) around the deformation points and output the deformation height of the affected pixels using an atomic minimum. This atomic minimum is necessary as several deformation points can affect overlapping pixels in the heightmap. Since the only unordered access view (UAV) types that allow atomic operations in DirectX 11 are 32-bit integer types, our deformation heightmap UAV is an `R32_UINT`.

18.5.2 Trail Elevation

What has been described thus far is sufficient to render snow trails with depression, but not trails with both depression and elevation (Figure 18.5). Elevation can occur when the deformation height exceeds the snow height, though using this difference alone is not enough. The foot height must also be taken into account, as a foot height greater than the snow height signifies no depression, and therefore no trail and no elevation (Figure 18.6). For this reason the foot height is also stored in the deformation texture using the least significant 16 bits (Figure 18.7). It is important that the deformation height remain in the most significant 16 bits for the atomic minimum used in the deformation shader. Should the snow shader sample a foot height that is above the vertex height, it early outs of the deformation and renders the snow untouched.

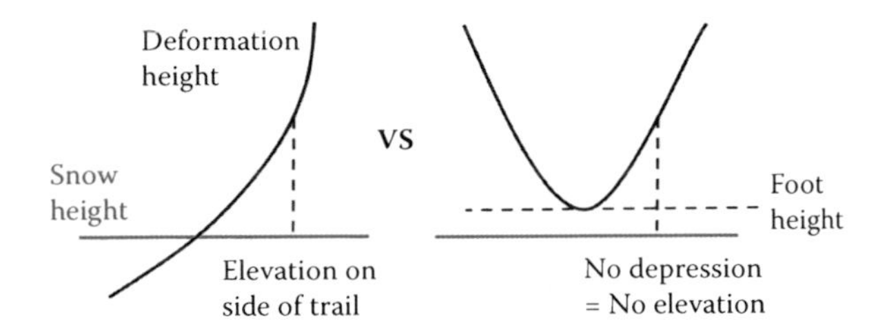

Figure 18.6. The deformation height alone is not enough to know if there is elevation. The foot height is also needed.

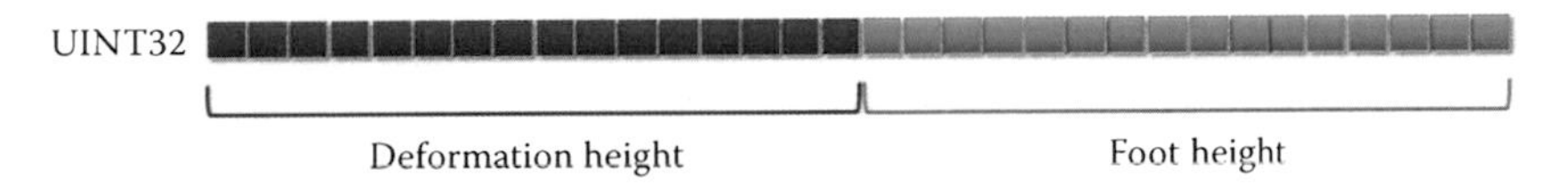

Figure 18.7. Bit allocation scheme used for the deformation heightmap.

18.5.3 Calculating the Elevation

Constructing the elevation first requires the elevation distance, i.e., the distance between the start of the elevation and the point being rendered (Figure 18.8). To calculate the elevation distance, the following variables are introduced (see Figure 18.9 for a detailed illustration):

- Depression distance: the distance between the center of the deformation and the end of the depression.
- Distance from foot: the distance between the center of the deformation and the point being rendered. Note that this is the sum of the depression distance and the elevation distance.

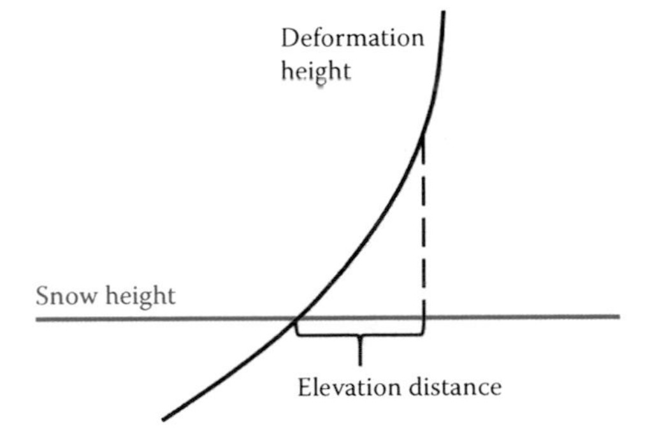

Figure 18.8. The elevation distance at a given point is the distance between that point and the start of the elevation.

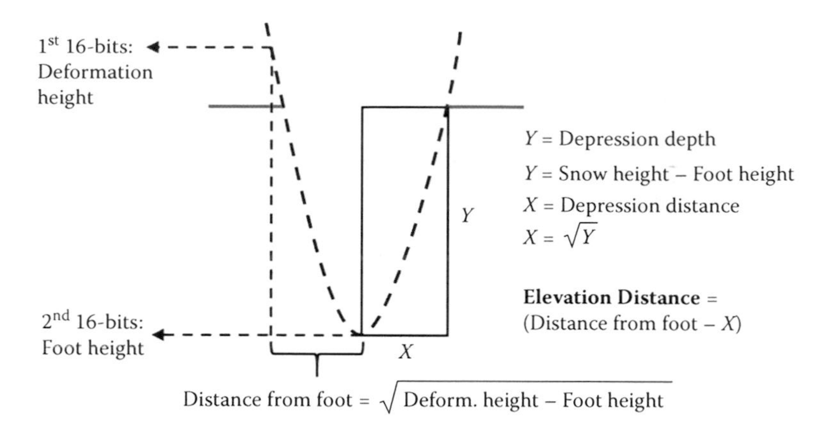

Figure 18.9. Illustration of the various calculations involved in rendering the elevation.

The above values are calculated as follows:

$$\text{depression distance} = \sqrt{\text{snow height} - \text{foot height}},$$

$$\text{distance from foot} = \sqrt{\text{deformation height} - \text{foot height}},$$

$$\text{distance from foot} = \text{depression distance} + \text{elevation distance},$$

$$\text{elevation distance} = \text{distance from foot} - \text{depression distance}.$$

The elevation should scale with the depth of the trail—deeper trails produce greater elevation along the trail edges because more snow is being displaced. The depression depth is therefore used to calculate a maximum elevation distance using a linear function with an artist-driven scale. Knowing the elevation distance and the maximum elevation distance, we can compute an elevation ratio and pass this ratio into a quadratic function to get a smooth, round elevation on the edges of the trail:

$$\text{ratio} = \frac{\text{elevation distance}}{\text{max. elevation distance}},$$

$$\text{height} = \text{max. elevation distance} \times \text{artist's scale},$$

$$\text{elevation} = ((0.5 - 2 \times \text{ratio})^2 + 1) \times \text{height}.$$

18.5.4 Texture Selection

To give deformed snow a more disorderly and chaotic look, different snow textures are applied to different parts of the deformation. A smooth texture selection value between 0 and 2 is generated to choose between the textures. (See Figure 18.10.) The value 0 corresponds to the center of the trail, the value 1 corresponds to the

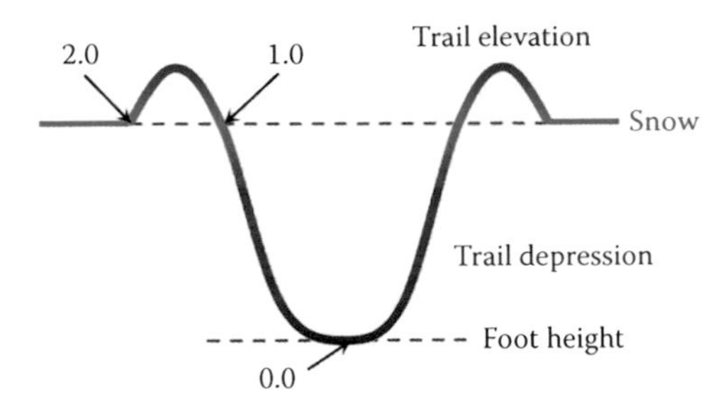

Figure 18.10. Different textures are used for the trail using a generated selection value between 0 and 2.

end of the depression and start of the elevation, and the value 2 corresponds to undeformed snow after the elevation. Artists fetch this value in *Rise of the Tomb Raider*'s shader node system and use it to dynamically select the desired textures. The texture selection variable is calculated using the depression distance, maximum elevation distance, and distance from foot variables.

18.6 Deformation Heightmap

The deformation heightmap is a 32-bit 1024 × 1024 texture (4 MB) with a resolution of 4 cm per pixel, covering an area of 40.96 m^2 centered on Lara. The texture holds two 16-bit values (deformation height and foot height). It is created as an `R16G16_TYPELESS` texture and given an `R16G16_UNORM` shader resource view (SRV) and an `R32_UINT` UAV (needed for the atomic minimum in the compute shader).

18.6.1 Sliding Window Heightmap

In order to keep the area of deformation centered on Lara, the deformation heightmap acts as a sliding window around her position. As Lara moves, the pixels of the heightmap that fall out of range are repurposed as the new pixels that have fallen into range. The implementation of this feature falls into two parts: reading and writing. In both cases, points out of range of the deformation area centered on Lara cause an early out for the shader to prevent tiling. (See also Figure 18.11.)

- Reading: In order to read from a sliding window texture, it is sufficient to scale world-space coordinates to match the texture resolution and then use them as UVs with a wrap sampler. Tiling is prevented with the early out mentioned above.

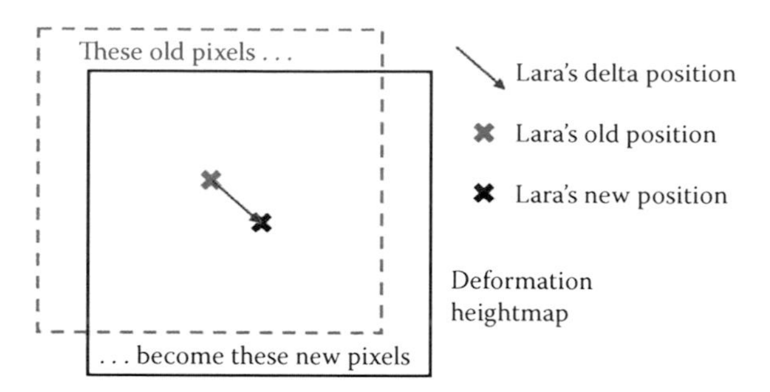

Figure 18.11. The deformation heightmap acts as a sliding window to keep the snow deformation area centered on Lara.

```
float2 Modulus(float2 WorldPos, float2 TexSize) {
  return WorldPos - (TexSize * floor(WorldPos/TexSize));
}
```

Listing 18.2. Modulus function for the compute shader.

- Writing: Writing to a sliding window heightmap is possible with the use of compute shaders and unordered access views. The deformation shader writes in 32×32 pixel areas around the deformation points, with the output pixels calculated at runtime. In order for the deformation shader to work with a sliding window texture, the calculations of these output pixels use the modulus function in Listing 18.2, which acts in the same way a wrap sampler would.

18.6.2 Overlapping Deformable Meshes

Despite the use of a single heightmap, deferred deformation allows for vertically overlapping snow meshes, i.e., deformable snow on a bridge and deformable snow under a bridge. This is accomplished by overriding the heightmap deformation in the deformation shader if the newly calculated deformation differs by more than a certain amount (in our case, 2 m), regardless of whether it is higher or lower than the existing deformation. The snow shader then early outs if the sampled foot height differs too greatly from the snow height (again 2 m in our case). Thus, snow deformation on a bridge will be ignored by snow under the bridge because the foot height is too high, and deformation under the bridge will be ignored by the snow on the bridge because the foot height is too low.

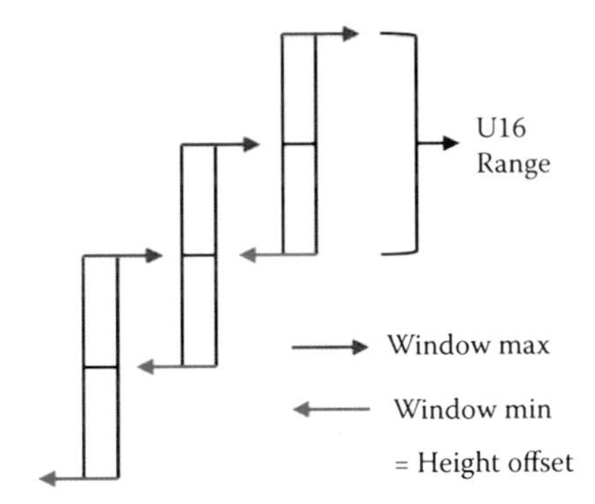

Figure 18.12. The vertical range of the snow deformation dynamically adjusts with Lara's position. This allows us to increase the precision of the deformation height and foot height stored in the deformation heightmap.

18.6.3 Vertical Sliding Window

Working with 16-bit values and extracting a sufficient level of precision from them means limiting their effective range. To overcome this problem, deferred deformation employs a vertical sliding window technique for the deformation heightmap (Figure 18.12). At any given time, the snow deformation has a minimum global height. This is used as an offset when the deformation heightmap is rendered and sampled. Whenever Lara goes below this height offset, the sliding window shifts down by half the window frame. Whenever she climbs above the window's maximum height, the sliding window shifts up by half a frame. The reason half a frame is used as the increment/decrement value is to avoid cases where minor changes in Lara's position will cause the window to switch back and forth. Whenever the sliding window shifts up or down, half the range is also added/subtracted to the global fill rate (Section 18.7) for that frame, bringing the heightmap values in accordance with the new height offset.

18.7 Filling the Trail over Time

Snow deformation in *Rise of the Tomb Raider* emulates blizzard-like conditions by filling the snow tracks over time. For this, a second compute shader called the *fill shader* is dispatched with 1024^2 threads to cover the entire heightmap. This fill shader increases the value of each pixel by a global fill rate. It is not sufficient, however, to only increase the deformation height, as this will cause the elevations on the trail edges to move inward, giving a weird and unnatural result. Separate fill rates for both the deformation height and the foot height are required to remedy this, with manual adjustments needed to attain convincing results.

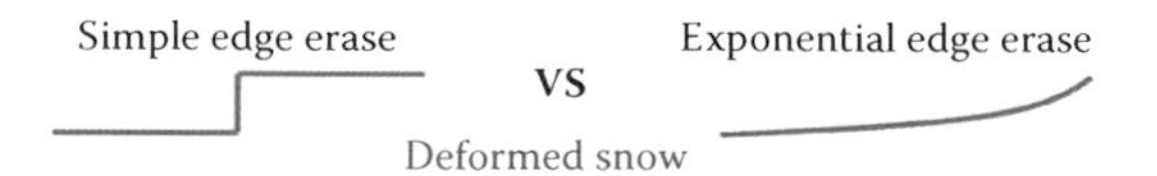

Figure 18.13. Exponential edge erase provides a much smoother finish to the trails over a more simple approach.

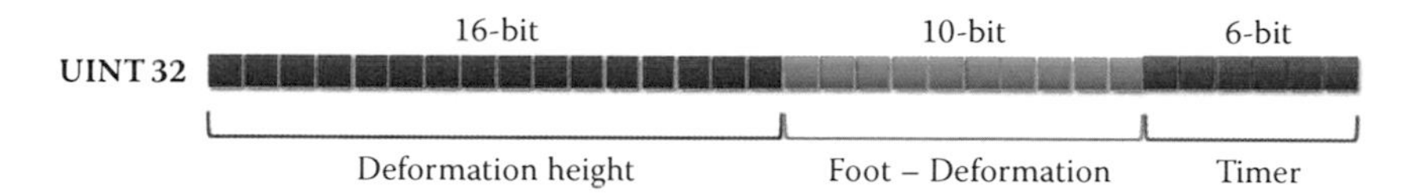

Figure 18.14. New bit allocation scheme used for the deformation heightmap.

18.7.1 Erasing the Edges of the Sliding Window

A key component in the sliding window functionality is erasing the pixels along the edge of the sliding window. A straightforward way to do this is by resetting the values of the pixels to `UINT32_MAX` along the row and the column of the pixels farthest away from Lara's position (use the `Modulus` function in Listing 18.2 to calculate this row and column). The downside to this approach is that it will create very abrupt lines in the snow trails along the edges of the sliding window, something the player will notice if they decide to backtrack.

Instead of erasing one row and one column, a better solution is to take eight rows and eight columns along the sliding window border and apply a function that exponentially increases the snow fill rate for these pixels. This will end the trails with a nice upward curve that looks far more natural (Figure 18.13).

18.7.2 Reset Timer

Filling trails over time conflicts with the ability to have vertically overlapping snow meshes. If a trail under a bridge fills over time, it will eventually create a trail on top of the bridge. However, this will only happen if the initial trail is filled for a long time. A per-pixel timer was therefore implemented to reset deformation after a set period. This period is long enough to allow for the deep tracks to fill completely and short enough to prevent overlapping snow meshes from interfering with each other. Once the timer reaches its maximum value, the pixel is reset to `UINT32_MAX`.

The implementation of this timer uses the least significant 6 bits of the foot height in the deformation heightmap (Figure 18.13). This leaves the foot height with only 10 bits (Figure 18.14). To compensate for the lost precision, the heightmap does not store the foot height but rather the deformation height minus the foot height. The foot height is then reconstructed in the snow shader.

Figure 18.15. High-poly snow mesh without tessellation. Normal pass: 3.07 ms. Composite pass: 2.55 ms.

18.8 Hardware Tessellation and Performance

With deferred deformation, the cost shifts from rendering the deformation heightmap to reconstructing the depression and elevation during the snow render pass. Because these calculations involve multiple square roots and divisions, the snow vertex shader's performance takes a significant hit. This makes statically tessellated, high-poly snow meshes prohibitively expensive (offscreen triangles and detail far from the camera are a big part of this cost). (See Figure 18.15.)

Much of this cost is alleviated with adaptive tessellation and a reduced vertex count on the snow meshes. The tessellation factors are computed in image space, with a maximum factor of 10. Frustum culling is done in the hull shader, though back-face culling is left out because the snow is mostly flat. Derivative maps [Mikkelsen 11] are used to calculate the normals in order to reduce the vertex memory footprint, which is crucial for fast tessellation. Further performance is gained by using Michał Drobot's ShaderFastMathLib [Drobot 14], without any noticeable decrease in quality or precision. (See Figure 18.16.)

The timings for the fill shader and deformation shader are 0.175 ms and 0.011 ms, respectively, on Xbox One.

18.9 Future Applications

Given that our deferred deformation technique does not care about the geometry it deforms, the same deformation heightmap can be repurposed for a wide variety of uses (for example, mud, sand, dust, grass, etc.). Moreover, if the desired deformation does not require any kind of elevation, the technique becomes all the more simple to integrate. We therefore hope to see this technique adopted, adapted, and improved in future AAA titles.

Figure 18.16. High-poly snow mesh with tessellation. Normal pass: 1.60 ms. Composite pass: 1.14 ms.

18.10 Acknowledgments

We would like to thank the guys from the Labs team at Eidos Montreal, Nixxes Software, and Crystal Dynamics for their help in implementing this feature in *Rise of the Tomb Raider*.

Bibliography

[Barré-Brisebois 14] Colin Barré-Brisebois. "Deformable Snow Rendering in *Batman: Arkham Origins*." Presented at Game Developers Conference 2014, San Francisco, CA, March 17–21, 2014.

[Drobot 14] Michał Drobot. "ShaderFastMathLib." *GitHub*, https://github.com/michaldrobot/ShaderFastLibs/blob/master/ShaderFastMathLib.h, 2014.

[Mikkelsen 11] Morten Mikkelsen. "Derivative Maps." *Mikkelsen and 3D Graphics*, http://mmikkelsen3d.blogspot.com/2011/07/derivative-maps.html, 2011.

[St-Amour 13] Jean-François St-Amour. "Rendering Assassin's Creed III." Presented at Game Developers Conference 2013, San Francisco, CA, March 25–29, 2013.

Catmull-Clark Subdivision Surfaces

Wade Brainerd

19.1 Introduction

Catmull-Clark subdivision surfaces, or SubDs, are smooth surfaces defined by bicubic B-spline patches extracted from a recursively subdivided control mesh of arbitrary topology [Catmull and Clark 78]. SubDs are widely used in animated film production and have recently been used in games [Brainerd 14]. They are valued for their intuitive authoring tools and the quality of the resultant surface (Figure 19.1).

In recent years, research has advanced rapidly with respect to rendering subdivision surfaces on modern GPUs. Stanford University's survey [Nießner et al. ar] gives a comprehensive picture of the state of the art.

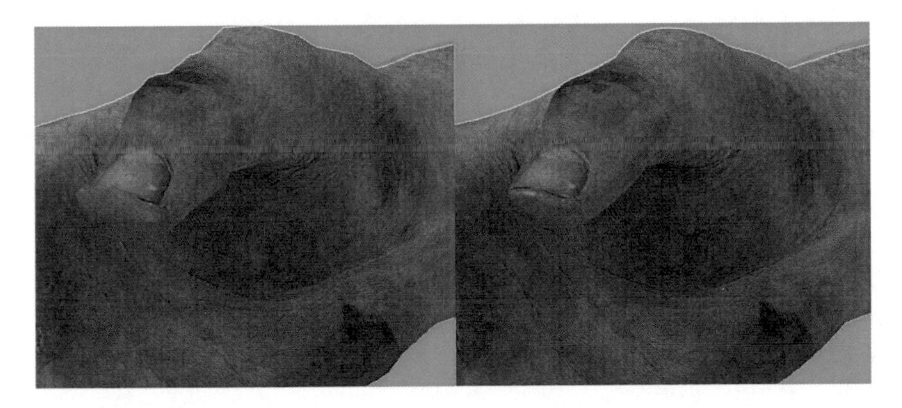

Figure 19.1. A hand modeled as a Catmull-Clark subdivision surface (right), with its corresponding control mesh on the left.

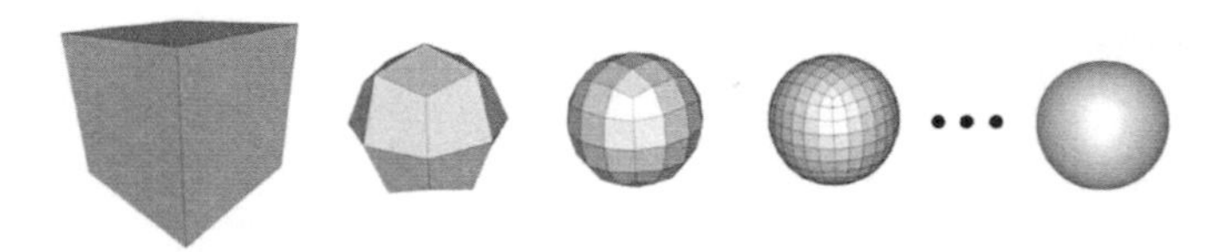

Figure 19.2. Several iterations of Catmull-Clark subdivision applied to a cube, with the eventual limit surface on the far right. Note that each corner is an extraordinary vertex.

In this chapter, we describe a real-time method for rendering subdivision surfaces that is utilized for key assets in *Call of Duty* titles, running in 1920×1080 resolution at 60 frames per second on Playstation 4 hardware. As long as the topology remains constant, our implementation allows the control mesh to deform and animate while being subdivided in real time and dynamically adapts the subdivision amount to the geometry curvature and the view.

19.1.1 Terminology

We provide definitions for a few basic and important terms. For a thorough survey of quad mesh terminology, refer to the 2013 survey by Bommes et al. [Bommes et al. 13].

control point A vertex that is used as a B-spline support.

control mesh A mesh consisting of control points, which defines a surface.

valence The number of edges around a face, or incident to a vertex.

regular vertex A vertex with valence 4.

extraordinary vertex A vertex with valence other than 4.

regular quad A quad in which all vertices are regular.

manifold mesh A mesh in which each edge is connected to either one or two faces, and the faces connected to each vertex form a continuous fan.

limit surface A smooth surface resulting from a Catmull-Clark subdivision.

19.1.2 Catmull-Clark Subdivision Surfaces

Catmull-Clark subdivision surfaces are a generalization of bicubic B-spline surfaces to arbitrary topology. Standard bicubic B-spline and NURBS surfaces require that control meshes be constructed from regular grids without extraordinary vertices. Careful stitching between grids is necessary to maintain smoothness. This can be inconvenient for artists, and the stitching does not always hold up well under animation.

Figure 19.3. A model rendered using feature adaptive subdivision. Red patches are regular in the control mesh, green patches after one subdivision, and so on. Tiny purple faces at the centers of rings are connected to the extraordinary vertices.

The method of Catmull and Clark solves this problem by finely subdividing the control mesh and extracting regular B-spline grids from the subdivided result. The subdivision rules are chosen to preserve the base B-spline surface for regular topology and to produce a smooth, aesthetically pleasing surface near extraordinary vertices. (See Figure 19.2.)

In theory, infinite subdivisions are required to produce a surface without holes at the extraordinary vertices. The result of infinite subdivisions is called the *limit surface*. In practice, the limit surface may be evaluated directly from the control mesh by exploiting the eigenstructure of the subdivision rules [Stam 98] or approximated by halting subdivision after some number of steps.

19.1.3 Feature Adaptive Subdivision

Feature adaptive subdivision [Nießner et al. 12] is the basis for many real-time subdivision surface renderers, such as OpenSubdiv [Pixar 15]. It is efficient, is numerically stable, and produces the exact limit surface.

In feature adaptive subdivision, a preprocessing step extracts bicubic B-spline patches from the control mesh where possible, and the remaining faces are subdivided. Extraction and subdivision are repeated until the desired subdivision level is reached. To weld T-junctions that cause surface discontinuities between subdivision levels, triangular *transition patches* are inserted along the boundaries. Finally, all the extracted patch primitives are rendered using hardware tessellation. With repeated subdivision, extraordinary vertices become isolated but are never eliminated, and after enough subdivision and extraction steps to reach the desired level of smoothness, the remaining faces are rendered as triangles. (See Figure 19.3.)

19.1.4 Dynamic Feature Adaptive Subdivision

In dynamic feature adaptive subdivision, Schäfer et al. extend feature adaptive subdivision to dynamically control the number of subdivisions around each extraordinary vertex [Schäfer et al. 15].

The subdivided topology surrounding each extraordinary vertex is extracted into a *characteristic map* of n subdivision levels. To render, a compute shader determines the required subdivision level $l \leq n$ for the patches incident to each extraordinary vertex and then uses the characteristic map as a guide to emit control points and index buffers for the patches around the vertex.

Dynamic feature adaptive subdivision reduces the number of subdivisions and patches required for many scenes and is a significant performance improvement over feature adaptive subdivision, but it does add runtime compute and storage costs.

19.2 The Call of Duty Method

Our method is a subset of feature adaptive subdivision; we diverge in one important regard: B-spline patches are only extracted from the first subdivision level, and the remaining faces are rendered as triangles. The reason is that patches that result from subdivision are small and require low tessellation factors, and patches with low tessellation factors are less efficient to render than triangles.

We render the surface using a mixture of hardware-tessellated patch geometry and compute-assisted triangle mesh geometry (Figure 19.4). Where the control mesh topology is regular, the surface is rendered as bicubic B-spline patches using hardware tessellation. Where the control mesh topology is irregular, vertices which approach the limit surface are derived from the control mesh by a compute shader and the surface is rendered as triangles.

Because we accept a surface quality loss by not adaptively tessellating irregular patches, and because small triangles are much cheaper than small patches, we also forgo the overhead of dynamic feature adaptive subdivision.

19.3 Regular Patches

A *regular patch* is a control mesh face that is a regular quad embedded in a quad lattice. More specifically, the face must be regular, and the faces in its *one ring neighborhood* (Figure 19.5) must all be quads. The control points of neighboring faces are assembled into a 4×4 grid, and the limit surface is evaluated using the bicubic B-spline basis functions in the tessellation evaluation shader.

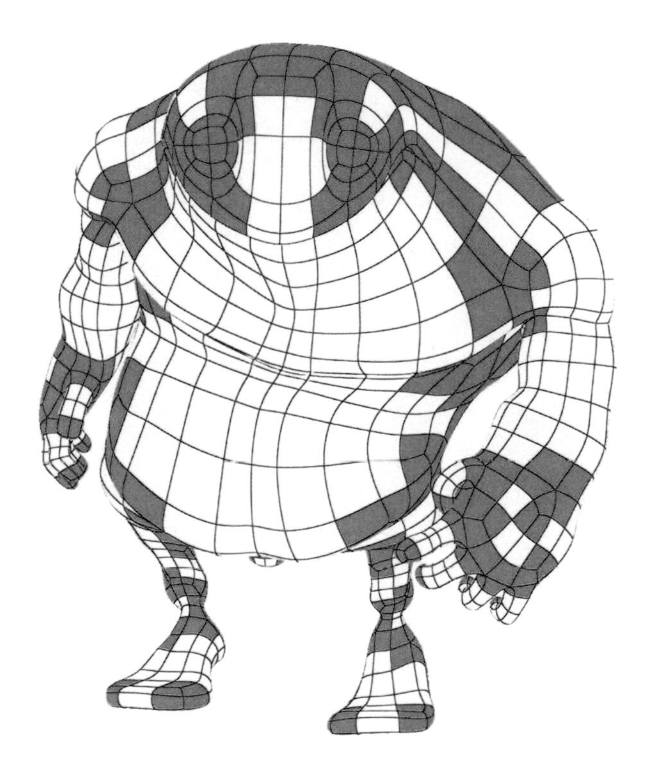

Figure 19.4. Bigguy model with regular patches in white and irregular patches in red.

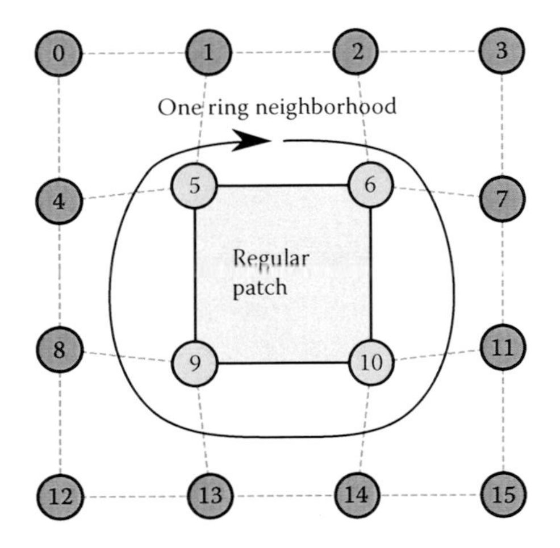

Figure 19.5. A regular patch with its one ring neighborhood faces and numbered control points.

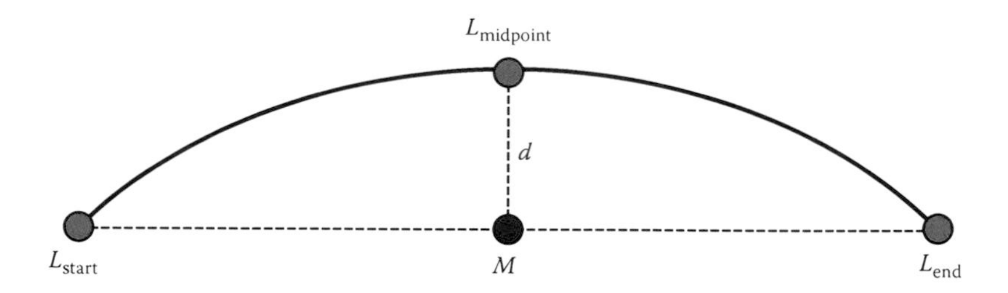

Figure 19.6. An edge with elements labeled related to the adaptive tessellation metric.

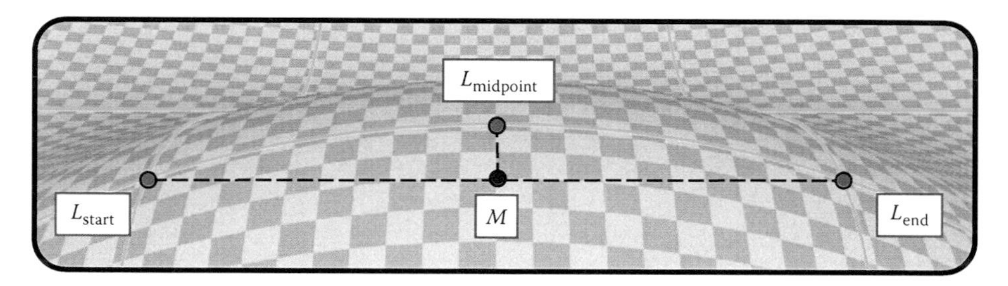

Figure 19.7. An edge with elements labeled related to the adaptive tessellation metric.

19.3.1 Adaptive Tessellation

A principal feature of the hardware tessellation pipeline is the ability to vary tessellation dynamically along each edge and in the interior of the patch. We utilize dynamic tessellation adaptively, to increase tessellation as needed to represent curvature and to decrease tessellation in flatter areas to reduce costs. Savings include vertex evaluation costs and also overshading costs caused by small or thin triangles being submitted to the rasterizer.

Our adaptive tessellation metric (Figures 19.6 and 19.7) requires evaluation of three limit surface points per patch edge: midpoint L_{midpoint} and endpoints L_{start} and L_{end}. Point L_{midpoint} is projected onto the line through L_{start} and L_{end} as M, and the square root of distance a between L_{midpoint} and M, multiplied by a constant quality factor k, becomes the tessellation factor f:

$$f \leftarrow k\sqrt{\left\|\texttt{bspline}(t{=}.5) - \frac{\texttt{bspline}(t{=}0) + \texttt{bspline}(t{=}1)}{2}\right\|}. \qquad (19.1)$$

The quality factor k is the reciprocal of the target distance between the limit surface and the rendered edge segments, in screen coordinates. The points are then projected to the screen space, to control tessellation in a view-dependent manner. After projection, depth should be clamped to $\epsilon \leq z$, and screen coordinates should be clamped to a guard band outside the view frustum $|x| \leq g$,

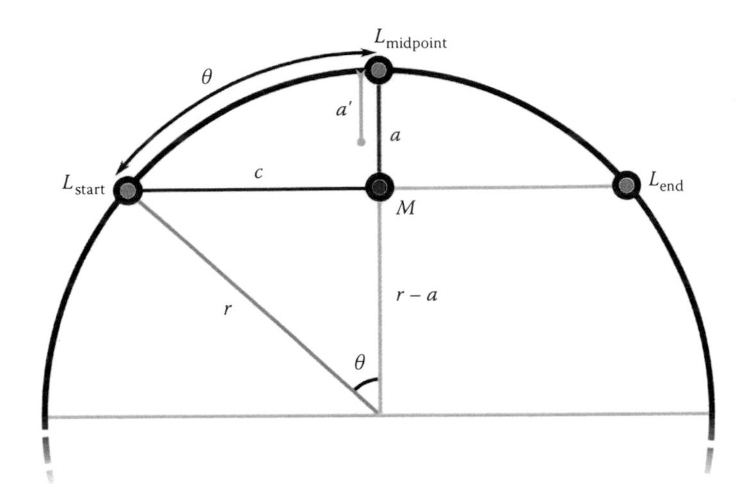

Figure 19.8. Curve segment approximated as an arc of a circle.

$|y| \leq g$, $g \approx 2.0$, to avoid over-tessellation of edges connected to vertices that cross outside near the view and develop large magnitudes.

Rationale We begin by approximating the curvature of the patch edge as a circular arc from L_{start} to L_{end}, intersecting L_{midpoint}, with radius r and angle 2θ. (See Figure 19.8.) Given half-segment $\vec{C}$ and distance a between midpoint M and L_{midpoint}, we can determine the radius r and angle θ of the half-arc. For the triangle with edges r, c, and $r - a$, length r can be determined from a and c as follows:

$$\begin{aligned}(r-a)^2 + c^2 &= r^2\\ r^2 - 2ra + a^2 + c^2 &= r^2\\ -2ra + a^2 + c^2 &= 0\\ a^2 + c^2 &= 2ra\\ \frac{(a^2+c^2)}{2a} &= r.\end{aligned} \tag{19.2}$$

For the angle θ between $\vec{A}$ and $\vec{R}$, note that $\angle CA$ is 90°:

$$\begin{aligned}\theta &= \arccos\left(\frac{r-a}{r}\right)\\ &= \arccos\left(1 - \frac{a}{r}\right).\end{aligned} \tag{19.3}$$

Consider an error threshold a' (similar to a) representing the maximum desired distance between segment $\vec{C}$ and the arc. If $a \leq a'$, no tessellation is needed.

Using a segment with length c' for the same curve ($r' = r$), given a' and r, without knowing c we can determine θ':

$$\theta' = \arccos\left(1 - \frac{a'}{r}\right). \tag{19.4}$$

If tessellation factor $f = 1$ represents the arc θ, we roughly need to subdivide θ into f segments of θ' that satisfy the error threshold. In terms of starting distance a, starting segment length c, and target distance a',

$$\begin{aligned} f &= \frac{\theta}{\theta'} \\ &= \frac{\arccos\left(1 - \frac{a}{r}\right)}{\arccos\left(1 - \frac{a'}{r}\right)}. \end{aligned} \tag{19.5}$$

For small values of x, we can approximate $\arccos x$:

$$\begin{aligned} \cos x &\approx 1 - \frac{x^2}{2} \\ \arccos\left(1 - \frac{x^2}{2}\right) &\approx x, \\ \text{let } y &= \frac{x^2}{2}, \\ \arccos(1-y) &\approx \sqrt{2y}. \end{aligned} \tag{19.6}$$

Thus, we can reasonably approximate the tessellation factor f in terms of a and a':

$$\begin{aligned} f &= \frac{\arccos\left(1 - \frac{a}{r}\right)}{\arccos\left(1 - \frac{a'}{r}\right)} \\ &\approx \frac{\sqrt{\frac{2a}{r}}}{\sqrt{\frac{2a'}{r}}} \\ &\approx \sqrt{\frac{a}{a'}}. \end{aligned} \tag{19.7}$$

The constant factor k in Equation (19.1) corresponds to $\frac{1}{a'}$, where a' is the screen-space distance threshold.

Results Our adaptive metric is fast and high quality, compared with global subdivision. In Figure 19.9, note that adaptive subdivision disables tessellation on straight edges and flat patches and maintains or increases tessellation in areas of high curvature.

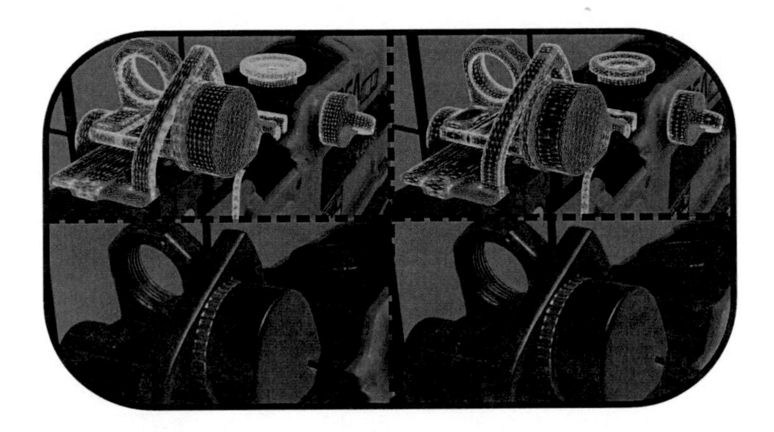

Figure 19.9. Our adaptive metric in practice; blue edges represent tessellated regular patches. Clockwise from top left: wireframe global subdivision, wireframe adaptive subdivision, shaded adaptive subdivision, and shaded global subdivision.

19.3.2 Rendering

Regular patches are rendered using the hardware tessellation pipeline. In OpenGL, this consists of the vertex shader, tessellation control shader, tessellation evaluation shader, and fragment shader stages. In Direct3D, it consists of the vertex shader, hull shader, domain shader, and fragment shader stages. For consistency with the sample code, we use OpenGL terms in this text.

The static index buffer contains the 16 control points per patch, and the vertex buffer holds the vertices of the control mesh (Algorithm 19.1).

Procedure `setup()`
  $vertexBuffer_{control} \leftarrow controlMesh$;
  $indexBuffer \leftarrow$ `extractControlPoints(`$mesh$`)`;
Procedure `render()`
  `setGpuControlPointsPerPatch(`*16*`)`;
  `drawPatches(`$vertexBuffer_{control}$, $indexBuffer$, $bsplineShaders$`)`

Algorithm 19.1. Render bicubic B-spline patches using hardware tessellation.

19.4 Irregular Patches

Control mesh faces that do not not meet the regular patch criteria are called *irregular patches*; these are recursively subdivided and rendered as triangles. To subdivide irregular patches, for each subdivision level, the following steps are taken (follow along with Figure 19.10):

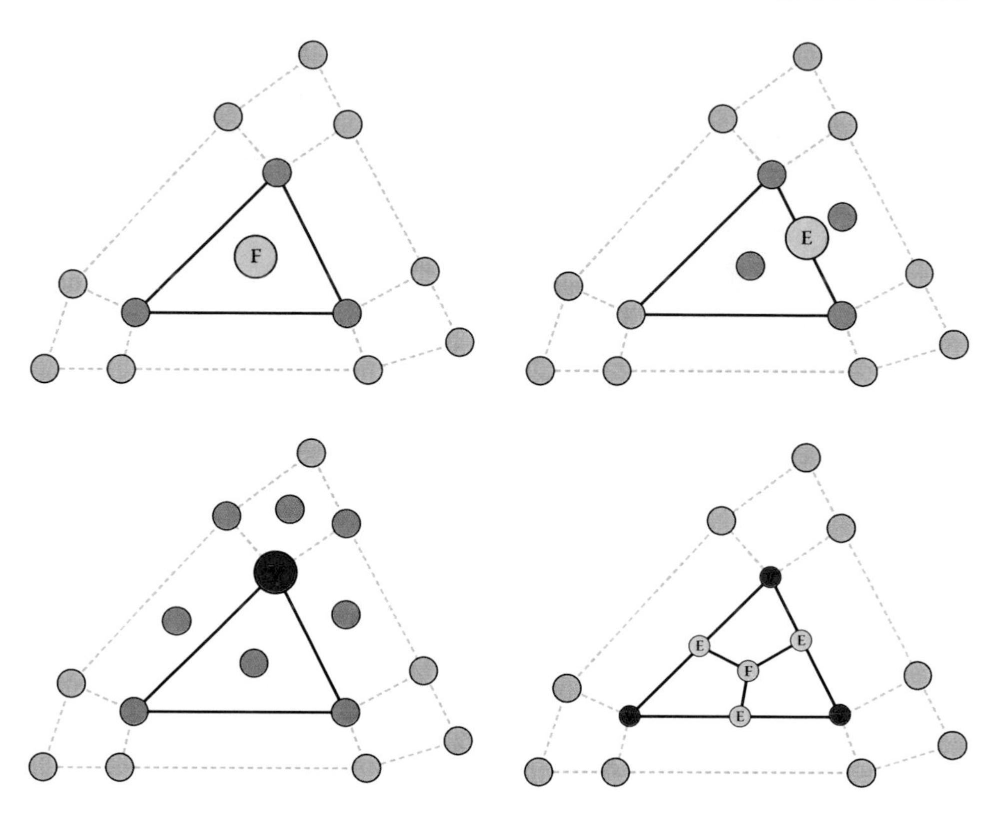

Figure 19.10. A triangle with its subdivided face point, edge point, and vertex point influences highlighted, followed by the subdivided quads.

- A new *face point* is added at the center of each control mesh face.
- A new *edge point* is added along each control mesh edge.
- A new *vertex point* replaces each control mesh vertex.

To subdivide each face, the face point is connected to an edge point, vertex point, and subsequent edge point to form a new quad. The process is repeated for every edge on the face. For a face of valence n, n quads are introduced. Note that after a single step of subdivision, only quads remain in the mesh. These quads are typically not planar.

Each Catmull-Clark subdivision produces a new, denser control mesh of the same subdivision surface. The face, edge, and vertex points of one control mesh become the vertex points of the next. With repeated subdivison, the control mesh becomes closer to the limit surface. In our experience with video game models, we have found it sufficient to stop after two subdivisions of irregular patches.

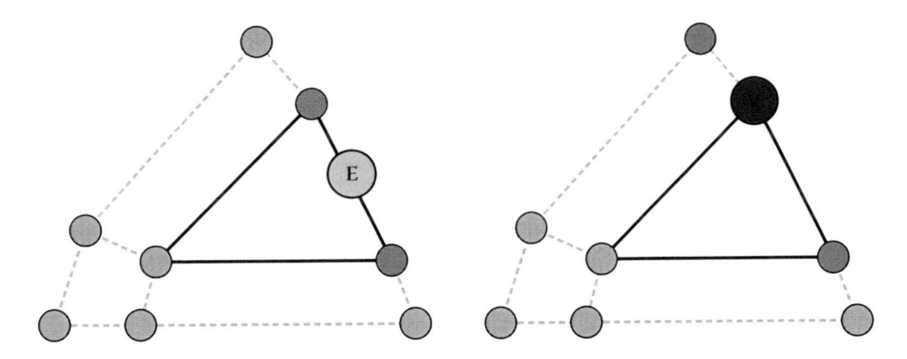

Figure 19.11. Influences for edge points and vertex points with border edges.

19.4.1 Face, Edge, and Vertex Point Rules

For a closed mesh, the Catmull-Clark rules for subdividing vertices are as follows (the influences are illustrated in Figure 19.10):

face point Average of the face's vertices: e.g., the centroid.

edge point Average of the edge's endpoints and its adjacent face points.

vertex point Weighted average of the vertex and its adjacent face and edge endpoints. The vertex weight is $\frac{n-2}{n}$, and the face and edge point weights are $\frac{1}{n^2}$, where n is the vertex valence.

Note that the edge point and vertex point calculations depend on the results of face point calculations; this has implications for parallel evaluation.

19.4.2 Borders and Corners

In the absence of neighboring faces and vertices, the weighted averages are altered. An edge connected to only one face is called a *border edge*. A vertex connected to two border edges is called a *border vertex* (Figure 19.11).

border edge point Average of the edge endpoints: e.g., the midpoint.

border vertex point Weighted average of the vertex and its two border edge endpoints. The vertex weight is $\frac{3}{4}$, the endpoint weights are $\frac{1}{8}$.

Rules are not given for non-manfiold, double-sided, bowtie, degenerate, and other problematic topology cases. Their subdivision surfaces are not well defined, and they should typically be rejected by the asset pipeline.

Corners and creases Extensions to the subdivision rules allow edges to be tagged as sharp creases [Hoppe et al. 94], and as semi-sharp creases with a fractional *sharpness* value [DeRose et al. 98]. Our pipeline does not explicitly support creases. Instead, we represent sharp creases by splitting vertices and relying on boundary rules. Semi-sharp creases may be emulated in smooth meshes by manually inserting edge loops and bevels. SubD implementations may also elevate border vertex points to *corner vertex points* based on their valence (see [Pixar 15]).

corner vertex point Remains pinned to its location.

Consistency There are several variations of the subdivision rules to choose from, each having subjective qualities with respect to the behavior of the limit surface. Ultimately, models need to be rendered consistently with how they were built, so these decisions need to take into account the behavior of the modeling software being used.

In our pipeline, we follow the default behavior of Autodesk Maya, which, although undocumented, is straightforward to reverse-engineer by construction of test cases.

19.4.3 Subdivision Tables

To render irregular patches using the GPU, we must first factor the subdivision rules into a form that can be processed by a compute shader. As the Catmull-Clark subdivision rules can all be defined as weighted averages of neighboring vertices, we generalize them into a table of weights and vertex indices, called a *subdivision table* (Figure 19.12)

The subdivision table is stored in a GPU buffer, which is processed by a *subdivision table compute shader* (Listing 19.1). For each table row, the compute shader accumulates the weighted influences from the control mesh vertices and writes the result to the vertex buffer. To allow a single index buffer to reference control mesh vertices and subdivision table outputs, the control mesh vertices C are prepended to the subdivision table output. Table row k therefore stores its weighted average at vertex buffer location $||C|| + k$.

19.4.4 Subdivision Table Factorizing

Each subdivision table row is a weighted average of the face, edge, and vertex points from the same or prior subdivision. These points are in turn weighted averages of face, edge, and vertex points from the prior subdivision. At the first subdivision, all influences are weighted averages of vertices in the control mesh.

To account for dependencies between table rows, the subdivision table evaluation must be partitioned into separate face, edge, and vertex dispatches per subdivision level, with GPU read/write fences in between. However, as every

kind	id	label	influences	weights
control mesh verts	0	c0		
	1	c1		
	2	c2		
	3	c3		
face points	4	f0	0 1 2 3	0.25...
edge points	5	e0	0 1 4 ?	0.25...
	6	e1	1 2 4 ?	0.25...
	7	e2	3 2 4 ?	0.25...
	8	e3	2 0 4 ?	0.25...
vertex points	9	v0	0 1 3 4 ? ? ? ?	0.5 0.625...
	10	v1	1 0 2 4 ? ? ? ?	0.5 0.625...
	11	v2	2 1 3 4 ? ? ? ?	0.5 0.625...
	12	v3	3 0 2 4 ? ? ? ?	0.5 0.625...

Figure 19.12. A subset of subdivision tables generated for the vertices connected to one quad. Each row of the table represents one subdivided vertex. Note that "?" is used to depict vertices that are present but not shown in the drawing. The control mesh vertices c_n are implicitly prepended to the table.

weighted average is a linear combination of its inputs, *all* subdivision tables may be factorized to depend only on the vertices of the control mesh.

Factorizing is accomplished by recursively replacing each weighted influence that is not from the control mesh with its own influences, appropriately weighted. Though this increases the average number of influences per table row, it eliminates dependencies between tables and therefore allows all subdivision tables to be evaluated in a single compute dispatch.

19.4.5 Rendering

Irregular patches are rendered using the standard triangle pipeline consisting of vertex shader and fragment shader. The static index buffer contains the triangulated subdivided quads, and the vertex buffer is filled by the subdivision table compute shader.

As the control mesh vertices are typically prepended to the subdivision table output vertex buffer, this is typically the same vertex buffer that is used to render regular patches. If any control mesh vertices have changed, the subdivision table compute shader is dispatched before rendering to update the subdivided vertex buffer (Algorithm 19.2).

19.5 Filling Cracks

19.5.1 Transition Points

While the B-spline evaluation performed by the tessellation hardware evaluates points on the limit surface, recursive subdivision evaluates points that merely

```
layout( local_size_x = 32, local_size_y = 1) in;

uniform uint baseVertCount;

layout( std430, binding = 0 ) buffer TablesBuffer {
    uint tables[];
};

layout( std430, binding = 1 ) buffer InfluencesBuffer {
    uint influences[];
};

layout( std430, binding = 2 ) buffer VertsBuffer {
    float verts[];
};

void main()
{
  uint index = gl_GlobalInvocationID.x;

  uint data = tables[index];

  uint first = data & 0xffffff;
  uint count = data >> 24;

  vec3 result = vec3( 0 );

  for ( uint i = first; i < first + count; i++ )
  {
    uint vertIn = influences[i * 2 + 0];

    float weight = uintBitsToFloat( influences[i * 2 + 1] );

    vec3 p = vec3(
      verts[vertIn * 3 + 0],
      verts[vertIn * 3 + 1],
      verts[vertIn * 3 + 2] );

    result += p * weight;
  }

  uint vertOut = baseVertCount + index;

  verts[vertOut * 3 + 0] = result.x;
  verts[vertOut * 3 + 1] = result.y;
  verts[vertOut * 3 + 2] = result.z;
}
```

Listing 19.1. Subdivision table compute shader.

approach the limit surface. Where a regular patch and an irregular patch share an edge, this manifests as a crack in the rendered surface.

The discrepancy can be resolved by using the B-spline basis functions in the subdivision table evaluation compute shader to evaluate the limit surface position for the irregular vertices along the edge. We call these limit surface points

```
Procedure setup()
    mesh ← controlMesh;
    vertexBuffer_control ← controlMesh;
    foreach i in subdivisions do
        facePoints ← extractFacePoints(mesh);
        edgePoints ← extractEdgePoints(mesh);
        vertexPoints ← extractVertexPoints(mesh);
        faces ← subdivideFaces(mesh);
        shaderBuffer_i ←
          factorizeTables(facePoints, edgePoints, vertexPoints);
        indexBuffer_i ← triangulate(faces);
        mesh ← (facePoints, edgePoints, vertexPoints, faces);
    end
Procedure render()
    i ← chooseSubdivisionLevel(camera);
    if control mesh vertices changed then
        vertexBuffer ←
          dispatchCompute(vertexBuffer_control, shaderBuffer_i, tableShader);
        waitForCompute();
    end
    drawTriangles(vertexBuffer, indexBuffer_i, standardShaders)
```

Algorithm 19.2. Render irregular patches as triangles.

transition points, and they are written to the vertex buffer by an alternate code path within the subdivision table evaluation compute shader.

Transition point tables are appended to the subdivision table and use largely the same data format. In place of influences, transition points store the 16 control points of the regular patch, and in place of weights, they store the domain location to be evaluated (Figure 19.13).

19.5.2 Tessellation Factor Synchronization

While hardware-tessellated regular patches can be divided into any number of segments, recursively subdivided irregular patches are limited to power-of-two divisions.

We resolve this inconsistency by flagging edges of regular patches that are shared with irregular patches. When a regular patch edge is flagged, it forgoes its adaptive metric and snaps to the global tessellation factor, which corresponds to the level of recursive subdivision (Figure 19.14). In the hardware tessellation pipeline, patch flags are stored in a static shader buffer that is bound to the tessellation control shader.

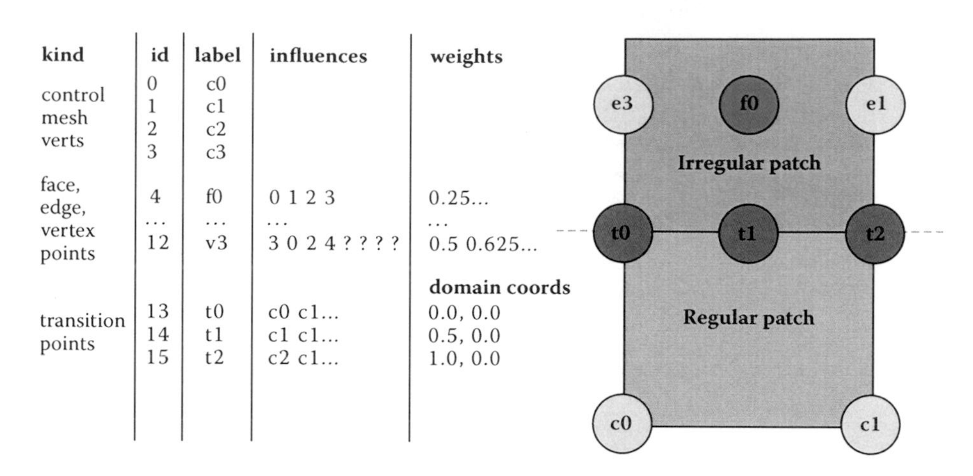

kind	id	label	influences	weights
control mesh verts	0	c0		
	1	c1		
	2	c2		
	3	c3		
face, edge, vertex points	4	f0	0 1 2 3	0.25...
	...	...	...	...
	12	v3	3 0 2 4 ? ? ? ?	0.5 0.625...
				domain coords
transition points	13	t0	c0 c1...	0.0, 0.0
	14	t1	c1 c1...	0.5, 0.0
	15	t2	c2 c1...	1.0, 0.0

Figure 19.13. A regular and irregular patch sharing an edge, with crack-welding transition points along the boundary.

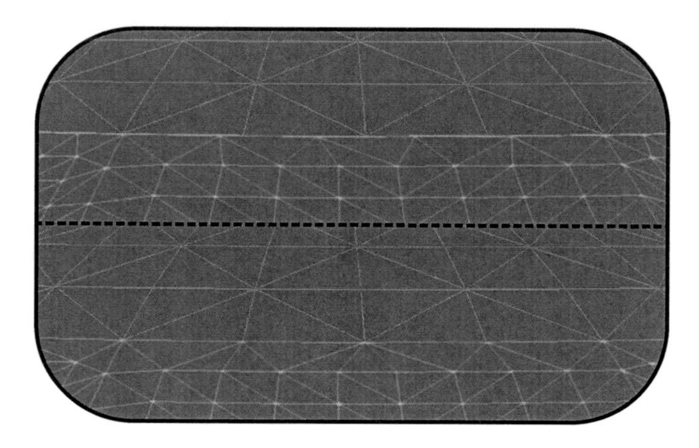

Figure 19.14. An edge shared by regular patches (blue) and irregular patches (red), shown with tessellation factors desynchronized (top) and synchronized (bottom).

19.5.3 Bireversal Invariant B-Spline Basis

When two regular patches share an edge, control mesh topology may dictate that they parameterize the edge in opposite directions. That is, the edge may interpolate A to B from 0 to 1 in one patch, and B to A from 0 to 1 in the other (Figure 19.15). Along these edges, numeric precision errors can introduce slight cracks at high tessellation factors.

To avoid cracking, we use a direction-invariant version of the B-spline basis from [Nießner et al. 12] that is mathematically equivalent when interpolating from

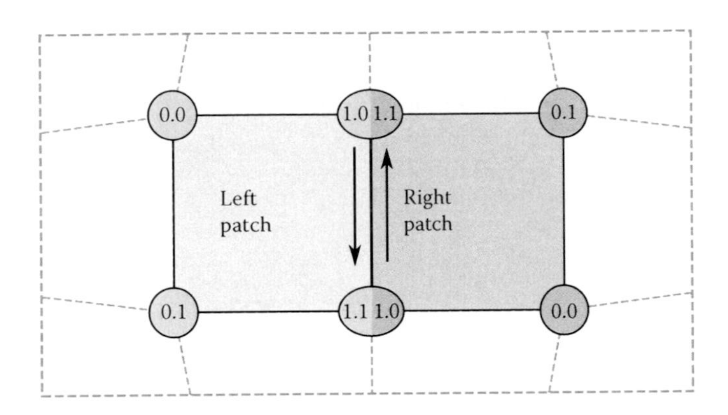

Figure 19.15. Two clockwise-oriented regular patches with opposite parameterization along the shared edge.

```
void EvaluateBSplineBasis(float u, out vec4 b)
{
  float s = 1.0 - u;
  float t = u;

  b.x = (s*s*s                              ) * 1.0/6.0;
  b.y = (4*s*s*s + t*t*t + 12*s*t*s + 6*t*s*t) * 1.0/6.0;
  b.z = (4*t*t*t + s*s*s + 12*t*s*t + 6*s*t*s) * 1.0/6.0;
  b.w = (t*t*t                              ) * 1.0/6.0;
}

void EvaluateBSpline(vec3 cp[16], vec2 uv,
                     out vec3 position)
{
  vec4 uBasis, vBasis;
  EvaluateBSplineBasis(uv.x, uBasis);
  EvaluateBSplineBasis(uv.y, vBasis);

  position = vec3(0);
  for (int i = 0; i < 4; i++)
  {
    position += vBasis[i] * (cp[i*4 + 0] * uBasis.x +
                             cp[i*4 + 1] * uBasis.y +
                             cp[i*4 + 2] * uBasis.z +
                             cp[i*4 + 3] * uBasis.w);
  }
}
```

Listing 19.2. Evaluating a B-spline patch using the bireversal invariant method from feature adaptive subdivision [Nießner et al. 12].

A to B by u, and when interpolating from B to A by $1 - u$. Listing 19.2 gives an implementation of bireversal invariant bicubic B-spline evaluation.

19.6 Going Further

We have thus far given an introduction to implementing Catmull-Clark subdivision surfaces in a game engine, but more work remains to complete the pipeline.

This section describes extensions that may be implemented or not, depending on individual game requirements.

19.6.1 Secondary Components

A vertex may have different values for secondary components, such as colors, texture coordinates, tangent basis, etc., for each incident face. With the exception of texture coordinates (see Section 19.6.3), it is usually acceptable to linearly interpolate secondary components across subdivided patches. When extracting control points and subdivision tables from the mesh topology, a *render vertex* must be generated that is a unique combination of the vertex and secondary components from the correct faces.

For regular patch control points that are strictly *supports* (not one of the interior four), render vertices may be generated without regard to the values of the secondary components. To avoid introducing extra render vertices, these components may be drawn from any face that is connected to the supporting control point.

19.6.2 Normals and Tangents

Normals and tangents may be evaluated directly by the shader, giving the true limit surface normal, or they may be treated as secondary components and interpolated. The choice is a tradeoff between quality and performance: limit normals and tangents give better shading but add calculation cost (Listing 19.3).

Note that if the same tangent-space normal map is applied to a SubD mesh and traditional LOD meshes, limit normals and tangents must be transferred to the LOD meshes to avoid using an inconsistent tangent basis. Additionally, normal map baking tools must render to a mesh with limit normals and tangents, to ensure that the normal map is encoded in the proper basis.

For regular patches, the limit surface normal is evaluated by the cross product of the patch tangent vectors. Irregular patches use the same method, but tangents are evaluated using *limit stencils* (see [Halstead et al. 93], Appendix A).

19.6.3 Texture Coordinate Smoothing

Because the subdivision rules weight a vertex by its neighbors without regard for their endpoints, the relative length of incident edges affects the subdivided position. For example, if a regular vertex is connected to three short edges and one long edge, the subdivided vertex will move toward the endpoint of the long edge. This can cause control mesh faces to change size and shape in the subdivided

```
void EvaluateBSplineBasis(float u, out vec4 b, out vec4 d)
{
  float s = 1.0 - u;
  float t = u;

  b.x = (s*s*s                                    ) * 1.0/6.0;
  b.y = (4*s*s*s + t*t*t + 12*s*t*s + 6*t*s*t) * 1.0/6.0;
  b.z = (4*t*t*t + s*s*s + 12*t*s*t + 6*s*t*s) * 1.0/6.0;
  b.w = (t*t*t                                    ) * 1.0/6.0;

  d.x = -s*s;
  d.y = -t*t - 4*s*t;
  d.z =  s*s + 4*s*t;
  d.w =  t*t;
}

void EvaluateBSpline( vec3 cp[16], float u, float v,
                      out vec3 position, out vec3 normal )
{
  vec4 uBasis, vBasis, uDeriv, vDeriv;
  EvaluateBSplineBasis(uv.x, uBasis, uDeriv);
  EvaluateBSplineBasis(vv.x, vBasis, vDeriv);

  position       = vec3(0);
  vec3 tangent   = vec3(0);
  vec3 bitangent = vec3(0);

  for (int i = 0; i < 4; i++)
  {
    vec3 positionBasis = (cp[i*4 + 0] * uBasis.x +
                          cp[i*4 + 1] * uBasis.y +
                          cp[i*4 + 2] * uBasis.z +
                          cp[i*4 + 3] * uBasis.w);

    vec3 positionDeriv = (cp[i*4 + 0] * uDeriv.x +
                          cp[i*4 + 1] * uDeriv.y +
                          cp[i*4 + 2] * uDeriv.z +
                          cp[i*4 + 3] * uDeriv.w);

    position  += vBasis[i] * positionBasis;
    tangent   += vBasis[i] * positionDeriv;
    bitangent += vDeriv[i] * positionBasis;
  }

  normal = normalize( cross( bitangent, tangent ) );
}
```

Listing 19.3. Bicubic B-spline evaluation shader extended to return normals.

mesh. If texture coordinates are linearly interpolated across the face, the texture parameterization will be distorted (Figure 19.16).

The solution employed by modeling packages such as Autodesk Maya is to construct a second topology from the *texture coordinates* of the control mesh and to smooth it in two dimensions using the Catmull-Clark subdivision rules. Smoothing the texture coordinates effectively inverts the distortion caused by smoothing the vertices.

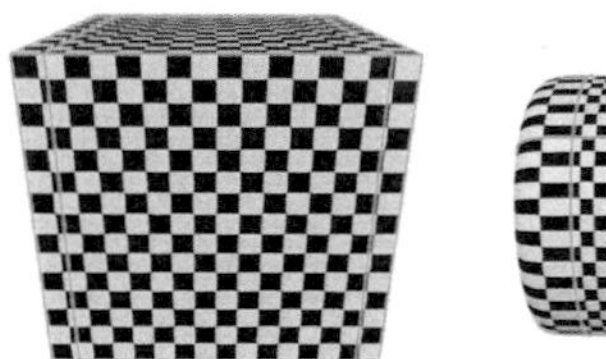  

Figure 19.16. From left to right: A textured cube control mesh with beveled caps, the surface with linearly interpolated texture coordinates, and the surface with smoothed texture coordinates.

To implement texture coordinate smoothing efficiently, we utilize *vertex-dominant topology*. Intuitively, texture-coordinate topology follows position topology but may introduce texture-only boundary edges where discontinuous parameterizations meet. More formally, the topology of secondary coordinates is embedded in the vertex topology with limitations: An edge that is a boundary in vertex topology must be a boundary in secondary topology, and an edge that is smooth in vertex topology either is a boundary in secondary topology or else connects the same two faces as in vertex topology.

When extracting regular patches from vertex-dominant topology, the texture-coordinate topology must be checked to ensure that it is also regular. If it is not, the entire patch is directed to recursive subdivision. Secondary component smoothing need not be specific to texture coordinates, but it is expensive and should be limited to where it has the greatest impact.

19.6.4 Regular Patch Extrapolation

To extract the 16 control points required to make a regular patch from a control mesh face, the face, its vertices, and its one ring neighborhood faces must all have valence 4. For some boundary and corner patches, it is possible to *extrapolate* missing control points and construct a B-spline boundary patch that evaluates to the Catmull-Clark limit surface (Figure 19.17).

We begin by defining some special cases of boundary vertices.

non-corner boundary vertex Vertex of valence 3 with two boundary edges.

convex corner boundary vertex Vertex of valence 2 with two boundary edges.

concave corner boundary vertex Vertex of valence 4 or more with two boundary edges.

Non-corner and convex corner boundary vertices may have their supporting control points extrapolated to form a valid B-spline patch:

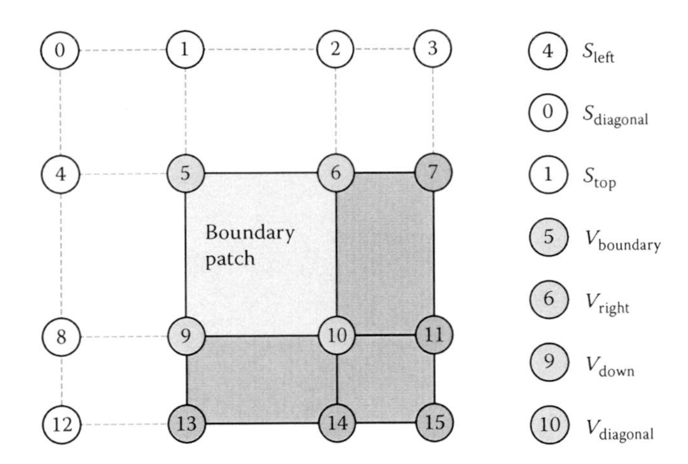

Figure 19.17. Extrapolated supporting control points for a convex corner boundary vertex.

- Non-corner boundary vertices require one extrapolated support:

$$S_{\text{edge}} = 2V_{\text{boundary}} - V_{\text{opposite}}. \tag{19.8}$$

- Convex corner boundary vertices require three extrapolated supports:

$$\begin{aligned} S_{\text{left}} &= 2V_{\text{boundary}} - V_{\text{right}}, \\ S_{\text{diagonal}} &= 4V_{\text{boundary}} - 2V_{\text{right}} - 2V_{\text{down}} + V_{\text{diagonal}}, \\ S_{\text{top}} &= 2V_{\text{boundary}} - V_{\text{down}}. \end{aligned} \tag{19.9}$$

- Concave corner boundary vertices may not be extrapolated and require recursive subdivision.

If all vertices of a control mesh face are regular or are borders that support extrapolation, the needed supports may be added as rows to the subdivision tables and the patch may be treated as regular. If texture coordinates are smoothed (Section 19.6.3), regular patch extrapolation may be applied to texture coordinates as well. Note that the extrapolation formulae are linear combinations of control mesh vertices and are therefore compatible with the subdivision table compute shader.

Causing more faces to render as regular patches in this manner improves image quality and reduces subdivision table size (Figure 19.18).

19.6.5 View Frustum Culling

In the hardware tessellation pipeline, it is possible to cheaply discard patches by setting the tessellation factor to 0. Additionally, the convex hull of the control points of a B-spline patch is a convex hull of the surface.

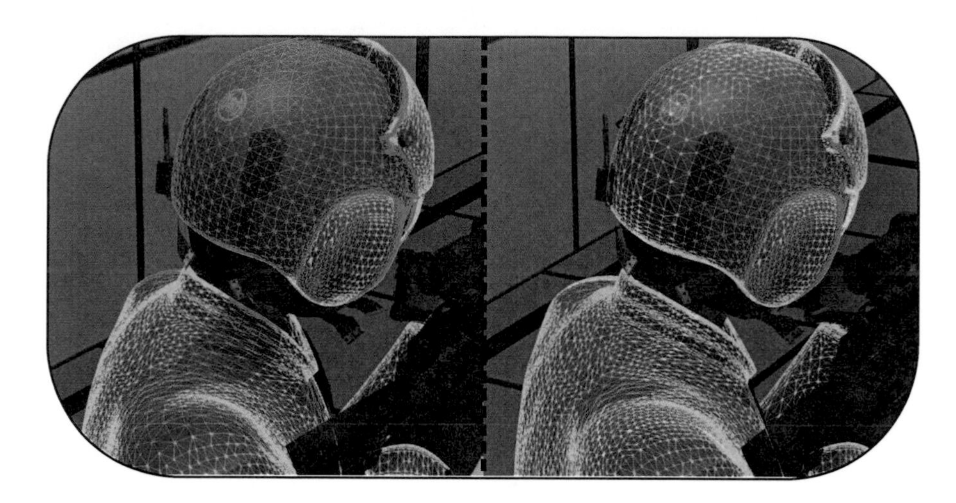

Figure 19.18. Comparison of models with (right) and without (left) regular patch extrapolation for texture coordinates. Blue wireframe represents regular patches, and yellow edges are extrapolated borders.

We can utilize these properties to add coarse view frustum culling of patches, saving surface evaluation and triangle setup costs. In the tessellation control shader, we transform each control point to clip space and test against the unit cube:

$$\begin{aligned} x_{\text{clip}} &> w_{\text{clip}}, \\ y_{\text{clip}} &> w_{\text{clip}}, \\ x_{\text{clip}} &< -w_{\text{clip}}, \\ y_{\text{clip}} &< -w_{\text{clip}}, \\ w_{\text{clip}} &\leq 0. \end{aligned} \tag{19.10}$$

If all control points pass one of the tests, the patch is discarded.

19.6.6 Back Patch Culling

Shirman and Abi-Ezzi describe a *cone of normals*: a region of space from which no part of a corresponding Bézier patch may be seen front facing [Shirmun and Abi-Ezzi 93].

Using the cone of normals, we can implement *back patch culling*, discarding entire patches without evaluating the surface or submitting triangles for rasterization. Once calculated, the cone test is extremely fast, consisting of a single dot product and comparison:

$$\hat{v} \cdot \hat{a} \leq \sin(\alpha). \tag{19.11}$$

The calculation of the cone is expensive, however, and requires converting the B-spline patch control points to the Bézier basis. Therefore, this test is reserved for control meshes that do not animate.

It is also possible to consider occlusion in the culling calculation. "Patch-Based Occlusion Culling for Hardware Tessellation" by Nießner and Loop [Nießner and Loop 12] describes a method for building and testing a hierarchal depth buffer in the tessellation control shader.

19.7 Conclusion

In this chapter, we have described a practical real-time implementation of Catmull-Clark subdivision surfaces that has been utilized in multiple AAA console games. It is hoped that the reader will come away with an appreciation for the opportunities presented by the tessellation hardware in modern GPUs and the knowledge that it is practical to implement SubDs in games today.

19.8 Acknowledgments

The author would like to thank Paul Allen Edelstein for improvements to the quality of the adaptive tessellation metric and for deriving its mathematical basis, and the reviewers for their feedback.

Bibliography

[Bommes et al. 13] David Bommes, Bruno Lévy, Nico Pietroni, Enrico Puppo, Claudio Silva, Marco Tarini, and Denis Zorin. "Quad-Mesh Generation and Processing: A Survey." *Computer Graphics Forum* 32:6 (2013), 51–76. Article first published online, March 4, 2013, DOI: 10.1111/cgf.12014, http://vcg.isti.cnr.it/Publications/2013/BLPPSTZ13a.

[Brainerd 14] Wade Brainerd. "Tessellation in Call of Duty: Ghosts." http://wadeb.com/siggraph_2014_tessellation_in_call_of_duty_ghosts.zip, 2014.

[Catmull and Clark 78] E. Catmull and J. Clark. "Recursively Generated B-Spline Surfaces on Arbitrary Topological Meshes." *Computer-Aided Design* 10:6 (1978), 350–355.

[DeRose et al. 98] Tony DeRose, Michael Kass, and Tien Truong. "Subdivision Surfaces in Character Animation." In *SIGGRAPH '98: Proceedings of the 25th Annual Conference on Computer Graphics and Interactive Techniques*, pp. 85–94. New York: ACM, 1998. Available online (http://graphics.pixar.com/library/Geri/).

[Halstead et al. 93] Mark Halstead, Michael Kass, and Tony DeRose. "Efficient, Fair Interpolation Using Catmull-Clark Surfaces." In *SIGGRAPH '93: Proceedings of the 20th Annual Conference on Computer Graphics and Interactive Techniques*, pp. 35–44. New York: ACM, 1993.

[Hoppe et al. 94] H. Hoppe, T. DeRose, T Duchamp, M. Halstead, H. Jin, J. McDonald, J. Schweitzer, and W. Stuetzle. "Piecewise Smooth Surface Reconstruction." In *SIGGRAPH '94: Proceedings of the 25th Annual Conference on Computer Graphics and Interactive Techniques*, pp. 295–302. New York: ACM, 1994.

[Nießner and Loop 12] Matthias Nießner and Charles Loop. "Patch-Based Occlusion Culling for Hardware Tessellation." Paper presented at Computer Graphics International, Poole, UK, June 12–15, 2012.

[Nießner et al. 12] M. Nießner, C. Loop, M. Meyer, and T. DeRose. "Feature-Adaptive GPU Rendering of Catmull-Clark Subdivision Surfaces." *ACM Transactions on Graphics (TOG)* 31:1 (2012), 6.

[Nießner et al. ar] Matthias Nießner, Benjamin Keinert, Matthew Fisher, Marc Stamminger, Charles Loop, and Henry Schäfer. "Real-Time Rendering Techniques with Hardware Tessellation." *Computer Graphics Forum.* First published online DOI: 10.1111/cgf.12714, September 21, 2015.

[Pixar 15] Pixar. "Subdivision Surfaces." *OpenSubdiv Documentation*, http://graphics.pixar.com/opensubdiv/docs/subdivision_surfaces.html, 2015.

[Schäfer et al. 15] Henry Schäfer, Jens Raab, Benjamin Keinert, and Matthias Nießner. "Dynamic Feature-Adaptive Subdivision." In *Proceedings of the ACM SIGGRAPH Symposium on Interactive 3D Graphics and Games*, pp. 31–38. New York: ACM, 2015.

[Shirmun and Abi-Ezzi 93] Leon A. Shirmun and Salim S. Abi-Ezzi. "The Cone of Normals Technique for Fast Processing of Curved Patches." *Computer Graphics Forum* 12:3 (1993), 261–272.

[Stam 98] Jos Stam. "Exact Evaluation Of Catmull-Clark Subdivision Surfaces at Arbitrary Parameter Values." In *SIGGRAPH '98: Proceedings of the 25th Annual Conference on Computer Graphics and Interactive Techniques*, pp. 395–404. New York: ACM, 1998.

About the Contributors

Daniel Bagnell is a software developer at Analytical Graphics, Inc., where he is working on Cesium, a WebGL virtual globe. He received his BS degrees in mathematics and computer science from Drexel University.

Xavier Bonaventura is currently a PhD student at the Graphics & Imaging Laboratory of University of Girona, researching on viewpoint selection. He developed his master's thesis on hardware tessellation at Budapest University of Technology and Economics, within the Erasmus program.

Tamy Boubekeur is an associate professor in computer science at Telecom ParisTech (formally ENST Paris), the telecommunication graduate school of the Paris Institute of Technology (France). He is a faculty member of LTCI, the CNRS laboratory for information processing and communication. His areas of research include 3D computer graphics, geometric modeling, real-time rendering, and interaction. More info at www.telecom-paristech.fr/~boubek.

Wade Brainerd is a principal technical director at Activision, where he enjoys all kinds of graphics programming and performance optimization. He lives in Portland, Maine, with his wife and two children.

Alan Chambers obtained an MSc with Distinction in computer science from the University of Wales, Swansea in 2002. Since then he has worked at Sony Computer Entertainment Europe on games such as *Formula 1* and *WipEout*. He is now the Lead Graphics Engineer at New Zealand's largest game development studio, Sidhe, and has a specific interest in engine programming and optimizations. Alan is also a keen pilot and can often be found cruising around the skies of New Zealand in his spare time.

Patrick Cozzi is coauthor of *3D Engine Design for Virtual Globes* (2011), coeditor of *OpenGL Insights* (2012), and editor of *WebGL Insights* (2015). At Analytical Graphics, Inc., he leads the graphics development of Cesium, an open source WebGL virtual globe. He teaches "GPU Programming and Architecture" at the University of Pennsylvania, where he received a masters degree in computer science.

Alex Dunn, as a developer technology engineer for NVIDIA, spends his days passionately working toward advancing real-time visual effects in games. A former graduate of Abertay University's Games Technology Course, Alex got his first taste of graphics programming on the consoles. Now working for NVIDIA, his time is spent working on developing cutting-edge programming techniques to ensure the highest quality and best player experience possible is achieved.

Jose I. Echevarria received his MS degree in computer science from the Universidad de Zaragoza, Spain, where he is currently doing research in computer graphics. His research fields range from real-time to off-line rendering, including appearance acquisition techniques.

Holger Gruen ventured into creating real-time 3D technology over 20 years ago writing fast software rasterizers. Since then he has worked for games middleware vendors, game developers, simulation companies, and independent hardware vendors in various engineering roles. In his current role as a developer technology engineer at NVIDIA, he works with games developers to get the best out of NVIDIA's GPUs.

Diego Gutierrez is a tenured associate professor at the Universidad de Zaragoza, where he got his PhD in computer graphics in 2005. He now leads his group's research on graphics, perception, and computational photography. He is an associate editor of three journals, has chaired and organized several conferences, and has served on numerous committees, including the SIGGRAPH and Eurographics conferences.

Graham Hemingway is a PhD student (who really hopes to graduate one day soon) at Vanderbilt University in Nashville, Tennessee. His research interests include CAD model visualization, formal methods for developing safety-critical embedded systems, and large-scale heterogeneous distributed simulation.

Jorge Jimenez is a real-time graphics researcher at the Universidad de Zaragoza, in Spain, where he received his BSc and MSc degrees, and where he is pursuing a PhD in real-time graphics. His passion for graphics started after watching old school demos in his brother's Amiga A1000. His interests include real-time photorealistic rendering, special effects, and squeezing rendering algorithms to be practical in game environments. He has numerous contributions in books and journals, including *Transactions on Graphics*, where his skin renderings made the front cover of the SIGGRAPH Asia 2010 issue. He loves challenges, playing games, working out in the gym, and more than anything, breaking in the street.

Benjamin Keinert is a PhD candidate in the DFG-funded Research Training Group "Heterogeneous Image Systems" at the University of Erlangen-Nuremberg, Germany. He received his MSc degree from the same university in 2013 after completing his thesis about "Dynamic Attributes for Hardware Tessellated Meshes."

His current research involves real-time rendering techniques as well as interactive global illumination approaches.

Denis Kravtsov graduated from Saint-Petersburg State Polytechnic University. He worked on the multiplatform title *TimeShift* at Saber Interactive. Denis is currently a PhD student at the National Centre for Computer Animation, Bournemouth University.

Sean Lilley is a senior studying digital media design at the University of Pennsylvania. He has interned at Electronic Arts and AMD, which reflects his passion for both video games and 3D graphics. Currently he is building a game engine with his brother Ian as part of their senior design project. He hopes to join the game industry after graduation.

Benjamin Mistal graduated with a degree in computer science from the University of Victoria in 1998, and he spent a number of years developing industrial software applications utilizing 2D and 3D laser cloud data. After joining the 3D Application Research Group at ATI (now part of AMD), he helped develop, and later lead the development of, the popular RenderMonkey shader development tool. He has been involved with Right Hemisphere's Deep Exploration and Deep Creator products, and he cofounded Esperient Corporation to further develop the Deep Creator (renamed to Esperient Creator) product. He is currently researching and developing programming languages and web-related graphics technologies for Mistal Research, Inc. (www.mistal-research.com) and is also developing products for Qualcomm Technologies Inc., highlighting their next generation of mobile graphics hardware. Often found in various places throughout the Rocky Mountains, he is an avid hiker, rock climber, and runner.

Anton Kai Michels, while studying computer science at Concordia University, entered a game design competition with fellow students Zoe Briscoe, Kirsty Beaton, Joel Daignault, and Nicolas Cattanéo. Their game *Panopticon* earned them all internships at Ubisoft Montreal, after which Anton was hired fulltime as a graphics programmer on *Rainbow Six: Siege*. He later left Ubisoft to join a talented research-and-development team at Eidos Montreal, where he devised novel rendering techniques for *Tomb Raider*, *Deus Ex*, and *Hitman*. He is now the rendering lead at DICE LA.

Aleksander Netzel is a masters degree student at the University of Wrocław. He is currently employed as a graphic programmer at Techland, where he adapts the latest cutting-edge graphic techniques for the current generation of gaming platforms. His goal is to develop new solutions for real-time graphics problems as part of a research team.

Matthias Nießner is a visiting assistant professor at Stanford University affiliated with the Max Planck Center for Visual Computing and Communication. Previous to his appointment at Stanford, he earned his PhD from the University of Erlangen-Nuremberg, Germany, under the supervision of Günther Greiner. His research focuses on different fields of computer graphics and computer vision, including real-time rendering, reconstruction of 3D scene environments, and semantic scene understanding.

Gustavo Bastos Nunes is a graphics engineer in the Engine team at Microsoft Turn 10 Studios. He received his BSc in computer engineering and MSc in computer graphics from Pontifícia Universidade Católica do Rio de Janeiro, Brazil. He has several articles published in the computer graphics field. He is passionate about everything real-time graphics related. Gustavo was part of the teams that shipped Microsoft Office 2013, Xbox One, *Forza Motorsport 5*, *Forza Horizon 2*, and *Forza Motorsport 6*.

Christopher Oat is a graphics lead at Rockstar Games, where he works on real-time rendering techniques used in Rockstar's latest titles. Previously he was the demo team lead for AMD's Game Computing Applications group. Christopher has published his work in various books and journals and has presented at graphics and game developer conferences worldwide. Many of the projects that he has worked on can be found on his website: www.chrisoat.com.

David Pangerl is the CEO of Actalogic, where he is working as a lead researcher and engine architecture designer. He has been involved in computer graphics and engine research for over a decade.

Emil Persson is the Head of Research at Avalanche Studios, where he is conducting forward-looking research, with the aim to be relevant and practical for game development, as well as setting the future direction for the Avalanche Engine. Previously, he was an ISV Engineer in the Developer Relations team at ATI/AMD. He assisted tier-one game developers with the latest rendering techniques, identifying performance problems and applying optimizations. He also made major contributions to SDK samples and technical documentation.

João Lucas Guberman Raza is a program manager at Microsoft's 343 Industries, where he works in the services cloud compute systems. Previously he was in the Windows Phone division, where he helped ship the SDK for game developers. An avid gamer, he has worked in the game industry for over five years. He holds a bachelor of computer science from Universidade Federal de São Carlos (UFSCar). He runs the blog www.versus-software.com, where he writes about his main interests in graphics, networking, and game design.

Christophe Riccio is a graphics programmer with a background in digital content creation tools, game programming, and GPU design research. He is also a keen supporter of real-time rendering as a new medium for art. He has an MSc degree in computer game programming from the University of Teesside. He joined e-on software to study terrain editing and to design a multi-threaded graphics renderer. He worked for Imagination Technologies on the PowerVR series 6 architecture. He is currently working for AMD doing some OpenGL gardening. For the past ten years, Christophe has been an active OpenGL community contributor, including contributions to the OpenGL specifications. Through G-True Creation, he writes articles to promote modern OpenGL programming. He develops tools, GLM and the OpenGL Samples Pack, which are part of the official OpenGL SDK.

Pawel Rohleder claims he has been interested in the computer graphics and game industry since he was born. He is keen on knowing how all things (algorithms) work, then trying to improve them and developing new solutions. He started programming games professionally in 2002; he has been 3D graphics programmer at Techland's ChromeEngine team for three years and since 2009 has been working as a Build Process Management Lead / R&D manager. He is also a PhD student in computer graphics (at Wroclaw University of Technology, since 2004).

Henry Schäfer is currently working toward a PhD degree in computer graphics at the University of Erlangen-Nuremberg. His research interests include realistic image synthesis, data-compression techniques, and real-time rendering.

Peter Sikachev graduated from Lomonosov Moscow State University in 2009, majoring in applied mathematics and computer science. He started his career game development in 2011 at Mail.Ru Games as a graphics programmer. He contributed to a range of rendering features of the *Skyforge* next-generation MMORPG. In 2013 he joined Eidos Montreal as a research-and-development graphics programmer. He helped ship *Thief* and *Rise of the Tomb Raider* and contributed to *Deus Ex: Universe*. Peter has been an author of multiple entries in the *GPU Pro* series and a speaker at ACM SIGGRAPH courses. He now holds a senior graphics programmer position at Rockstar Toronto.

Marc Stamminger is a professor for computer graphics at the University of Erlangen-Nuremberg, Germany, since 2002. After finishing his PhD thesis on finite element methods for global illumination in 1999, he was a post doctorate at MPI Informatics in Saarbrücken, Germany, and at the INRIA Sophia-Antipolis, France. In his research he investigates novel algorithms to exploit the power of current graphics hardware for rendering, geometry processing, and medical visualization. He participates in the program committees of all major computer graphics conferences and was program co-chair of Eurographics 2009 and the Eurographics Rendering Symposium 2010. Since 2012, he is head of the DFG-funded Research Training Group "Heterogeneous Image Systems."